EXPERIMENTAL CHARACTERIZATION OF ADVANCED COMPOSITE MATERIALS

FOURTH EDITION

EXPERIMENTAL CHARACTERIZATION OF ADVANCED COMPOSITE MATERIALS

FOURTH EDITION

Leif A. Carlsson • Donald F. Adams • R. Byron Pipes

CRC Press
Taylor & Francis Group
Boca Raton London New York

CRC Press is an imprint of the
Taylor & Francis Group, an **informa** business

CRC Press
Taylor & Francis Group
6000 Broken Sound Parkway NW, Suite 300
Boca Raton, FL 33487-2742

Printed on acid-free paper
Version Date: 20130731

International Standard Book Number-13: 978-1-4398-4858-6 (Hardback)

Library of Congress Cataloging-in-Publication Data

Carlsson, Leif.
Experimental characterization of advanced composite materials / authors, Leif Carlsson, Donald F. Adams, and R. Byron Pipes. -- 4th edition.
pages cm
Revised edition of: Experimental characterization of advanced composite materials / Donald F. Adams, Leif A. Carlsson, R. Byron Pipes. 2003. Carlsson's name appears first on earlier editions.
Includes bibliographical references and index.
ISBN 978-1-4398-4858-6
1. Composite materials--Testing. I. Adams, Donald Frederick, 1935- II. Pipes, R. Byron. III. Title.

TA418.9.C6A263 2014
620.1'180287--dc23 2013029839

Visit the Taylor & Francis Web site at
http://www.taylorandfrancis.com

and the CRC Press Web site at
http://www.crcpress.com

Contents

Preface

As new composite systems have been developed and new applications found, new requirements for test method development have continually evolved. In the early days of advanced composite materials (1960s), the primary structural composite material available to compete with metals consisted of boron and carbon fibers in an epoxy matrix of low toughness. Matrix-controlled toughness as a property was not considered by the composite materials community. Demanding commercial and military applications, however, emerged in the 1980s, promoting the development of many new matrix materials (e.g., toughened epoxies and high-temperature thermoplastics) and resulting in the need to develop test methods for determining the toughness and damage tolerance of composite materials. Soon, many test methods not previously applied to composites or any other materials were proposed, including mode I, mode II, mode III, and mixed-mode fracture mechanics tests, plate impact, compression after impact, and open-hole tension and compression.

This evolution of test methods to meet new demands has continued over the years as additional aspects have risen in importance (e.g., elevated temperature, absorption of moisture and solvents, and other factors that might affect performance and durability). Also, improvements in fiber-matrix interfacial bonding, the introduction of organic fibers (e.g., aramid, polyethylene, liquid crystal polymer), natural fibers (e.g., hemp and jute), and ultrahigh-modulus inorganic fibers (e.g., ceramics) have occurred. Likewise, new classes of matrix materials such as bismalimides, polyimides, vinylesters, and many others have necessitated still more test methods or revisions of existing ones.

As we are more than a decade into the twenty-first century, applications of all types of composite materials to commercial products are being emphasized. In anticipation of this development, the 1990s were a period of test method consolidation and attempts to better understand those being used. Thus, the present text comes at an opportune time, that is, when the evolution of test methods is in a relatively stable period and definitive recommendations can be made. This, then, is the goal of this text: to present primarily those mechanical test methods that have achieved some consensus as the "best presently available," recognizing that "best" is often subjective.

The primary audience for this text will be university and technical school graduate and undergraduate students taking a course in experimental mechanics of composite materials, this being the required text. However, this text also addresses a much larger audience. Quite frequently, engineers and technicians in industry and government laboratories are assigned composite material testing responsibilities but have little or no prior experience with such materials. These individuals are associated with a wide range of organizations, including corporate research, federal laboratories, university research, material suppliers, contract design organizations, and custom fabrication shops. They need to choose among competing test methods to perform or supervise the performance of mechanical testing and then interpret the experimental data obtained. In this sense, this text complements ASTM (American Society for Testing and Materials) and other standards. This text is sufficiently straightforward and concise to appeal to this group of individuals who need a "quick start."

Another potential audience is those who attend composite material characterization short courses and tutorials. The present text, because of its concise wording and numerous figures and tables, will serve as both a set of course notes and a permanent reference source of topics covered.

The 19 chapters of the text are organized to meet the class laboratory schedule needs of a one-semester or one-quarter course. Specific topics (chapters) can be deleted as required to fit the actual time available. The text is intended to be self-contained, with no reference texts required.

The first chapter provides an introduction to the special terminology and conventions that have evolved related to composite materials, Chapter 2 summarizes the unique analysis methods and data reduction formulas required, and Chapter 3 presents sufficient laminate-processing information to permit the reader to fabricate his or her own composite panels required to determine mechanical properties of the specific composite of interest. Chapter 4 details the preparation of test specimens and explains testing equipment required for the mechanical tests. Chapter 5 introduces cure shrinkage and residual stress testing of polymer resins.

Chapters 6 and 7 cover single-fiber testing, matrix characterization, and fiber-matrix interface testing. Chapters 8 through 10 cover the most basic lamina tests (viz., tension, compression, and shear). Flexure and off-axis testing of the lamina is explained in Chapters 11 and 12. Principles of lamina and laminate thermoelastic testing are outlined in Chapter 13. Through-thickness tension and compression testing of composite laminates is covered in Chapter 14. Specific analysis and mechanical tests for laminates are discussed in Chapter 15. Test procedures related to damage tolerance are described in Chapters 16 through 18. Sandwich panel and core testing are discussed in the final Chapter 19.

Appendices A–D cover stiffness transformations, examples of test panel dimensions, a sample laboratory report, and unit conversions, respectively. The sample report is intended to serve as a guide for the student in the preparation of an acceptable laboratory report.

L.A.C.
Boca Raton, Florida

D.F.A.
Salt Lake City, Utah

R.B.P.
West Lafayette, Indiana

Acknowledgments

We are indebted to many people who have contributed to and supported the four editions of this text. Seija Carlsson typed the manuscript for the first edition. Rosemarie Chiucchi and Teresa Perez typed the revised manuscript for the second edition, and Teresa also typed several chapters for the third edition. Laura Thornton and Melissa Mancao assisted us with typing of the fourth edition. We would also like to thank Touy Thiravong for help in clarifying many technical details related to testing and preparation of specimens; Dr. John W. Gillespie, Jr., Dale W. Wilson, and William A. Dick for reviewing the manuscript for the first edition; and Dr. Anthony Smiley, Dr. William Sanford, and Rod Don, who shared their knowledge of processing of thermoplastic and thermoset composites. We acknowledge Dr. Pascal Hubert and Prof. Anoush Poursatip and his research students at the University of British Columbia for their contributions to Chapter 3 of the third and fourth editions.

We would also like to thank the late Woody Snyder, who supplied many photographs for the first edition of this book. Thanks are due to Judy Joos and Mark Deshon, Shawn Pennell, Sherri VonHartman, and Jeffrey Weinraub for the artwork. Dr. James R. Reeder of NASA Langley Research Center kindly supplied information of the MMB test, and Dr. Shaw Ming Lee of Hexcel Corporation contributed much useful information on the ECT specimen. Dr. Xiaoming Li of Bell Helicopter provided test data generated on the ECT specimen. Students who scrutinized the first edition of the text and suggested improvements are Robert Rothschilds, Bruce Trethewey, Gary Becht, Ellen Brady, and James Newill. We thank the many students for providing the test results presented in numerous graphs throughout all four editions of the text. Of those, we single out Robert Jurf, Thomas Chapman, David Adkins, Richard Givler, Robert Wetherhold, Richard Walsh, Nicolass Ballityn, Bruce Yost, James York, Yong-Zhen Chen, Uday Kashalikar, Mark Cirino, Felipe Ramirez, Grant Browning, Alex Figliolini. Michael Dale, Elio Saenz, Johnathan Goodsell, Oleksandr Kravchenko, Josh Dustin, and Andrew Ritchey. Finally, many thanks go to Allison Shatkin, editor, CRC, and her colleagues for their support and patience.

1

Introduction

Most of the subject matter in this text has been taught for years to advanced undergraduate and first-year graduate students at the University of Delaware, Florida Atlantic University, the University of Wyoming, and Purdue University. During this time, we realized there was no textbook offering a concise treatment of the experimental characterization of composite materials. Most current textbooks deal only with the analysis of composite materials. If the present text appears to emphasize advanced composite materials, it is only because these materials often present the greatest challenges to experimental characterization. These also are the materials most often used in structural applications, where accurate characterization is most important. Interestingly, today, many high-performance designs demand the use of advanced composite materials.

The objective of this textbook is to present processing techniques, cure kinetics for thermosets, specimen preparation, analyses of test methods, test procedures, and data reduction schemes to determine mechanical properties, thermal expansion coefficients, and fracture and strength data for composite materials. Emphasis is placed on practical matters such as preparation and testing of specimens and data reduction methodology. Many of the test methods presented are American Society for Testing and Materials (ASTM) or other national or international standards. Others, although originating within an individual organization and sometimes continuing to be refined in terms of test specimen and fixture geometries, test procedures, and data reduction schemes, are being widely used within the composites testing community.

No attempt is made to present a detailed review of composite mechanics or fracture mechanics. Such a treatment has been presented in many other textbooks, to which references are made. Only a brief elementary outline of the mechanics is provided in Chapter 2. Moreover, no attempt is made to present an overview of all test methods; such reviews are available elsewhere and are referenced. The methods presented here are deemed the most appropriate and widely accepted at present. Additional developments can, however, be expected in this evolving field.

This text was prepared for students who have an interest in experimental aspects of composite materials. It will also be useful for engineers in industrial or government laboratories who desire to extend their expertise into experimental characterization of anisotropic materials.

1.1 Background

Composite materials, in the context of high-performance materials for structural applications, have been used increasingly since the early 1960s, although materials such as glass fiber-reinforced polymers were already being studied 20 years earlier. Initially, conventional test methods, originally developed for determining the physical and mechanical properties of metals and other homogeneous and isotropic construction materials, were used. It was soon recognized, however, that these new materials, which are nonhomogeneous and anisotropic (orthotropic), require special consideration for determining physical and mechanical properties.

During this initial period, composite materials technology was developed primarily within the aerospace community. Because composite material test methods were not standardized, each airframe manufacturer tended to develop its own procedures. Although these procedures were not usually proprietary, there was little incentive to adopt common test methods, particularly because few methods had emerged as clearly superior to others of the same type. The problem was further complicated by the continuous emergence of new materials (e.g., boron and carbon fibers in the mid-1960s and Kevlar® fibers in the early 1970s, along with new epoxies, polyimides, and other matrix materials, including ceramics and metals). A specific test method, which may have performed reasonably well for the types of composite materials of the past, was not necessarily adequate for the material evaluated at that time. That is, there was little possibility of standardization. As a result, many diverse test methods were developed for measuring the same properties. Some were easy to use but provided only limited results or data of questionable quality. Others were complex, operator dependent, and perhaps also of questionable quality.

In the United States, the federal government sponsored much of the early development work in composite materials, primarily through agencies such as the Department of Defense and the National Aeronautics and Space Administration. The problems associated with the lack of standards were recognized, and attempts were made to identify general test methods, to generate a database for comparison purposes, and to establish standards. These attempts were largely unsuccessful, primarily because newer composite materials did not necessarily behave in the same manner as the prior generation of materials around which the test methods had been established.

Today, almost six decades after these initial attempts, complete standards for testing composite materials still do not exist, perhaps still for the same practical reasons. That is, as new generations of composite materials are developed, existing test methods have to be modified to accommodate them. Rigid standardization would not permit this. On the other hand, consensus organizations such as ASTM have done much to maintain a degree of uniformity and an awareness of the general problem of achieving standardization.

As additional industries (e.g., automotive, sporting goods, electronics, machine tool, and civil infrastructure) have moved toward the more extensive use of composite materials in their products, this general lack of standardization has become particularly disturbing to them. Most of the more traditional industries are accustomed to following specific design standards, purchasing materials to standards, and testing to standards. Thus, acceptance of the general lack of test method standards has become part of the indoctrination of newer industries into this relatively new technology.

This lack of standardization in composite materials testing is not necessarily a negative aspect, although it may often be inconvenient for the new user. That is, the term *composite material* does not define a specific class of fabrication materials, but rather a broad spectrum of materials of widely varying properties. Thus, it can be expected that different test methods will be required for different classes of composite materials. This philosophy is no different from that associated with using a different test method for testing low-carbon steel than for testing a ceramic.

With this general background and philosophy in mind, current composite material characterization methods are discussed and evaluated in the following chapters. Not every known method is introduced, however. Some methods that were previously popular are now rapidly fading from use. Thus, although these names are familiar to many, and are frequently quoted in the literature, particularly in the older literature, they are becoming obsolete and need not be discussed here. See also the texts of Tarnopolskii and Kincis (1985) and Hodgkinson (2000).

1.2 Laminate Orientation Code

Typically, the basic building block of a composite material structural component is a unidirectional lamina, that is, a thin layer consisting of reinforcing fibers all oriented in the same direction and embedded in a matrix such as a polymer. Alternatively, the reinforcement can be in the form of fibers woven to form a layer of fabric, a thin mat of randomly oriented fibers, or some similar form. All of these laminae are typically characterized experimentally using the test methods described in this text.

However, in the actual structural design process, these individual laminae are stacked and processed together to form a laminate of the desired properties. Such a laminate can be made as complex as required to satisfy the specified design criteria by adding more and more plies of arbitrary orientations, reinforcement forms, and material types. Until the early 1970s, there was no unified system for defining the layup patterns of composite laminates. As composite materials moved from the research laboratory to the production shop, the need for a common terminology became obvious.

1.2.1 Standard Laminate Code

The Air Force Flight Dynamics Laboratory included a Laminate Orientation Code in the third edition of its *Advanced Composites Design Guide* (1973). This code, established by general consensus of the aerospace industry at the time, has survived to the present with minimal modification and continues to be used almost universally by the composites community. Thus, it is important for the reader to know at least its general features.

The Standard Laminate Code, used to describe a specific laminate uniquely, is most simply defined by the following detailed descriptions of its features:

1. The plies are listed in sequence from one laminate face to the other, starting with the first ply laid up, with square brackets used to indicate the beginning and end of the code.
2. A subscript capital T following the closing square bracket should be used to indicate that the total laminate is shown. Although it is not good practice, as shown subsequently, the T is often omitted. For a symmetric laminate (see Chapter 2), only the plies on one side of the midplane are shown, and a subscript capital S follows the closing bracket. A subscript capital Q is also defined in the code to designate an antisymmetric laminate (however, improperly termed a quasi-symmetric laminate). It must be pointed out that antisymmetric laminates are not commonly used.
3. Each ply within the laminate is denoted by a number representing its orientation in degrees as measured from the geometric x-axis of the laminate to the lamina principal material coordinate direction (1-axis). Material and geometric coordinate axis systems are described in Chapter 2. Positive angles are defined as clockwise when looking toward the layup tool surface. Note that this convention is consistent with the definition of a positive angle in Figure 2.3, because there the view is away from the layup tool surface.
4. When two or more plies of identical properties and orientation are adjacent to each other, a single number representing the angular orientation, with a numerical subscript indicating the number of identically oriented adjacent plies, is used.

   For example, a laminate consisting of just three –45° plies would be designated as $[-45_3]_T$. The notation $[-45]_{3T}$ is also acceptable and in fact is more commonly used.
5. If the angles of otherwise-identical adjacent plies are different, or if the angles are the same but the materials are different, the plies are separated in the code by a slash.

For example, a two-ply laminate consisting of a +45°-ply and a −30°-ply of the same material would be expressed as $[45/-30]_T$. Note that the first ply listed in the code is always the first ply to be laid up in the fabrication process. Note also that the plus sign is not used unless omitting it would create an ambiguity.

A six-ply symmetric laminate consisting of identical plies oriented at +45, 0, −30, −30, 0, and +45° would be expressed as $[45/0/-30]_S$.

When a symmetric laminate contains an odd number of plies of the same material, such as −30, 90, 45, 90, and −30°, the center ply is designated with an overbar, for instance, $[-30/90/\overline{45}]_S$.

6. When adjacent plies are at angles of the same magnitude but of opposite sign, the appropriate use of plus and minus signs is employed. Each plus or minus sign represents one ply and supersedes the use of the numerical subscript, which is used only when the directions are identical (as in Item 4).

   For example, a four-ply laminate consisting of plies oriented at 20, 20, −30, and 30° would be designated as $[20_2/\mp 30]_T$. Note that ∓ and not ± is used here to preserve the intended order.

7. Repeating sequences of plies are called sets and are enclosed in parentheses. A set is coded in accordance with the same rules that apply to a single ply.

   For example, a six-ply 45, 0, 90, 45, 0, and 90° laminate would be designated as $[(45/0/90)_2]_T$ or alternatively as $[(45/0/90)]_{2T}$. As in item 4, the latter form is no more correct but is more commonly used.

8. If a laminate contains plies of more than one type of material or thickness, a distinguishing subscript (or superscript) is used with each ply angle to define the characteristics of that ply, for example, $[0_g/90_k/45_c]_S$ for a glass, Kevlar, and carbon/fiber laminate.

1.2.2 Basic Condensed Code

When the exact number of plies need not be specified (as in a preliminary design), the Basic Condensed Code can be used. The plies are written in the order of ascending angle (magnitude) with only the relative proportions expressed by whole-number subscripts.

For example an actual $[30_2/0_6/-45_2/90_4]_T$ laminate would be expressed using the Basic Condensed Code as $[0_3/30/-45/90_2]$. An actual 30-ply $[90/\pm(0/45)]_{3S}$ laminate would be expressed as $[0_2/\pm 45/90]$. In both examples, the lack of a subscript after the closing bracket indicates that it is a Basic Condensed Code.

1.2.3 Specific Condensed Code

When the total number of plies and their orientations need be preserved, but not their order (stacking sequence) within the laminate, the Specific Condensed Code is used. This code is useful at that point in preliminary design when the laminate is being sized (i.e., when the required total number of plies is being specified). It is also particularly useful to the materials purchasing group because the scrap losses during cutting of the plies, and thus the amount of material that must be ordered, depends on the orientation of each ply in the laminate.

Using the Specific Condensed Code, the actual 30-ply $[90/\pm(0/45)]_{3S}$ laminate used in the previous example would be expressed as $[0_2/\pm45/90]_{6C}$. Note that a full 30-ply laminate is still expressed, the subscript C indicating, however, that the stacking sequence of the plies has not been retained.

1.2.4 Summary

Although the Laminate Orientation Code may appear complicated at first, it is systematically constructed and is as concise as possible. For simple laminates, the code reduces to a simple form and is easily and quickly written. Yet, the most complex laminate can be coded with equal conciseness.

1.3 Influence of Material Orthotropy on Experimental Characterization

The individual lamina (i.e., layer or ply) of a composite material is often the basic building block from which high-performance composite structures are designed, analyzed, and fabricated. Unless stated otherwise, the lamina material is usually assumed to exhibit linearly elastic material response, as, for example, in the analyses presented in Chapter 2. In many cases, this is a reasonable assumption. However, there are exceptions, particularly in terms of shear response, that sometimes must be accounted for.

1.3.1 Material and Geometric Coordinates

Each composite lamina typically possesses some degree of material symmetry; that is, principal material coordinate axes can be defined. These lamina are then oriented within a multiple-lamina composite (the laminate) at specified angles with respect to some general geometric coordinate system.

For example, in designing or analyzing the stresses in an automobile, it may be logical to define the x-axis as the forward direction of the vehicle, the y-axis as the lateral direction, and the z-axis as the vertical direction, maintaining a right-handed coordinate system. This coordinate system is termed *geometric* (or global) because its directions correspond to the geometry of the body to which it is attached.

Because the stiffness and strength properties of a lamina (ply) of composite material are typically not isotropic (e.g., the material is typically orthotropic), it is convenient to define these material properties in terms of directions coinciding with any material symmetries that exist. The corresponding coordinate system is termed a material (or ply) coordinate system.

For analysis purposes, it is necessary to express the properties of all plies of the laminate in terms of a common (global) coordinate system, the logical choice being the geometric coordinate system. Thus, it is necessary to transform the material properties of each ply from its own material coordinate system to the global (geometric) coordinate system. These transformation relations (similar to the Mohr's circle transformations for stress and strain) must therefore be developed and are presented in Chapter 2.

1.3.2 Stress–Strain Relations for Anisotropic Materials

The number of independent material constants relating stresses to strains, or strains to stresses, is dependent on the extent of material symmetry that exists (see Jones 1999). If the components of stress are expressed in terms of components of strain, these constants are called stiffnesses. If the components of strain are expressed in terms of components of stress, these constants are called compliances. Defying simple logic, the symbol C is customarily used to represent stiffness, and the symbol S is used to represent compliance. Literal translations from the non-English language of the original developers account for this confusion. These notations are now seemingly too ingrained to reverse, despite the novice's desire to do so.

In the most general case of a fully anisotropic material (i.e., no material symmetries exist), a total of 21 material constants must be experimentally determined. As material symmetry is introduced, it can be shown that certain of the stiffness terms (and the corresponding compliance terms) become zero, thus reducing the number of independent material constants. Some examples of practical interest are indicated in Table 1.1.

The last entry in Table 1.1, that of isotropic material behavior, is a familiar one. In this case, the material properties are the same in all directions; that is, an infinite number of planes of symmetry exist, and only two stiffness constants are required to fully define the stress–strain response of the material. Engineers commonly utilize E, ν, and G, termed the Young's modulus, Poisson's ratio, and shear modulus, respectively. However, the elastic constants must mutually satisfy the isotropic relation (Beer and Johnston 1992):

$$G = \frac{E}{2(1+\nu)} \tag{1.1}$$

Thus, only two of the three quantities can be independently prescribed.

TABLE 1.1

Number of Independent Material Constants as a Function of Material Symmetry

Type of Symmetry	Number of Independent Material Constants
None (triclinic material)	21
One plane of symmetry (monoclinic)	13
Three planes of symmetry (orthotropic)	9
Transversely isotropic (one plane of isotropy)	5
Infinite planes of symmetry (isotropic)	2

Source: Jones, R.M., 1999. *Mechanics of Composite Materials*, 2nd ed., Taylor & Francis, Philadelphia.

In Chapter 2, an orthotropic material is chosen because it is the material symmetry of major interest. For example, it is representative of a unidirectional composite lamina, as well as most other composite material forms.

1.4 Typical Unidirectional Composite Properties

A very large number of different fiber–matrix combinations have been developed over the years. Nevertheless, the general classes of polymer–matrix composites can be characterized by a few representative materials. In the examples presented in Table 1.2, all properties are normalized to a common fiber volume of 60%.

The columns are ordered from left to right in terms of increasing composite axial stiffness, as primarily dictated by the fiber type. Spectra® is a polyethylene fiber developed by Honeywell (Petersburg, VA). Its relative inability to bond with polymer matrices accounts for the low transverse normal, longitudinal shear, and axial compressive strengths indicated for the unidirectional composite. Note that its highly oriented polymer structure also results in extreme values of coefficients of thermal expansion. Another polymeric fiber is Kevlar 49, an aramid fiber produced by E.I. du Pont de Nemours and Company. While the compressive and transverse properties of this composite are generally better than those of the Spectra polyethylene fiber composite, they are still low relative to most of the other composites. AS4, IM6, and GY70 are low-, medium-, and high-modulus carbon fibers. The first two were produced by Hercules Corporation, and the third was produced by the Celanese Corporation. In all cases, the epoxy matrix indicated is Hercules 3501-6 or a similar polymer. This is an epoxy with a high structural performance, but

TABLE 1.2

Typical Properties of Various Types of Polymer Matrix Unidirectional Composites (Nominal 60% Fiber Volume)

Composite Property	Spectra/ Epoxy	E-Glass/ Epoxy	S2-Glass/ Epoxy	Kevlar 49/ Epoxy	AS4/ PEEK	AS4/ Epoxy	IM6/ Epoxy	Boron/ Epoxy	GY70/ Epoxy
E_1 (GPa)	31	43	52	76	134	138	172	240	325
E_2 (GPa)	3.4	9.7	11.7	5.5	10.1	10.3	10.0	18.6	6.2
G_{12} (GPa)	1.4	6.2	7.6	2.1	5.9	6.9	6.2	6.6	5.2
ν_{12}	0.32	0.26	0.28	0.34	0.28	0.30	0.29	0.23	0.26
X_1^T (MPa)	1100	1070	1590	1380	2140	2275	2760	1590	760
X_2^T (MPa)	8	38	41	30	80	52	50	60	26
X_1^C (MPa)	83	870	1050	275	1105	1590	1540	2930	705
X_2^C (MPa)	48	185	234	138	200	207	152	200	70
S_6 (MPa)	24	72	90	43	120	131	124	108	27
$\alpha_1 (10^{-6}/°C)$	−11	6.4	6.2	−2.0	−0.1	−0.1	−0.4	4.5	−0.5
$\alpha_2 (10^{-6}/°C)$	120	16	16	57	29	18	18	20	18
$\beta_1 (10^{-4}/\%M)$	1.0	1.3	1.1	1.9	0.5	0.4	0.3	0.2	0.2
$\beta_2 (10^{-3}/\%M)$	3.2	3.0	3.0	3.5	3.2	3.1	3.1	3.2	3.2
ρ (g/cm³)	1.13	2.00	2.00	1.38	1.57	1.55	1.60	2.02	1.59

brittle, resulting in strong but not highly impact resistant composites. The PEEK matrix is polyetheretherketone, a high-temperature thermoplastic. It is included here along with the brittle epoxy matrix to permit, in particular, a direct comparison with the AS4/epoxy composite system.

The rows in Table 1.2 indicate unidirectional composite in-plane material properties (i.e., stiffnesses, strengths, and hygrothermal properties). As discussed in greater detail in Chapter 2, the subscript 1 indicates the axial direction (the fiber direction), and subscript 2 the in-plane transverse direction. Standard symbols are used and are defined as follows:

E_1 Axial stiffness
E_2 Transverse stiffness
G_{12} In-plane shear stiffness
ν_{12} Major Poisson's ratio
X_1^T Axial tensile strength
X_2^T Transverse tensile strength
X_1^C Axial compressive strength
X_2^C Transverse compressive strength
S_6 Inplane shear strength
α_1 Axial coefficient of thermal expansion
α_2 Transverse coefficient of thermal expansion
β_1 Axial coefficient of moisture expansion
β_2 Transverse coefficient of moisture expansion
ρ Density

It is immediately obvious from Table 1.2 that the various unidirectional composite materials are highly orthotropic in terms of all of their mechanical and physical properties. The axial stiffness varies by an order of magnitude. Many composites have negative axial coefficients of thermal expansion, some much more negative than others. Yet, all of the materials have positive transverse coefficients of thermal expansion, although some are almost an order of magnitude higher than others. Many similar observations can be made by studying this table.

Overall, the use of composite materials offers the designer tremendous design flexibility and potential. However, because the strengths are also highly orthotropic, being very low in transverse tension and compression, and in shear, special care must be taken to design properly with them.

2

Analysis of Composite Materials

2.1 Constitutive Relations

Laminated composites are typically constructed from orthotropic plies (laminae) containing unidirectional fibers or woven fabric. Generally, in a macroscopic sense, the lamina is assumed to behave as a homogeneous orthotropic material. The constitutive relation for a linear elastic orthotropic material in the material coordinate system (Figure 2.1) is (Daniel and Ishai 2006, Gibson 2007, Herakovich 1998, Hyer 2009, Jones 1999, Reddy 1997)

$$\begin{bmatrix} \varepsilon_1 \\ \varepsilon_2 \\ \varepsilon_3 \\ \gamma_{23} \\ \gamma_{13} \\ \gamma_{12} \end{bmatrix} = \begin{bmatrix} S_{11} & S_{12} & S_{13} & 0 & 0 & 0 \\ S_{12} & S_{22} & S_{23} & 0 & 0 & 0 \\ S_{13} & S_{23} & S_{33} & 0 & 0 & 0 \\ 0 & 0 & 0 & S_{44} & 0 & 0 \\ 0 & 0 & 0 & 0 & S_{55} & 0 \\ 0 & 0 & 0 & 0 & 0 & S_{66} \end{bmatrix} \begin{bmatrix} \sigma_1 \\ \sigma_2 \\ \sigma_3 \\ \tau_{23} \\ \tau_{13} \\ \tau_{12} \end{bmatrix} \quad (2.1)$$

where the stress components σ_i, τ_{ij} are defined in Figure 2.1, and the S_{ij} are elements of the compliance matrix. The engineering strain components ε_i, γ_{ij} are defined as implied in Figure 2.2.

In a thin lamina, a state of plane stress is commonly assumed by setting

$$\sigma_3 = \tau_{23} = \tau_{13} = 0 \quad (2.2)$$

For Equation (2.1), this assumption leads to

$$\varepsilon_3 = S_{13}\sigma_1 + S_{23}\sigma_2 \quad (2.3a)$$

$$\gamma_{23} = \gamma_{13} = 0 \quad (2.3b)$$

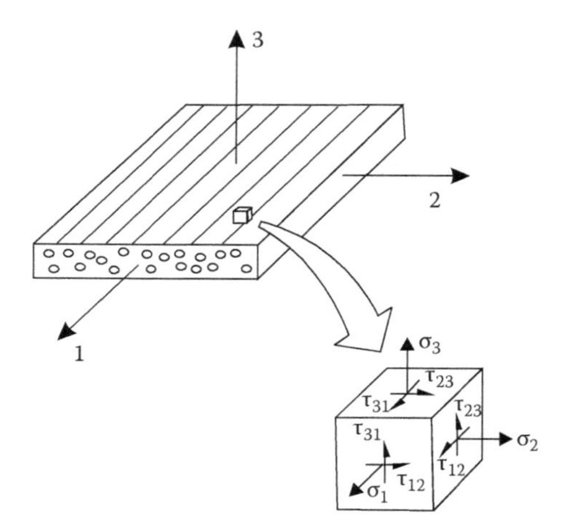

FIGURE 2.1
Definitions of principal material directions for an orthotropic lamina and stress components.

Thus, for plane stress the through-the-thickness strain ε_3 is not an independent quantity and does not need to be included in the constitutive relationship. Equation (2.1) becomes

$$\begin{bmatrix} \varepsilon_1 \\ \varepsilon_2 \\ \gamma_{12} \end{bmatrix} = \begin{bmatrix} S_{11} & S_{12} & 0 \\ S_{12} & S_{22} & 0 \\ 0 & 0 & S_{66} \end{bmatrix} \begin{bmatrix} \sigma_1 \\ \sigma_2 \\ \tau_{12} \end{bmatrix} \tag{2.4}$$

The compliance elements S_{ij} may be related to the engineering constants (E_1, E_2, G_{12}, ν_{12}, ν_{21}),

$$S_{11} = 1/E_1 \quad S_{12} = -\nu_{12}/E_1 = -\nu_{21}/E_2 \tag{2.5a}$$

$$S_{22} = 1/E_2, \quad S_{66} = 1/G_{12} \tag{2.5b}$$

The engineering constants are average properties of the composite ply. The quantities E_1 and ν_{12} are the elastic stiffness and Poisson's ratio, respectively, corresponding to stress σ_1 (Figure 2.2a):

$$\varepsilon_1 = \sigma_1/E_1 \tag{2.6a}$$

$$\nu_{12} = -\varepsilon_2/\varepsilon_1 \tag{2.6b}$$

E_2 and ν_{21} correspond to stress σ_2 (Figure 2.2b):

$$E_2 = \sigma_2/\varepsilon_2 \tag{2.7a}$$

$$\nu_{21} = -\varepsilon_1/\varepsilon_2 \tag{2.7b}$$

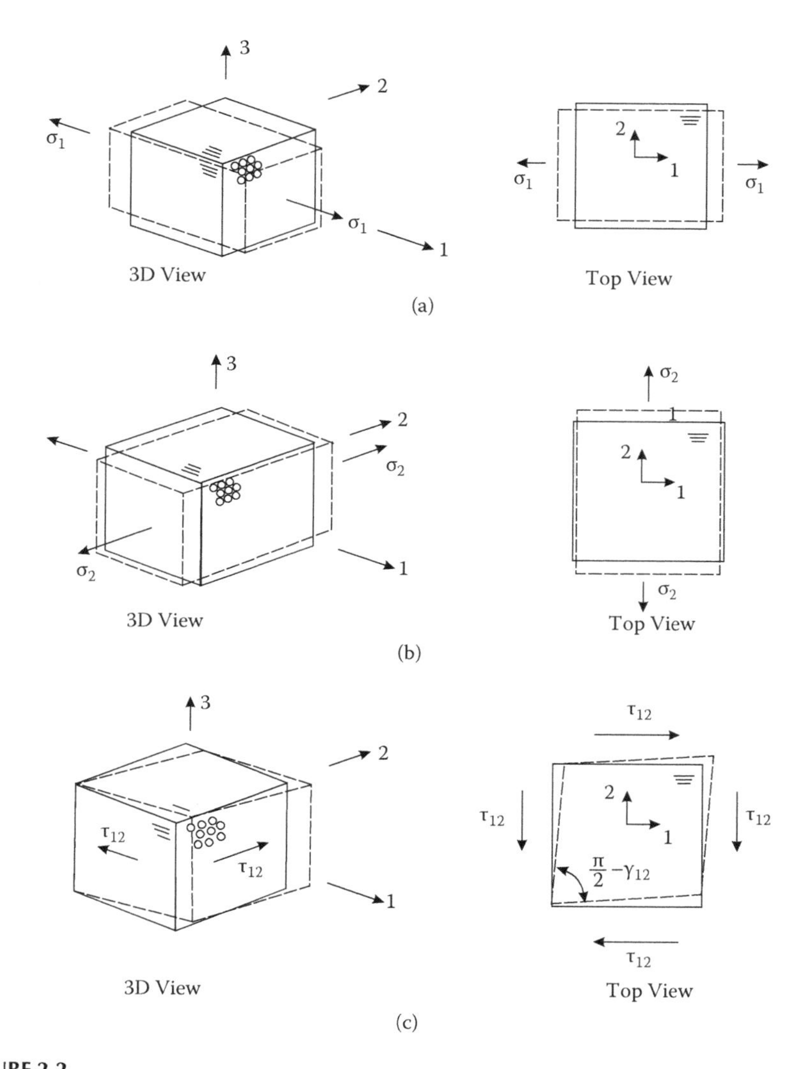

FIGURE 2.2
Illustration of deformations of an orthotropic material due to (a) stress σ_1, (b) stress σ_2, and (c) stress τ_{12}.

For a unidirectional composite, E_2 is much less than E_1, and ν_{21} is much less than ν_{12}. For a balanced fabric composite, $E_1 \approx E_2$ and $\nu_{12} \approx \nu_{21}$. The Poisson ratios ν_{12} and ν_{21} are not independent [see Equation (2.5a)]:

$$\nu_{21} = \nu_{12} E_2 / E_1 \tag{2.8}$$

The in-plane shear stiffness G_{12} is defined as (Figure 2.2c)

$$G_{12} = \tau_{12} / \gamma_{12} \tag{2.9}$$

It is often convenient to express stresses as functions of strains. This is accomplished by inversion of Equation (2.4):

$$\begin{bmatrix} \sigma_1 \\ \sigma_2 \\ \tau_{12} \end{bmatrix} = \begin{bmatrix} Q_{11} & Q_{12} & 0 \\ Q_{12} & Q_{22} & 0 \\ 0 & 0 & Q_{66} \end{bmatrix} \begin{bmatrix} \varepsilon_1 \\ \varepsilon_2 \\ \gamma_{12} \end{bmatrix} \tag{2.10}$$

where the reduced stiffnesses Q_{ij} can be expressed in terms of the engineering constants:

$$Q_{11} = E_1/(1-\nu_{12}\nu_{21}) \tag{2.11a}$$

$$Q_{12} = \nu_{12}E_1/(1-\nu_{12}\nu_{21}) = \nu_{21}E_1/(1-\nu_{12}\nu_{21}) \tag{2.11b}$$

$$Q_{22} = E_2/(1-\nu_{12}\nu_{21}) \tag{2.11c}$$

$$Q_{66} = G_{12} \tag{2.11d}$$

2.1.1 Transformation of Stresses and Strains

For a lamina whose principal material axes (1,2) are oriented at an angle θ with respect to the x,y coordinate system (Figure 2.3), the stresses and strains can be transformed. It may be shown (see, e.g., Hyer 2009) that both the stresses and strains transform according to

$$\begin{bmatrix} \sigma_1 \\ \sigma_2 \\ \tau_{12} \end{bmatrix} = [T] \begin{bmatrix} \sigma_x \\ \sigma_y \\ \tau_{xy} \end{bmatrix} \tag{2.12}$$

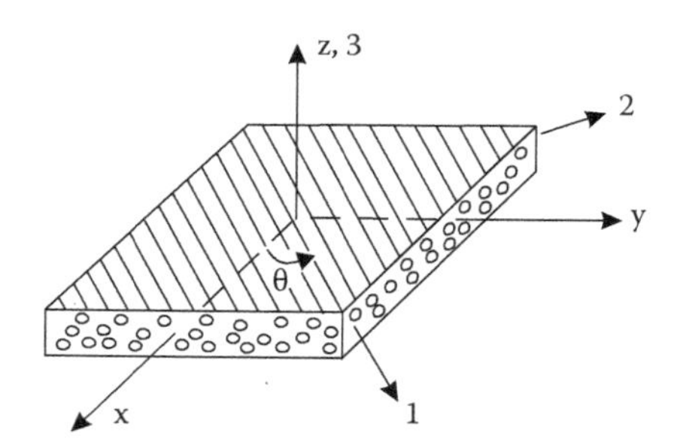

FIGURE 2.3
Positive (counterclockwise) rotation of principal material axes (1,2) from arbitrary x,y-axes.

and

$$\begin{bmatrix} \varepsilon_1 \\ \varepsilon_2 \\ \gamma_{12}/2 \end{bmatrix} = [T] \begin{bmatrix} \varepsilon_x \\ \varepsilon_y \\ \gamma_{xy}/2 \end{bmatrix} \tag{2.13}$$

where the transformation matrix is (Hyer 2009)

$$[T] = \begin{bmatrix} m^2 & n^2 & 2mn \\ n^2 & m^2 & -2mn \\ -mn & mn & m^2 - n^2 \end{bmatrix} \tag{2.14}$$

and

$$m = \cos\theta,\ n = \sin\theta \tag{2.15a,b}$$

From Equations (2.12) and (2.13), it is possible to establish the lamina strain–stress relations in the (x,y) coordinate system:

$$\begin{bmatrix} \varepsilon_x \\ \varepsilon_y \\ \gamma_{xy} \end{bmatrix} = \begin{bmatrix} \bar{S}_{11} & \bar{S}_{12} & \bar{S}_{16} \\ \bar{S}_{12} & \bar{S}_{22} & \bar{S}_{26} \\ \bar{S}_{16} & \bar{S}_{26} & \bar{S}_{66} \end{bmatrix} \begin{bmatrix} \sigma_x \\ \sigma_y \\ \tau_{xy} \end{bmatrix} \tag{2.16}$$

The $\bar{S}_{ij}$ terms are the transformed compliances defined in Appendix A. Similarly, the lamina stress–strain relations become

$$\begin{bmatrix} \sigma_x \\ \sigma_y \\ \tau_{xy} \end{bmatrix} = \begin{bmatrix} \bar{Q}_{11} & \bar{Q}_{12} & \bar{Q}_{16} \\ \bar{Q}_{12} & \bar{Q}_{22} & \bar{Q}_{26} \\ \bar{Q}_{16} & \bar{Q}_{26} & \bar{Q}_{66} \end{bmatrix} \begin{bmatrix} \varepsilon_x \\ \varepsilon_y \\ \gamma_{xy} \end{bmatrix} \tag{2.17}$$

where the overbars denote transformed reduced stiffness elements, defined in Appendix A.

2.1.2 Hygrothermal Strains

If fibrous composite materials are processed at elevated temperatures, thermal strains are introduced during cooling to room temperature, leading to residual stresses and dimensional changes. Figure 2.4 illustrates dimensional changes of a composite subjected to a temperature increase of ΔT from the reference temperature T. Furthermore, polymer matrices are commonly hygroscopic, and moisture absorption leads to swelling of the material. The analysis of moisture expansion strains in composites is mathematically equivalent to that for thermal strains.

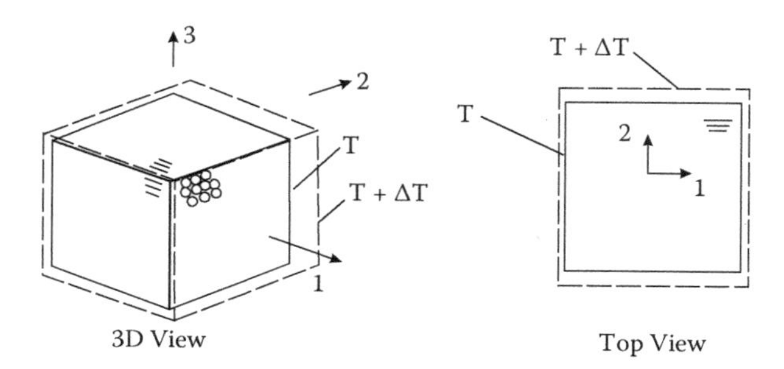

FIGURE 2.4
Deformation of a lamina subject to temperature increase.

The constitutive relationship, when it includes mechanical-, thermal-, and moisture-induced strains, takes the following form (Hyer 2009):

$$\begin{bmatrix} \varepsilon_1 \\ \varepsilon_2 \\ \gamma_{12} \end{bmatrix} = \begin{bmatrix} S_{11} & S_{12} & 0 \\ S_{12} & S_{22} & 0 \\ 0 & 0 & S_{66} \end{bmatrix} \begin{bmatrix} \sigma_1 \\ \sigma_2 \\ \tau_{12} \end{bmatrix} + \begin{bmatrix} \varepsilon_1^T \\ \varepsilon_2^T \\ 0 \end{bmatrix} + \begin{bmatrix} \varepsilon_1^M \\ \varepsilon_2^M \\ 0 \end{bmatrix} \tag{2.18}$$

where superscripts T and M denote temperature- and moisture-induced strains, respectively. Note that shear strains are not induced in the principal material system by a temperature or moisture content change (Figure 2.4). Equation (2.18) is based on the superposition of mechanical-, thermal-, and moisture-induced strains. Inversion of Equation (2.18) gives

$$\begin{bmatrix} \sigma_1 \\ \sigma_2 \\ \tau_{12} \end{bmatrix} = \begin{bmatrix} Q_{11} & Q_{12} & 0 \\ Q_{11} & Q_{12} & 0 \\ 0 & 0 & Q_{66} \end{bmatrix} \begin{bmatrix} \varepsilon_1 & -\varepsilon_1^T & -\varepsilon_1^M \\ \varepsilon_2 & -\varepsilon_2^T & -\varepsilon_2^M \\ & \gamma_{12} & \end{bmatrix} \tag{2.19}$$

Consequently, the stress-generating strains are obtained by subtraction of the thermal- and moisture-induced strains from the total strains. The thermal- and moisture-induced strains are often approximated as linear functions of the changes in temperature and moisture concentration,

$$\begin{bmatrix} \varepsilon_1^T \\ \varepsilon_2^T \end{bmatrix} = \Delta T \begin{bmatrix} \alpha_1 \\ \alpha_2 \end{bmatrix} \tag{2.20}$$

$$\begin{bmatrix} \varepsilon_1^M \\ \varepsilon_2^M \end{bmatrix} = \Delta M \begin{bmatrix} \beta_1 \\ \beta_2 \end{bmatrix} \tag{2.21}$$

where ΔT and ΔM are the temperature change and moisture concentration change, respectively, from the reference state.

The transformed thermal expansion coefficients α_x, α_y, α_{xy} are obtained from those in the principal system using Equation (2.13). Note, however, that in the principal material coordinate system, there is no shear deformation induced, that is, $\alpha_{16} = \beta_{16} = 0$,

$$\alpha_x = m^2\alpha_1 + n^2\alpha_2 \tag{2.22a}$$

$$\alpha_y = n^2\alpha_1 + m^2\alpha_2 \tag{2.22b}$$

$$\alpha_{xy} = 2mn(\alpha_1 - \alpha_2) \tag{2.22c}$$

The moisture expansion coefficients β_x, β_y, β_{xy} are obtained by replacing α with β in Equations (2.22).

The transformed constitutive relations for a lamina, when incorporating thermal- and moisture-induced strains, are

$$\begin{bmatrix} \varepsilon_x \\ \varepsilon_y \\ \gamma_{xy} \end{bmatrix} = \begin{bmatrix} \bar{S}_{11} & \bar{S}_{12} & \bar{S}_{16} \\ \bar{S}_{12} & \bar{S}_{22} & \bar{S}_{26} \\ \bar{S}_{16} & \bar{S}_{26} & \bar{S}_{66} \end{bmatrix} \begin{bmatrix} \sigma_x \\ \sigma_y \\ \tau_{xy} \end{bmatrix} + \begin{bmatrix} \varepsilon_x^T \\ \varepsilon_y^T \\ \gamma_{xy}^T \end{bmatrix} + \begin{bmatrix} \varepsilon_x^M \\ \varepsilon_y^M \\ \gamma_{xy}^M \end{bmatrix} \tag{2.23}$$

$$\begin{bmatrix} \sigma_x \\ \sigma_y \\ \tau_{xy} \end{bmatrix} = \begin{bmatrix} \bar{Q}_{11} & \bar{Q}_{12} & \bar{Q}_{16} \\ \bar{Q}_{12} & \bar{Q}_{22} & \bar{Q}_{26} \\ \bar{Q}_{16} & \bar{Q}_{26} & \bar{Q}_{66} \end{bmatrix} \begin{bmatrix} \varepsilon_x - \varepsilon_x^T - \varepsilon_x^M \\ \varepsilon_y - \varepsilon_y^T - \varepsilon_y^M \\ \gamma_{xy} - \gamma_{xy}^T - \gamma_{xy}^M \end{bmatrix} \tag{2.24}$$

2.2 Micromechanics

As schematically illustrated in Figure 2.5, micromechanics aims to describe the moduli and expansion coefficients of the lamina from properties of the fiber and matrix, the microstructure of the composite, and the volume fractions of the constituents. Sometimes, it is also necessary to consider the small transition region between bulk fiber and bulk matrix (i.e., interphase) (Drzal et al. 2000). Much fundamental work has been devoted to the study of the distributions of strain and stress in the constituents and the formulation of appropriate averaging schemes to allow definition of macroscopic engineering constants. Most micromechanics analyses have focused on unidirectional continuous-fiber composites (e.g., Hashin 1983, Christensen 1979,

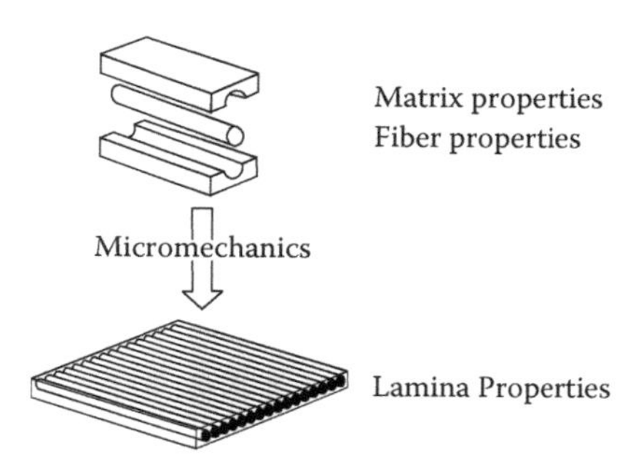

FIGURE 2.5
Role of micromechanics.

Hyer and Waas 2000), although properties of composites with woven fabric reinforcements can also be predicted with reasonable accuracy (Byun and Chou 2000).

The objective of this section is not to review the numerous micromechanics developments. The interested reader can find ample information in the previously referenced review articles. In this section, we limit the presentation to some commonly used simple estimates of the stiffness constants E_1, E_2, ν_{12}, ν_{21}, and G_{12} and thermal expansion coefficients α_1 and α_2 required for describing the small strain response of a unidirectional lamina under mechanical and thermal loads (see Section 2.1). Such estimates may be useful for comparison to experimentally measured quantities.

2.2.1 Stiffness Properties of Unidirectional Composites

Although most matrices are isotropic, many fibers, such as carbon and Kevlar (E.I. du Pont de Nemours and Company, Wilmington, DE) have directional properties because of molecular or crystal plane orientation effects (Daniel and Ishai 2006). As a result, the axial stiffness of such fibers is much greater than the transverse stiffness. The thermal expansion coefficients along and transverse to the fiber axis also are quite different. It is common to assume cylindrical orthotropy for fibers with axisymmetric microstructure. The stiffness constants of the fibers required for plane stress analysis of a composite are E_L, E_T, ν_{LT}, and G_{LT}, where L and T denote the longitudinal and transverse directions of a fiber, respectively. The corresponding thermal expansion coefficients are α_L and α_T, respectively.

The mechanics of materials approach (Hyer 2009) yields

$$E_1 = E_{Lf} V_f + E_m V_m \tag{2.25a}$$

$$E_2 = \frac{E_{Tf} E_m}{E_{Tf} V_m + E_m V_f} \tag{2.25b}$$

$$\nu_{12} = \nu_{LTf} V_f + \nu_m V_m \tag{2.25c}$$

$$G_{12} = \frac{G_{LTf} G_m}{G_{LTf} V_m + G_m V_f} \tag{2.25d}$$

where subscripts f and m represent fiber and matrix, respectively, and the symbol V represents volume fraction. Note that once E_1, E_2, and ν_{12} are calculated from Equation (2.25a), ν_{21} is obtained from Equation (2.8). Equations (2.25a) and (2.25c) provide good estimates of E_1 and ν_{12}. Equations (2.25b) and (2.25d), however, substantially underestimate E_2 and G_{12}. More realistic estimates of E_2 and G_{12} are provided in the work of Hyer (2009) and Rosen and Hashin (1987).

Simple, yet reasonable, estimates of E_2 and G_{12} may also be obtained from the Halpin-Tsai equations (Halpin and Kardos 1976):

$$P = \frac{P_m(1 + \xi\chi V_f)}{1 - \chi V_f} \tag{2.26a}$$

$$\chi = \frac{P_f - P_m}{P_f + \xi P_m} \tag{2.26b}$$

where P is the property of interest (E_2 or G_{12}), and P_f and P_m are the corresponding fiber and matrix properties, respectively. The parameter ξ is called the reinforcement efficiency; $\xi(E_2) = 2$ and $\xi(G_{12}) = 1$, for circular fibers.

2.2.2 Expansion Coefficients

Thermal expansion (and moisture-swelling) coefficients can be defined by considering a composite subjected to a uniform increase in temperature (or moisture content) (Figure 2.4).

The thermal expansion coefficients α_1 and α_2 of a unidirectional composite consisting of transversely isotropic fibers in an isotropic matrix may be determined using the mechanics of materials approach (Hyer 2009):

$$\alpha_1 = \frac{\alpha_{Lf} E_{Lf} V_f + \alpha_m E_m V_m}{E_{Lf} V_f + E_m V_m} \tag{2.27a}$$

$$\alpha_2 = \alpha_{Tf} V_f + \alpha_m V_m \tag{2.27b}$$

Predictions of α_1 using Equation (2.27a) are accurate, whereas Equation (2.27b) underestimates the actual value of α_2. An expression derived by Hyer and Waas (2000) provides a more accurate prediction of α_2:

$$\alpha_2 = \alpha_{Tf} V_f + \alpha_m V_m + \frac{(E_{Lf}\nu_m - E_m \nu_{LTf})}{E_{Lf} V_f + E_m V_m}(\alpha_m - \alpha_{Lf}) V_f V_m \tag{2.28}$$

2.3 Laminated Plate Theory

Structures fabricated from composite materials rarely utilize a single composite lamina because this unit is typically very thin and highly orthotropic. To achieve a thicker cross section and more balanced properties, plies of prepreg or fiber mats are stacked in specified directions. Such a structure is called a laminate (Figure 2.6). Most analyses of laminated structures are limited to flat panels (see, e.g., Daniel and Ishai 2006 and Hyer 2009). Extension to curved laminated shell structures may be found in the work of Herakovich (1998), Hyer (2000), and Vinson and Sierakowsky (2002).

In this section, attention is limited to a flat laminated plate under in-plane and bending loads. The classical theory of such plates is based on the assumption that a line originally straight and perpendicular to the middle surface remains straight and normal to the middle surface, and that the length of the line remains unchanged during deformation of the plate. These assumptions lead to the vanishing of the out-of-plane shear and extensional strains:

$$\gamma_{xz} = \gamma_{yz} = \varepsilon_z = 0 \tag{2.29}$$

where the laminate coordinate system (x, y, z) is indicated in Figure 2.6. Consequently, the laminate strains are reduced to ε_x, ε_y, and γ_{xy}. The assumption that the cross sections undergo only stretching and rotation leads to the following strain distribution (Hyer 2009):

$$\begin{bmatrix} \varepsilon_x \\ \varepsilon_y \\ \gamma_{xy} \end{bmatrix} = \begin{bmatrix} \varepsilon_x^0 \\ \varepsilon_y^0 \\ \gamma_{xy}^0 \end{bmatrix} + z \begin{bmatrix} \kappa_x \\ \kappa_y \\ \kappa_{xy} \end{bmatrix} \tag{2.30}$$

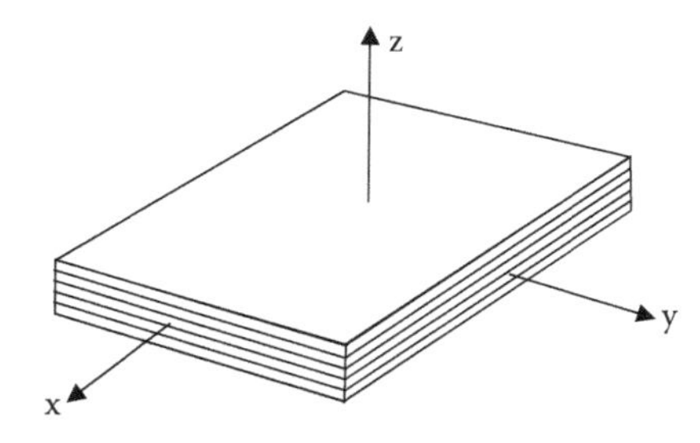

FIGURE 2.6
Laminate coordinate system.

where $[\varepsilon_x^0, \varepsilon_y^0, \gamma_{xy}^0]$ and $[\kappa_x, \kappa_y, \kappa_{xy}]$ are the midplane strains and curvatures, respectively, and z is the distance from the midplane.

Force and moment resultants, $[N_x, N_y, N_{xy}]$ and $[M_x, M_y, M_{xy}]$, respectively, are obtained by integration of the stresses in each layer over the laminate thickness h:

$$\begin{bmatrix} N_x \\ N_y \\ N_{xy} \end{bmatrix} = \int_{-h/2}^{h/2} \begin{bmatrix} \sigma_x \\ \sigma_y \\ \tau_{xy} \end{bmatrix}_k dz \tag{2.31}$$

$$\begin{bmatrix} M_x \\ M_y \\ M_{xy} \end{bmatrix} = \int_{-h/2}^{h/2} \begin{bmatrix} \sigma_x \\ \sigma_y \\ \tau_{xy} \end{bmatrix}_k zdz \tag{2.32}$$

where the subscript k represents the kth ply in the laminate. Combination of Equations (2.24) with Equations (2.30–2.32) leads to the following constitutive relationships among forces and moments and midplane strains and curvatures:

$$\begin{bmatrix} N_x + N_x^T + N_x^M \\ N_y + N_y^T + N_y^M \\ N_{xy} + N_{xy}^T + N_{xy}^M \end{bmatrix} = \begin{bmatrix} A_{11} & A_{12} & A_{16} \\ A_{12} & A_{22} & A_{26} \\ A_{16} & A_{26} & A_{66} \end{bmatrix} \begin{bmatrix} \varepsilon_x^0 \\ \varepsilon_y^0 \\ \gamma_{xy}^0 \end{bmatrix} + \begin{bmatrix} B_{11} & B_{12} & B_{16} \\ B_{12} & B_{22} & B_{26} \\ B_{16} & B_{26} & B_{66} \end{bmatrix} \begin{bmatrix} \kappa_x \\ \kappa_y \\ \kappa_{xy} \end{bmatrix} \tag{2.33}$$

$$\begin{bmatrix} M_x + M_x^T + M_x^M \\ M_y + M_y^T + M_y^M \\ M_{xy} + M_{xy}^T + M_{xy}^M \end{bmatrix} = \begin{bmatrix} B_{11} & B_{12} & B_{16} \\ B_{12} & B_{22} & B_{26} \\ B_{16} & B_{26} & B_{66} \end{bmatrix} \begin{bmatrix} \varepsilon_x^0 \\ \varepsilon_y^0 \\ \gamma_{xy}^0 \end{bmatrix} + \begin{bmatrix} D_{11} & D_{12} & D_{16} \\ D_{12} & D_{22} & D_{26} \\ D_{16} & D_{26} & D_{66} \end{bmatrix} \begin{bmatrix} \kappa_x \\ \kappa_y \\ \kappa_{xy} \end{bmatrix} \tag{2.34}$$

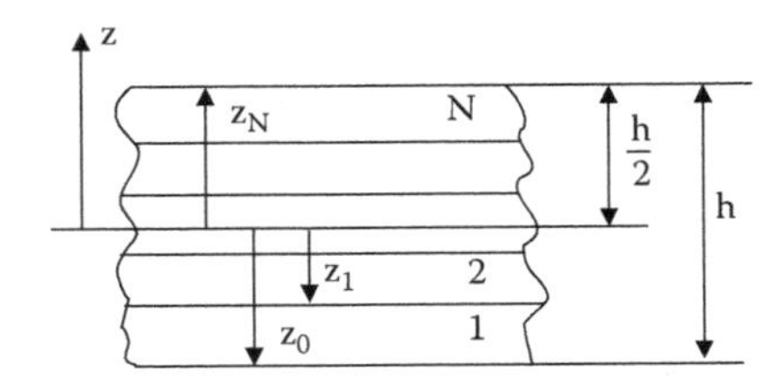

FIGURE 2.7
Definition of ply coordinates z_k.

where the A_{ij}, B_{ij}, and D_{ij} are called extensional stiffnesses, coupling stiffnesses, and bending stiffnesses, (Hyer 2009), given by

$$A_{ij} = \sum_{k=1}^{N} (\bar{Q}_{ij})_k (z_k - z_{k-1}) \tag{2.35a}$$

$$B_{ij} = \frac{1}{2} \sum_{k=1}^{N} (\bar{Q}_{ij})_k \left(z_k^2 - z_{k-1}^2 \right) \tag{2.35b}$$

$$D_{ij} = \frac{1}{3} \sum_{k=1}^{N} (\bar{Q}_{ij})_k \left(z_k^3 - z_{k-1}^3 \right) \tag{2.35c}$$

The so-called ply coordinates z_k ($k = 1, 2 \ldots$N), where N is the number of plies in the laminate, indicate the location of the ply boundaries (see Figure 2.7). z_k may be calculated from the following recursion formula:

$$z_0 = -h/2 \qquad k = 0 \tag{2.36a}$$

$$z_k = z_{k-1} + h_k \qquad k = 1, 2, \ldots, N \tag{2.36b}$$

in which h_k is the ply thickness of the kth ply. Normally, all plies are of the same thickness.

For situations involving a temperature change ΔT, it is common to consider only the steady-state condition at which the temperature is uniform throughout the laminate. However, in a transient situation, the transfer of heat by conduction (Ozisik 1980) or moisture diffusion (Jost 1960) has to be considered (see Pipes et al. 1976). For steady state, the thermal force resultants are determined from

$$\begin{bmatrix} N_x^T \\ N_y^T \\ N_{xy}^T \end{bmatrix} = \sum_{k=1}^{N} \begin{bmatrix} \bar{Q}_{11} & \bar{Q}_{12} & \bar{Q}_{16} \\ \bar{Q}_{12} & \bar{Q}_{22} & \bar{Q}_{26} \\ \bar{Q}_{16} & \bar{Q}_{26} & \bar{Q}_{66} \end{bmatrix}_k \begin{bmatrix} \alpha_x \\ \alpha_y \\ \alpha_{xy} \end{bmatrix}_k (z_k - z_{k-1}) \Delta T \tag{2.37}$$

The moisture-induced force resultants $[N_x^M, N_y^M, N_{xy}^M]$ are obtained in the same manner as the thermal force resultants, by replacing $[\alpha_x, \alpha_y, \alpha_{xy}]$ with $[\beta_x, \beta_y, \beta_{xy}]$ and ΔT with ΔM in Equation (2.37).

The thermal moment resultants $[M_x^T, M_y^T, M_{xy}^T]$ are determined from

$$\begin{bmatrix} M_x^T \\ M_y^T \\ M_{xy}^T \end{bmatrix} = \frac{1}{2}\sum_{k=1}^{N} \begin{bmatrix} \bar{Q}_{11} & \bar{Q}_{12} & \bar{Q}_{16} \\ \bar{Q}_{12} & \bar{Q}_{22} & \bar{Q}_{26} \\ \bar{Q}_{16} & \bar{Q}_{26} & \bar{Q}_{66} \end{bmatrix}_k \begin{bmatrix} \alpha_x \\ \alpha_y \\ \alpha_{xy} \end{bmatrix}_k \left(z_k^2 - z_{k-1}^2\right)\Delta T \quad (2.38)$$

The moisture-induced moment resultants $[M_x^M, M_y^M, M_{xy}^M]$ are obtained by replacing the α values with β values and ΔT with ΔM in Equation (2.38).

For laminates with the plies consisting of different materials, the moisture concentration may vary through the thickness in a stepwise manner. At steady state, this is simply incorporated into the analysis by letting $\Delta M = (\Delta M)_k$ (Carlsson 1981).

Equations (2.33) and (2.34) may conveniently be written as

$$\begin{bmatrix} N \\ M \end{bmatrix} = \begin{bmatrix} A & B \\ B & D \end{bmatrix} \begin{bmatrix} \varepsilon^0 \\ \kappa \end{bmatrix} \quad (2.39)$$

where $[N]$ and $[M]$ represent the left-hand side of Equations (2.33) and (2.34), that is, the sum of mechanical and hygrothermal forces and moments, respectively. Equation (2.39) represents the stiffness form of the laminate constitutive equations.

Sometimes, it is more convenient to express the midplane strains and curvatures as a function of the forces and moments. This represents the compliance form of the laminate constitutive equations, which is obtained by inversion of Equation (2.39):

$$\begin{bmatrix} \varepsilon^0 \\ \kappa \end{bmatrix} = \begin{bmatrix} a & b \\ c & d \end{bmatrix} \begin{bmatrix} N \\ M \end{bmatrix} \quad (2.40)$$

Expressions for the compliance matrices $[a]$, $[b]$, $[c]$, and $[d]$ are given in Appendix A.

2.4 St. Venant's Principle and End Effects in Composites

In the testing and evaluation of material properties, it is generally assumed that load introduction effects are confined to a region close to the grips or loading points, and a uniform state of stress and strain exists within the test section. The justification for such a simplification is usually based on the St. Venant principle, which states that the difference between the stresses caused by statically equivalent load systems is insignificant at distances greater than the largest dimension of the area over which the loads are acting (Timoshenko and Goodier 1970). This estimate, however, is based on isotropic material properties. For orthotropic composite materials, Horgan (1972, 1982), Choi and Horgan (1977), and Horgan and Carlsson (2000) showed that the application of St. Venant's principle for plane elasticity problems involving orthotropic materials is not justified in general. For the particular problem of a rectangular strip made of highly orthotropic material and loaded at the ends, Choi and Horgan (1977) demonstrated that the stress approached the uniform St. Venant solution much more slowly than the corresponding solution for an isotropic material.

The size of the region where end effects influence the stresses in a rectangular strip loaded with tractions at the ends may be estimated by

$$\lambda \approx \frac{b}{2\pi}(E_1/G_{12})^{1/2} \tag{2.41}$$

where b is the maximum dimension of the cross section, and E_1 and G_{12} are the longitudinal elastic and shear stiffnesses, respectively.

In this equation, λ is defined as the distance over which the self-equilibrated stress decays to $1/e$ of its value at the end $(e = 2.718)$. When the ratio E_1/G_{12} is large, the decay length is large, and end effects are transferred a considerable distance along the gage section. Testing of highly orthotropic materials thus requires special consideration of load introduction effects. Arridge et al. (1976), for example, found that a very long specimen with an aspect ratio ranging from 80 to 100 was needed to avoid the influence of clamping effects in tension testing of highly orthotropic, drawn polyethylene film. Several other cases are reviewed by Horgan and Carlsson (2000).

2.5 Lamina Strength Analysis

When any material is considered for a structure, an important task for the structural engineer is to assess the load-carrying ability of the particular material/structure combination. Prediction of the strength of composite materials has been an active area of research since the early work of Tsai (1968). Many failure theories have been suggested, although no universally accepted failure criterion exists (Hinton et al. 2004). As pointed out by Hyer (2009), however, no single criterion could be expected to accurately predict failure of all composite materials under all loading conditions. Popular strength criteria are maximum stress, maximum strain, and Tsai-Wu criteria

TABLE 2.1

Basic Strengths of Orthotropic Plies for Plane Stress

Direction/Plane	Active Stress	Strength	Ultimate Strain
1	σ_1	X_1^T, X_1^C	e_1^T, e_1^C
2	σ_2	X_2^T, X_2^C	e_2^T, e_2^C
1,2	τ_{12}	S_6	e_6

Note: All strengths and ultimate strains are defined by their magnitudes.

(Sun 2000). These criteria are phenomenological in the sense that they do not rely on physical modeling of the failure process. The reason for their popularity is that they are based on failure tests on simple specimens in tension, compression, and shear (Chapters 8–10) and are able to predict load levels required to fail more complicated structures under combined stress loading.

In the following presentation, failure of the lamina is first examined, and then failure of the laminate is briefly considered. It is assumed that the lamina, being unidirectional or a woven fabric ply, can be treated as a homogeneous orthotropic ply with known, measured strengths in the principal material directions. Furthermore, the shear strength in the plane of the fibers is independent of the sign of the shear stress. The presentation is limited to plane stress in the plane of the fibers. Table 2.1 lists the five independent failure stresses and strains corresponding to plane stress.

Notice here that superscripts T and C denote tension and compression, respectively, and that strengths and ultimate strains are defined as positive; that is, the symbols indicate their magnitudes. For example, a composite ply loaded in pure negative shear $(\tau_{12} < 0)$ would fail at a shear stress $\tau_{12} = -S_6$ and shear strain $\gamma_{12} = -e_6$.

2.5.1 Maximum Stress Failure Criterion

The maximum stress failure criterion assumes that failure occurs when any one of the in-plane stresses σ_1, σ_2, or τ_{12} attains its limiting value independent of the other components of stress. If the magnitudes of the stress components are less than their values at failure, failure does not occur, and the element or structure is considered safe. For determining the failure load, any of the following equalities must be satisfied at the point when failure occurs:

$$\sigma_1 = X_1^T \tag{2.42a}$$

$$\sigma_1 = -X_1^C \tag{2.42b}$$

$$\sigma_2 = X_2^T \tag{2.42c}$$

$$\sigma_2 = -X_2^C \tag{2.42d}$$

$$\tau_{12} = S_6 \tag{2.42e}$$

$$\tau_{12} = -S_6 \tag{2.42f}$$

For unidirectional and fabric composites, Equations (2.42a) and (2.42b) indicate failure of fibers at quite high magnitudes of stress, whereas Equations (2.42c)–(2.42f) indicate matrix or fiber–matrix interface-dominated failures at much lower magnitudes of stress for unidirectional composites. For fabric composites, however, Equations (2.42c) and (2.42d) indicate failure of the fibers oriented along the 2-direction.

2.5.2 Maximum Strain Failure Criterion

The maximum strain criterion assumes that failure of any principal plane of the lamina occurs when any in-plane strain reaches its ultimate value in uniaxial tension, compression, or pure shear.

$$\varepsilon_1 = e_1^T \tag{2.43a}$$

$$\varepsilon_1 = -e_1^C \tag{2.43b}$$

$$\varepsilon_2 = e_2^T \tag{2.43c}$$

$$\varepsilon_2 = -e_2^C \tag{2.43d}$$

$$\gamma_{12} = e_6 \tag{2.43e}$$

$$\gamma_{12} = -e_6 \tag{2.43f}$$

In these expressions, the symbol e represents the magnitude of the ultimate strain. If any of these conditions becomes satisfied, failure is assumed to be triggered by the same mechanism as uniaxial loading or pure shear loading. Similar to the maximum stress criterion, the maximum strain criterion has the ability of predicting the failure mode.

2.5.3 Tsai-Wu Failure Criterion

Tsai and Wu (1971) proposed a second-order tensor polynomial failure criterion for prediction of biaxial strength, which takes the following form for plane stress:

$$F_1\sigma_1 + F_2\sigma_2 + F_{11}\sigma_1^2 + F_{22}\sigma_2^2 + F_{66}\tau_{12}^2 + 2F_{12}\sigma_1\sigma_2 = 1 \tag{2.44}$$

Failure under combined stress is assumed to occur when the left-hand side of Equation (2.44) is equal to or greater than one. All of the parameters of the Tsai-Wu criterion, except F_{12}, can be expressed in terms of the basic strengths (Table 1.1),

$$
\begin{aligned}
F_1 &= 1/X_1^T - 1/X_1^C \qquad & F_{11} &= 1/\left(X_1^T X_1^C\right) \\
F_2 &= 1/X_2^T - 1/X_2^C \qquad & F_{22} &= 1/\left(X_2^T X_2^C\right) \\
& & F_{66} &= 1/S_6^2
\end{aligned}
\tag{2.45}
$$

F_{12} is a strength interaction parameter that has to be determined from a biaxial experiment involving stresses σ_1 and σ_2. Such experiments are, unfortunately, expensive and difficult to properly conduct. As an alternative, Tsai and Hahn (1980) suggested that F_{12} be estimated from the following relationship:

$$F_{12} = -\frac{1}{2}\sqrt{F_{11}F_{22}} \tag{2.46}$$

The Tsai-Wu criterion has found widespread applicability in the composite industry because of its versatility and that it provides reasonable predictions of strength. It does not, however, predict the mode of failure. Furthermore, it must be pointed out that the World-Wide Failure Exercise found that none of the current failure models provides accurate predictions (Hinton et al. 2004).

2.6 Laminate Strength Analysis

Analysis of failure and strength of laminated composites is quite different from the analysis of strength of a single ply. Failure of laminates often involves delamination (i.e., separation of the plies), which is discussed in Chapters 14 and 17. This failure mode is influenced by the three-dimensional state of stress that develops near free edges in laminated specimens (Pipes et al. 1973). Furthermore, multidirectional composite laminates are commonly processed at elevated temperatures, and the mismatch in thermal expansion between the plies leads to residual stresses in the plies on cooling (Hahn and Pagano 1975, Weitsman 1979, Jeronimidis and Parkyn 1988). Exposure of the laminate to moisture will also influence the state of residual stress in the laminate (Springer 1981).

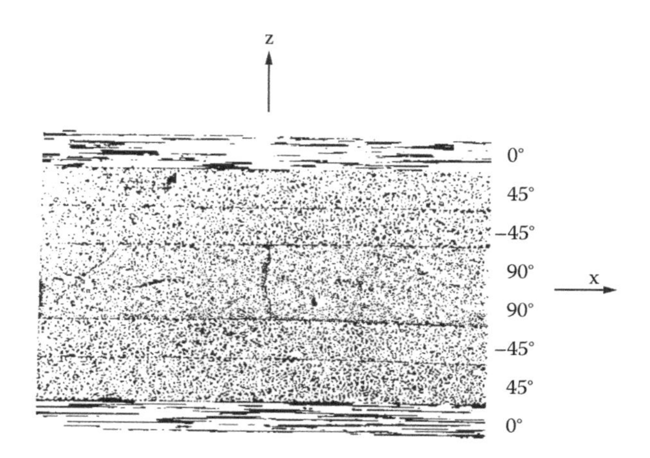

FIGURE 2.8
Matrix crack in a unidirectional ply in a laminate (first-ply failure).

A common failure mode in laminates containing unidirectional plies is matrix cracking, which is failure of the matrix and fiber–matrix interface in a plane perpendicular to the fiber direction (Figure 2.8). Such a failure is called first-ply failure and occurs because of the presence of a weak plane transverse to the fiber axis in such composites. In fabric composites, no such weak planes exist perpendicular to the xy plane, and failure initiates locally in fiber tows and matrix pockets before ultimate failure occurs (Alif and Carlsson 1997). At any instant, local failures tend to arrest by constraint of adjacent layers or tows in the laminate before the occurrence of catastrophic failure of the laminate. Wang and Crossman (1980), Crossman et al. (1980, 1982), and Flaggs and Kural (1982) found a very large constraint effect in composite laminates with unidirectional plies. They examined matrix cracking in a set of laminates containing unidirectional 90° plies grouped together and found that the in situ strength depends strongly on the number of plies of the same orientation grouped together and on the adjacent ply orientations. Consequently, there are a host of mechanisms influencing failure of laminates, and as a result, accurate failure prediction is associated with severe difficulties.

Various methods to predict ply failures and ultimate failure of composite laminates are reviewed by Sun (2000). A common method to analyze laminate failure is to determine the stresses and strains in plies in the laminate using laminated plate theory (Section 2.3) and then examine the loads and strains corresponding to the occurrence of first-ply failure as predicted by the failure criterion selected. The ply failure mode is then identified. Swanson and Trask (1989), and Swanson and Qian (1992), performed biaxial tension–tension and tension–compression testing on several carbon/epoxy laminate cylinders made from unidirectional plies. Ply failures were identified using the strength criteria mentioned in Section 2.5. Final failure of the cylinders was predicted by using a ply property reduction method (ply-by-ply discount method) by which failed plies are identified, and the transverse and shear stiffnesses E_2 and G_{12}, respectively, of the failed plies are assigned numbers very close to zero. The laminate with reduced stiffness is then again analyzed for stresses and strains. Comparison of the predictions with measured ultimate failure data of the cylinders revealed good agreement for all criteria. It was concluded that ultimate failure predictions based on the maximum stress and maximum strain criteria are quite insensitive to variations in the ply transverse failure strengths X_2^T and S_6. This is an advantage because, as discussed, these strengths are difficult to determine in situ.

2.7 Fracture Mechanics Concepts

The influence of defects and cracks on the strength of a material or structure is the subject of fracture mechanics. The object of fracture mechanics analysis is the prediction of the onset of crack growth for a body containing a flaw of a given size. To calculate the critical load for a cracked composite, a common assumption is that the size of the plastic zone at the crack tip is small compared to the crack length. Linear elastic fracture mechanics has been found useful for certain types of cracks in composites (i.e., interlaminar cracks) (Wilkins et al. 1982) or matrix cracks in unidirectional brittle matrix composites (Wang and Crossman 1980).

The equilibrium of an existing crack may be judged from the intensity of elastic stress around the crack tip. Solutions of the elastic stress field in homogeneous isotropic and orthotropic materials show that stress singularities are of the $r^{-1/2}$ type, where r is the distance from the crack tip (see Anderson 2005). Stress intensity factors may be determined for crack problems in which the crack plane is in any of the planes of orthotropic material symmetry (Sih et al. 1965). It is possible to partition the crack tip loading into the three basic modes of crack surface displacement shown in Figure 2.9. Mode I refers to opening of the crack surfaces, mode II refers to sliding, and mode III refers to tearing.

It has, however, become common practice to investigate interlaminar cracks using the strain energy release rate G. This quantity is based on energy considerations and is mathematically well defined and measurable in experiments. The energy approach, which stems from the original Griffith (1920) treatment is based on a thermodynamic criterion for fracture by considering the energy available for crack growth of the system on one hand and the surface energy required to extend an existing crack on the other hand. An elastic potential for a cracked body may be defined as

$$H = W - U \tag{2.47}$$

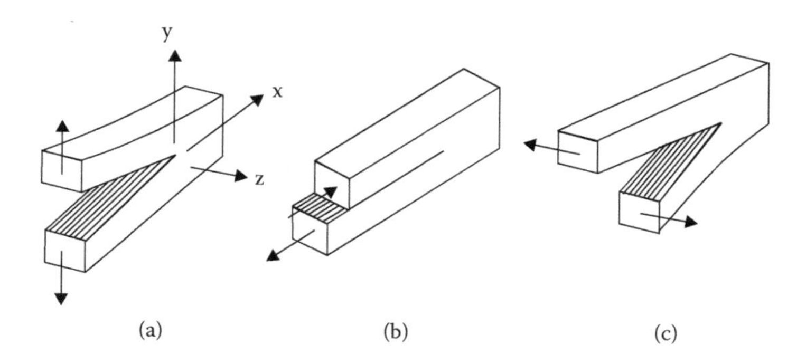

FIGURE 2.9
Modes of crack surface displacements. (a) mode I (opening), (b) mode II (sliding), and (c) mode III (tearing).

where W is the work supplied by the movement of the external forces, and U is the elastic strain energy stored in the body. If G_c is the work required to create a unit crack area, it is possible to formulate a criterion for crack growth:

$$\delta H \geq G_c \, \delta A \tag{2.48}$$

where δA is the increase in crack area.

A critical condition occurs when the net energy supplied just balances the energy required to grow the crack; that is,

$$\delta H = G_c \, \delta A \tag{2.49}$$

Equilibrium becomes unstable when the net energy supplied exceeds the required crack growth energy,

$$\delta H > G_c \, \delta A \tag{2.50}$$

The strain energy release rate G is defined as

$$G = \frac{\partial H}{\partial A} \tag{2.51}$$

In terms of G, the fracture criterion may thus be formulated as

$$G \geq G_c \tag{2.52}$$

This concept is illustrated for a linear elastic body containing a crack of original length a. Figure 2.10 shows the load P versus loading point

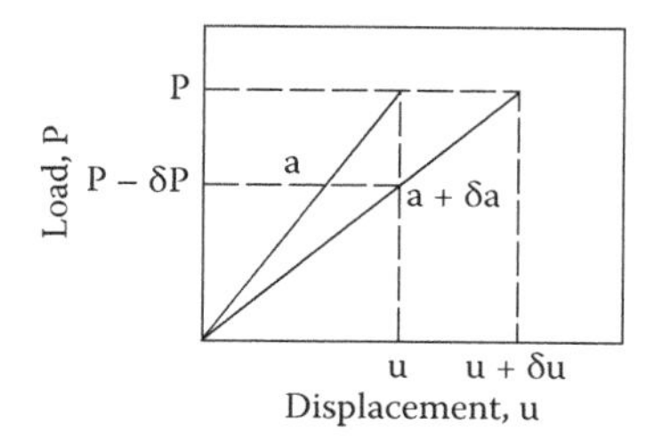

FIGURE 2.10
Load-displacement behavior for a cracked body at crack lengths a and $a + \delta a$.

displacement u for the cracked body where crack growth is assumed to occur either at constant load (fixed load) or at constant displacement (fixed grip).

For the *fixed-load* case,

$$\delta U = \frac{P\delta u}{2} \tag{2.53a}$$

$$\delta W = P\delta u \tag{2.53b}$$

Equation (2.47) gives

$$\delta H = P\delta u - P\delta u/2 = P\delta u/2 \tag{2.54}$$

and Equation (2.51) gives

$$G = \frac{P}{2}\frac{\partial u}{\partial A} \tag{2.55}$$

For the *fixed-grip* case, the work term in Equation (2.47) vanishes, and

$$\delta U = \frac{u\delta P}{2} \tag{2.56}$$

Note that δP is negative because of the loss in stiffness followed by crack extension, and G is

$$G = -\frac{u}{2}\frac{\partial P}{\partial A} \tag{2.57}$$

For a linear elastic body, the relationship between load and displacement may be expressed as

$$u = CP \tag{2.58}$$

where C is the compliance of the specimen. Substitution into Equation (2.55) (fixed load) gives

$$G = \frac{P^2}{2}\frac{\partial C}{\partial A} \tag{2.59}$$

For the fixed-grip case, substitution of $P = u/C$ into Equation (2.57) gives

$$G = \frac{u^2}{2C^2}\frac{\partial C}{\partial A} = \frac{P^2}{2}\frac{\partial C}{\partial A} \tag{2.60}$$

Consequently, both fixed-load and fixed-grip conditions give the same expression. This expression is convenient for the experimental determination of G and is employed in Chapter 17 for derivation of expressions for G for various delamination fracture specimens.

For a crack in a principal material plane, it is possible to decompose G into components associated with the three basic modes of crack extension illustrated in Figure 2.9:

$$G = G_I + G_{II} + G_{III} \tag{2.61}$$

Theoretically, the mode separation is based on Irwin's (1958) contention that if the crack extends by a small amount Δa, the energy absorbed in the process is equal to the work required to close the crack to its original length. For a polar coordinate system with the origin at the crack tip extended an amount Δa (Figure 2.9), the components of the energy release rate are

$$G_I = \lim_{\Delta a \to 0} \frac{1}{2\Delta a} \int_0^{\Delta a} \sigma_y(\Delta a - r)\bar{v}(r,\pi)\,dr \tag{2.62a}$$

$$G_{II} = \lim_{\Delta a \to 0} \frac{1}{2\Delta a} \int_0^{\Delta a} \tau_{xy}(\Delta a - r)\bar{u}(r,\pi)\,dr \tag{2.62b}$$

$$G_{III} = \lim_{\Delta a \to 0} \frac{1}{2\Delta a} \int_0^{\Delta a} \tau_{yz}(\Delta a - r)\bar{w}(r,\pi)\,dr \tag{2.62c}$$

where r is the radial distance from the crack tip; σ_y, τ_{xy}, and τ_{yz} are the normal and shear stresses near the crack tip; and $\bar{v}$, $\bar{u}$, and $\bar{w}$ are the relative opening and sliding displacements between points on the crack faces, respectively. π is the angle (in radians) in the polar coordinate system. These expressions form the basis for the virtual crack closure (VCC) method for separation of the fracture modes using finite element solutions of crack problems (Rybicki and Kanninen 1977, Krueger 2004).

2.8 Strength of Composite Laminates Containing Holes

Structures made from composite laminates containing cutouts or penetrations such as fastener holes (notches) offer a special challenge to the designer because of the stress concentration associated with the geometric discontinuity. The strength can be substantially reduced compared to the strength of the unnotched specimen. In laminates containing notches, a complex fiber-bridging zone develops near the notch tip (Mandell et al. 1975, Aronsson 1984). On the microscopic level, the damage appears in the form of fiber pull-out, matrix microcracking, and fiber–matrix interfacial failure. The type of damage and its growth depend strongly on the laminate stacking sequence, type of resin, and the fiber. As a consequence of the damaged material, the assumptions of a small process zone and self-similar growth of a single crack, inherent in linear elastic fracture mechanics, break down.

Because of the complexity of the fracture process for notched composite laminates, the methods developed for prediction of strength are semiempirical. Awerbuch and Madhukar (1985) review strength models for laminates containing cracks or holes loaded in tension. In this text, only the technically important case of a laminate containing a circular hole is considered.

A conservative estimate of the strength reduction is based on the stress concentration factor at the hole edge for a composite laminate containing a circular hole,

$$\frac{\sigma_N}{\sigma_0} = \frac{1}{K} \tag{2.63}$$

where σ_N and σ_0 are the notched and unnotched ultimate strengths of the laminate, respectively, and K is the stress concentration factor. The stress distribution can be obtained in closed form only for infinite, homogeneous, orthotropic plates containing an open hole (Lekhnitskii 1968). The stress concentration factor K_∞ for an infinite plate containing a circular hole loaded in uniaxial tension (Figure 2.11) is given in terms of the effective orthotropic engineering constants of the plate (Lekhnitskii 1968):

$$K_\infty = 1 + \sqrt{2\left(\sqrt{E_x/E_y} - \nu_{xy} + E_x/\left(2G_{xy}\right)\right)} \tag{2.64}$$

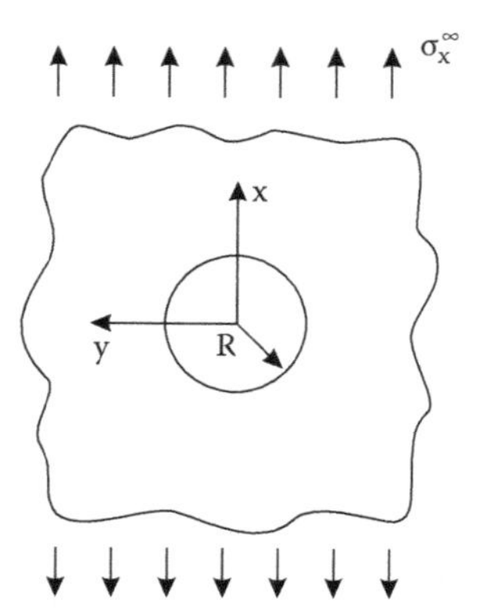

FIGURE 2.11
Infinite plate containing a circular hole under remote uniform tension.

Notice here that the stress concentration factor is defined as the stress at the hole edge divided by the applied far-field stress. Notice further that x and y are coordinates along and transverse to the loading direction (Figure 2.11). The stress concentration factor for finite-width plates containing holes K is larger than K_∞ (see, e.g., Konish and Whitney 1975, Gillespie and Carlsson 1988). Plates where the width and length exceed about six hole diameters, however, may be considered as infinite, and Equation (2.64) holds to a good approximation.

It can be easily verified from Equation (2.64) that the stress concentration factor for an isotropic material is 3. For highly orthotropic composites, the stress concentration factor is much greater (up to 9 for unidirectional carbon/epoxy).

3

Processing of Composite Laminates

3.1 Introduction

The processing of polymer matrix composite laminates has been the subject of considerable research during the last several decades (see Lee et al. 1982, Loos and Springer 1983, Bogetti and Gillespie 1991, and Khoun et al. 2010). Multiple physical and chemical phenomena must occur simultaneously and in the proper sequence to achieve desired laminate properties. There are several routes to achieve full consolidation and minimize void content of a polymeric matrix with a reinforcing fiber in volume fractions (50–65%) appropriate for structural applications. The most widely accepted approach is by impregnation of unidirectional fibers, textile fabrics, or random fiber arrays to create a thin sheet or tape. If the polymer is a thermoset, it is often advanced in its curing state to the "B" stage (a state of cure of the matrix that is incomplete but provides high room temperature viscosity). Known as *prepreg* in this form, it may be stored at low temperature (below freezing) to greatly reduce the rate of cure and thus increase the storage life. After being warmed to room temperature, these prepreg sheets or tapes may then be assembled into a laminate and subjected to a cure cycle.

It is also possible to assemble dry fibers into an appropriate geometric form and then impregnate the entire laminate in a single step. This approach is known as resin transfer molding (RTM) or resin infusion, and there are several variations, such as vacuum-assisted resin transfer molding (VARTM). The weaving of a fabric from reinforcing fibers is a widely accepted approach to creating the fiber preform, although there are other techniques designed to avoid fiber crimp and develop microstructures typical of that achieved with prepreg tape.

For prepreg, heat and pressure are first applied to the laminate to reduce the viscosity of the polymer matrix and achieve full densification of the laminate and coalescence of the laminae through matrix flow. The application of heat to the laminate is governed by the laws of heat transfer and is therefore a time-dependent phenomenon. Further, the pressure in the laminate is shared by the polymeric matrix and the fibers. For thermosetting polymers, the kinetic process to achieve gelation and vitrification is a themochemical process that is often exothermic. The decrease in polymer viscosity with temperature and its increase with degree of cure for thermosets requires that the necessary flow be achieved prior to gelation or vitrification. For thermoplastic polymers, the process involves both viscosity changes and changes in the polymer morphology (degree of crystallinity). Thermoplastic crystalline polymers will exhibit varying degrees of crystallinity depending on their thermal history (Velisaris and Seferis 1986).

The instantaneous degree of cure of a thermoset polymer is measured by the fraction of total heat generated at a given time divided by the total heat of reaction. Thus, the degree of cure ranges from 0 to 1.0 and can be measured using differential scanning calorimetry (DSC), which determines the heat of reaction as a function of time. As the reaction progresses and the macromolecular network forms, the rate-controlling phenomenon changes from kinetic to diffusion because of the reduction in polymer free volume (a measure of the internal space available within a polymer). An accompanying reduction in molecular mobility occurs because of molecular weight increase and cross-link formation.

Uneven distribution of resin may result from nonuniform flow of the polymer through the fiber reinforcement. This is particularly pronounced for laminates with curvilinear geometry and tapered thickness, in which local pressure gradients occur. The velocity of flow of a polymer through a porous medium such as a fiber array has been shown to be proportional to the pressure gradient and inversely proportional to the polymer viscosity. The proportionality constant is known as the fiber array permeability.

3.2 Characterization of Thermoset Matrices

The characterization of thermoset matrices for polymer composites requires the development of cure kinetic models to describe the state of cure within the polymer after the composite system has been subjected to a thermal history. Many of the performance and processing properties of the polymer matrix depend directly on the state of resin cure, as indicated, for example, by glass transition temperature, viscosity, cure shrinkage, stiffness, and thermal expansion. In the following, models and processes for determining these properties are discussed in the context of characterization. This is not to say that the processing phenomena are not important in determining efficient and effective means of manufacturing composite material systems, but the focus of the current chapter is meant to give an understanding of those processing conditions that can significantly affect the polymer composite properties.

Polymer matrices go through several phase transformations during the curing cycle. The prepreging operation leaves the thermosetting matrix in a state of cure that is appropriate to handling and to subsequent consolidation, but that is not significantly advanced. When the temperature of the thermoset matrix is raised to begin the curing cycle, the polymer assumes a fluid state with significantly reduced viscosity appropriate for flow necessary for complete penetration of the fiber array. As the cross-linking reaction advances, the physical characteristics of the thermoset resin change from a liquid to a rubbery state and then to a glassy state. *Gelation* is a term that defines the degree of cure wherein the polymer is transformed from the liquid state to the rubbery state by the presence of the cross-link network. Vitrification marks the transformation of this polymer gel to the glassy state after sufficient advance of the cure process. After vitrification and on subsequent heating, the polymer with a given extent of cure passes from the glassy back to the rubbery state at its glass transition temperature T_g. As will be discussed further, the glass transition temperature is a function of the extent of cure of the thermoset polymer and can be strongly influenced by the thermal curing history. Measurement of the glass transition temperature yields important information regarding the extent of cure as well as the upper use temperature of the composite system.

3.2.1 Phenomenological Cure Kinetic Coefficients

The curing reaction of a thermoset polymer transforms it into an infusible solid through formation of a cross-link network. To understand the properties of the solid, the cure kinetics must be established. Cure kinetics of a thermoset polymer such as the epoxies utilized in polymeric composites involve the determination of the extent of cure as measured by the amount of heat released by the exothermic reaction, ΔH_R, divided by the total heat of reaction released for complete cure ΔH_T. Thus, the degree of cure α varies between 1.0 (fully cured) and 0 (no cure). Composed of epoxide groups, these systems are typically cured with primary and secondary amines. When a common epoxy resin (bifunctional with two epoxide groups) reacts with a diamine (chemical compounds containing two amino groups), the network formation (cross-linking) can advance quickly. While there are many approaches to modeling the cure state in a thermoset polymer, a widely accepted approach utilizes an Arrhenius-type equation, which is a function of the temperature and the activation energy. The classic rate equation utilized in much of the phenomenological modeling of epoxies is given in the following, where the rate of the reaction is equal to the product of a function of temperature multiplied by 1 minus the extent of cure, raised to the nth power:

$$\frac{d\alpha}{dt} = K(T)(1-\alpha)^n \tag{3.1}$$

When the temperature increases linearly over time, it is appropriate to express the rate equation as a function of temperature.

$$T(t) = mt + T_0 \tag{3.2a}$$

$$\frac{dt}{dT} = \frac{1}{m} \tag{3.2b}$$

Where m is the heating rate, Equation (3.1) can be rewritten in terms of temperature.

$$\frac{d\alpha}{dT} = \frac{K(T)}{m}(1-\alpha)^n \tag{3.2c}$$

The objective of the experimental characterization of the cure kinetics of the thermosetting matrix is therefore the determination of the rate constants, $K(T)$ and n.

Equation (3.1) may be expressed in logarithmic form and plotted as ln (da/dt) versus the ln of the terms on the right-hand side. The linear relation in ln-ln space shows that the two constants n and ln K, respectively, are the slope and intercept of the line at ln $(1 - \alpha) = 0$.

$$\ln\left(\frac{d\alpha}{dt}\right) = n\ln(1-\alpha) + \ln K(T) \tag{3.3}$$

The second rate constant $K(T)$ can be expressed in a typical Arrhenius form as shown in Equation (3.4), where T is expressed in Kelvin; R is the universal gas constant, $R = 8.3145$ J/K; and E_a is the activation energy (J/mol):

$$K(T) = Ae^{-E_a/RT} \tag{3.4}$$

By expressing this equation in ln-ln format, it is possible to determine the preexponential factor A and the activation energy E_a from the intercept and slope of the resulting linear relationship of ln K versus $1/T$.

$$\ln K = \ln A - \frac{E_a}{RT} \tag{3.5}$$

Consider the following isothermal procedure for determining the parameters of the reaction. First, the residual heat of reaction is determined by DSC

TABLE 3.1

Isothermal Heat of Reaction Data for Typical Epoxy

Temp, °C	Time, s	ΔH_R, J/g	$\alpha = 1 - \Delta H_R/\Delta H_T$
25	720	140	0.362
25	1800	82.2	0.624
25	3240	48.3	0.780
25	4680	31.6	0.856
50	360	148	0.329
50	900	91.3	0.585
50	1620	55.3	0.749
50	2340	37.1	0.832
100	180	136	0.381
100	450	78.1	0.645
100	810	44.6	0.797
100	1670	28.8	0.869
150	90	137	0.379
150	225	78.5	0.643
150	405	44.9	0.796
150	835	29.1	0.868

measurements after curing a resin sample for the given time and under a specific isothermal condition. This process is repeated for several isothermal conditions. To calculate the parameters with any graphical method, it is first necessary to determine conversion (degree of reaction) as a function of time. To accomplish this goal, it is necessary to construct a table with data of residual heat as a function of time and temperature and transform into conversion, α as a function of time and temperature. A typical set of results is shown in Table 3.1, where each value for the heat of reaction ΔH_R corresponds to an individual experiment.

After cure of a resin sample at 50°C for an hour, the residual heat determined by the DSC experiment performed afterward was 2.65 J/g. Therefore, all the possible heat of reaction from the sample, except the remaining 2.65 J/g, was evolved during the isothermal hold prior to the DSC analysis. The following equation is used to calculate the corresponding conversion of the resin after 1 h at 50°C:

$$\alpha = 1 - \frac{\Delta H_R}{\Delta H_T} \tag{3.6}$$

Note that in an earlier experiment the total heat of reaction ΔH_T for this material was determined as 220 J/g.

The reduction of the data in Table 3.1 is accomplished by approximating the time derivative with a forward difference approximation:

$$\frac{d\alpha}{dt} = \frac{\alpha_2 - \alpha_1}{t_2 - t_1} \tag{3.7}$$

As an example, for the previous 25°C data, Table 3.2 can be constructed.

When isothermal cure is carried out at four temperatures (25, 50, 100, 150°C), the data can be plotted according to Equation (3.3) and presented as shown in Figure 3.1, where R^2 is the measure of fit.

As shown in Equation (3.5), the Arrhenius rate constant A must also be determined from the data. From Figure 3.1, the values of $K(T)$ extracted are given in Table 3.3.

TABLE 3.2

Isothermal Reaction Data Reduction (25 °C)

Time, s	α	ln(1 – α)	$d\alpha/dt$ x 10^{-4}	ln($d\alpha/dt$)
720	0.362	–0.449	2.45	–8.33
1800	0.624	–0.978	1.08	–9.13
3240	0.780	–1.514	0.527	–9.85
4680	0.856	–1.938		

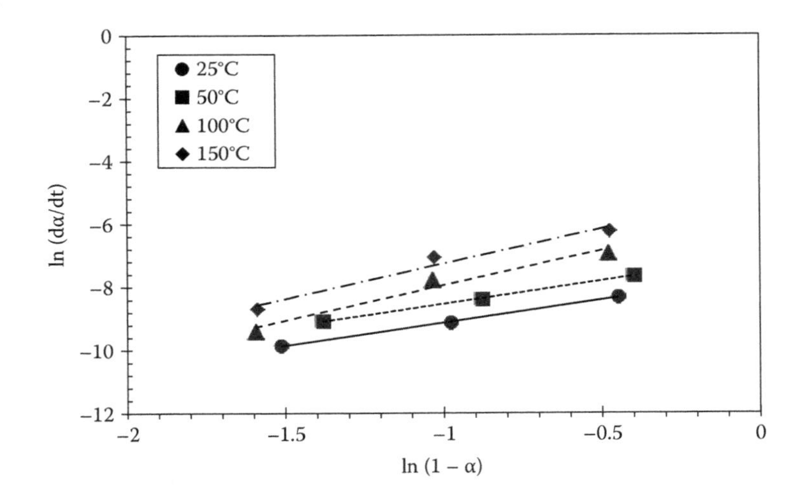

FIGURE 3.1
Typical isothermal cure data.

TABLE 3.3
Arrhenius Rate Constant Data

T, °C	T, K	$1/RT \times 10^{-4}$	$K(T)$
25	298	4.04	–7.698
50	323	3.72	–7.094
100	373	3.22	–6.263
150	423	2.84	–5.574

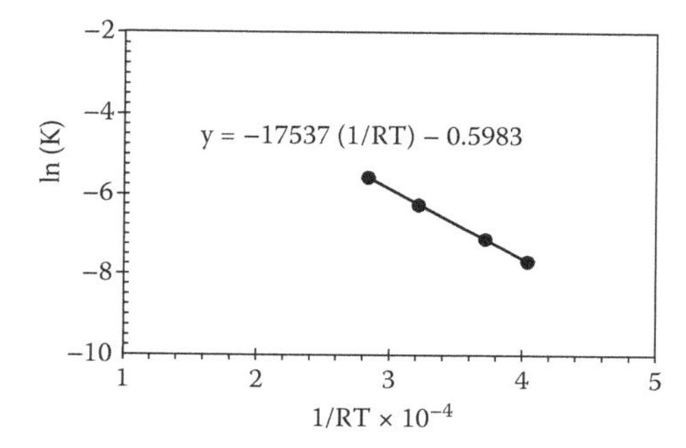

FIGURE 3.2
Typical kinetic parameter data.

The data in Table 3.3 are presented in Figure 3.2 for the determination of the rate constant A according to Equation (3.5) ($A = 0.564$).

The cure kinetic parameters determined by the four isothermal test conditions are summarized in Table 3.4. The data in Table 3.4 completely define the reaction kinetics of the polymer system under study and can be used in Equations (3.1) and (3.2) to determine the polymer cure state resulting from a specified thermal history.

3.2.2 Gelation of the Thermoset Polymer

Gelation is the transformation of the polymer from the liquid to rubbery state, and it is important to relate the extent of cure that corresponds to the onset

TABLE 3.4
Kinetic Parameters of the Reaction

Term	Value
n	1.43
E_a (kJ/mol)	17.6
A (s^{-1})	0.564

of gelation during the curing process. Using the fundamentals of gelation theory, the relationship in Equation (3.8) has been developed, wherein the extent of cure at gelation α_g is related to the molar ratio of reacting groups r, the fraction of amine hydrogen s, and the functionality of cross-linking groups f (Flory 1953).

$$\alpha_g = \frac{1}{[r + rs(f-2)]^{1/2}} \tag{3.8}$$

When an epoxy resin is cured with an aromatic diamine, the number of cross-linking groups is four ($f = 4$) because each amine group can react with two epoxide groups. For equimolar conditions, $r = 1$, and for this functionality of cross-linking groups, $s = 1$. This leads to a prediction of a degree of cure at gelation of 0.58. The test method to determine gel time (time for a 120-ml sample at 23°C to reach the gel point) previously described in ASTM (American Society for Testing and Materials) D2471 (2008) was withdrawn and is currently under review.

As mentioned, the glass transition temperature of a thermosetting polymer matrix is a strong function of the extent of cure due to the continued growth of the cross-linking network with cure. The DiBenedetto equation (1987) gives a convenient relationship between the glass transition temperature at a given degree of cure T_g, the glass transition temperature of the un-cross-linked polymer T_{go}, the ratio of lattice energies of cross-linked and partially cross-linked polymer E_x/E_m, and the ratio of segmental mobilities at the glass transition temperature F_x/F_m.

$$\frac{T_g - T_{go}}{T_{go}} = \frac{[E_x/E_m - F_x/F_m]\alpha}{1-(1-F_x/F_m)\alpha} \tag{3.9}$$

Expression (3.9) can be simplified by defining the glass transition temperature at full conversion as $T_{g\infty}$ to yield the following form:

$$\frac{T_g - T_{go}}{T_g^{\infty} - T_{go}} = \frac{\lambda\alpha}{1-(1-\lambda)\alpha} \tag{3.10}$$

where λ is typically 0.4–0.5. The dependence of T_g versus degree of cure is illustrated in Figure 3.3.

3.2.3 Viscosity and Cure Kinetics

The characterization of cure kinetics with rheological characteristics is based on the viscosity change of the thermosetting polymer as a function of its molecular weight and cross-link density. However, as the temperature of a thermoset resin is increased, two primary phenomena influence its viscosity. At first, the viscosity is reduced significantly by an increase in temperature,

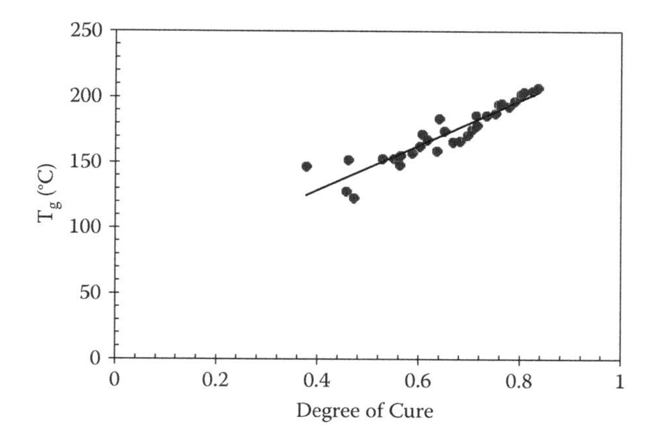

FIGURE 3.3
T_g versus degree of cure.

but as the reaction progresses, the viscosity increases exponentially due to the cross-link network formation. Typical viscosity versus time for a linear increase in temperature is shown in Figure 3.4.

While a number of phenomenological models have been developed for the relationship between viscosity, extent of cure, and temperature, the Castro-Macosko model (1982) has been chosen here for its versatility. As shown in Equations (3.11) and (3.12), the viscosity η_s of the thermoset polymer of

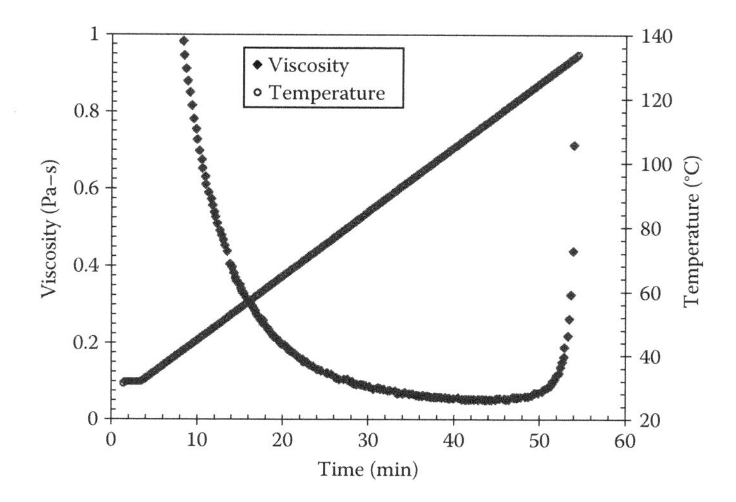

FIGURE 3.4
Viscosity versus time–temperature.

incomplete cure is related to the degree of cure at gelation α_g and the degree of cure of the polymer α.

$$\frac{\eta_c}{\eta_0(T)} = \left(\frac{\alpha_g}{\alpha_g - \alpha}\right)^{(A+B\alpha)} \tag{3.11}$$

The reference viscosity $\eta_0(T)$ follows an Arrhenius temperature dependence.

$$\eta_0(T) = C\exp\left(\frac{E_\mu}{R(T - T_0)}\right) \tag{3.12}$$

where E_a denotes the activation energy of the reaction, R is the universal gas constant, T is the absolute temperature of the system, and T_0 is the reference temperature.

Measurement of the viscosity of a thermoset presents challenges to the experimentalist because the material may pass from liquid to gel to vitrified states during the test. Some of the rheological equipment used in these measurements cannot accept these conditions, and care must be taken in the instrument choice. The ASTM D4473 (2008) test method specifies the use of a parallel plate rheometer. The system typically monitors the dynamic shear moduli of the polymer, consisting of the storage modulus and loss modulus of the system. Sweeps in frequency at constant temperature and temperature at constant frequency are the two testing options available. In the former, the change in viscosity under isothermal conditions can be determined through direct measurement, and in the latter, viscosity change with temperature can be observed. The typical frequency of the test ranges from quasi static to 100 Hz. The flat parallel plate geometry does not provide a uniform shear strain rate in the specimen, and for this reason, conical plate geometries are often utilized. By plotting the inverse of viscosity with time, it is possible to determine the point at which the viscosity approaches infinity and thereby gelation.

Consider the example viscosity data for an epoxy system in Table 3.5. Combining these data with the relationships given in Equations (3.11) and (3.12) provides guidance in determining the onset of gelation during the cure

TABLE 3.5

Viscosity Data for Typical Epoxy System

Term	Value
E_μ	76.5 kJ/gmol
C	3.45×10^{-10} Pa^{-s}
A	3.8
B	2.5
α_g	0.47

of the resin when the thermal cure cycle is specified and thereby illustrates the need for the determination of these properties for a given resin system.

3.3 Processing of Thermoset Composites

The development of an interlocking network during the cure of a thermoset polymer is illustrated in Figure 3.5. As temperature and time increase, the network interconnectivity grows according to the steps illustrated: (a) The prepolymer and curing agents are interspersed; (b) polymer molecular weight (size) increases; (c) gelation occurs, and a continuous network is achieved; and (d) cure is complete (see the time–temperature transformation diagram, Figure 3.6). After the polymer approaches vitrification (i.e., the polymer changes from a rubbery to a glassy state), the rate of conversion decreases significantly. Should vitrification occur before completion of the cure reaction, polymer properties will not be fully achieved, and voids may form in the laminate. These phenomena must be considered in the development of an appropriate cure cycle.

Figure 3.7 illustrates the flow and compaction phenomena during the curing and consolidation steps. Initially, the increase in temperature serves to decrease the viscosity of the polymer, and the polymer carries the applied pressure. As the laminate is vented and flow begins, the fibers deform and act as an elastic spring in assuming a portion of the applied pressure (Figure 3.7). Volatiles produced in the chemical reaction or trapped gases will then escape from the laminate. Finally, the total pressure is carried by the fully consolidated composite panel.

Given that composite laminates are often processed in an autoclave, wherein heat transfer is achieved with a pressurizing medium (normally nitrogen, an inert gas), it is important to recognize that the instantaneous temperature within the laminate may not be equal to that of the autoclave.

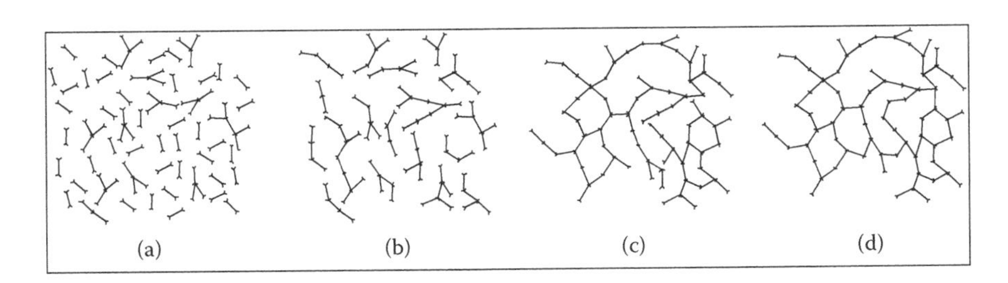

FIGURE 3.5
Dynamics of thermoset gelation and vitrification. (From L.A. Berglund and J.M. Kenny, *SAMPE J.*, 27(2), 1991. With permission.)

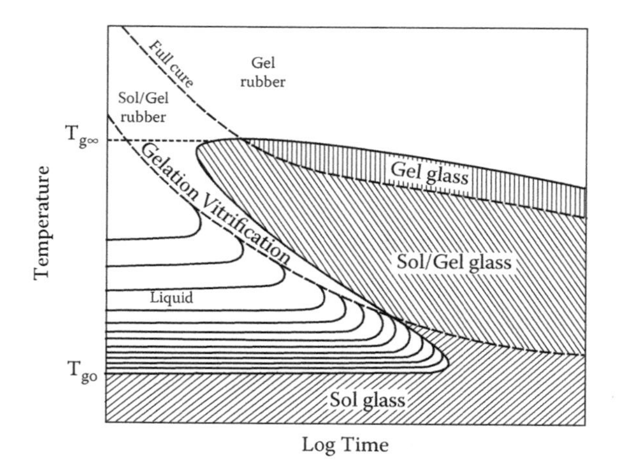

FIGURE 3.6
Time–temperature transformation diagram. (From L.A. Berglund and J.M. Kenny, *SAMPE J.*, 27(2), 1991. With permission.)

Figure 3.8 illustrates a typical thermal cycle and shows that the temperature of the laminate can differ from top surface to interior (center) to tool surface. Thus, the dynamics of heat transfer must be considered when an appropriate cure cycle is developed.

Consider the typical cure cycle shown in Figure 3.9, where internal composite temperature lags autoclave temperature. Initially, the autoclave temperature is

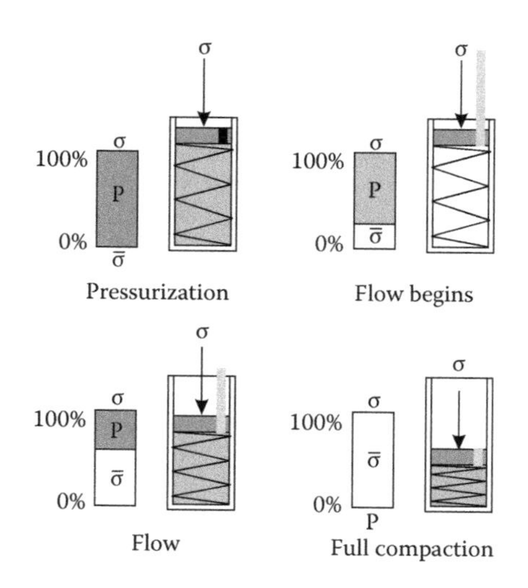

FIGURE 3.7
Polymer and perform pressurization and flow. (From P. Hubert, Ph.D. thesis, University of British Columbia, 1996. With permission.)

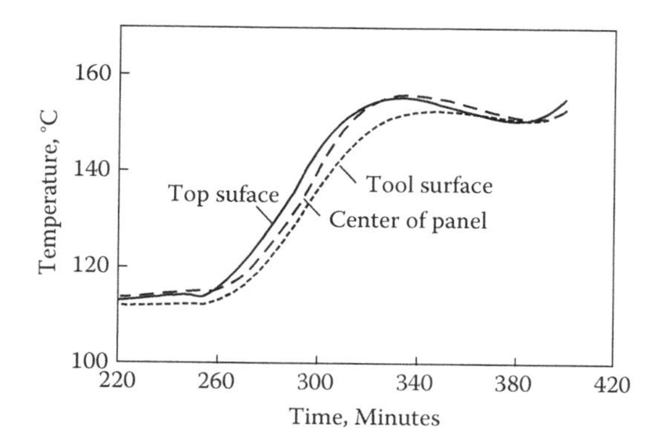

FIGURE 3.8
Heat transfer through laminate thickness. (From P. Hubert, University of British Columbia Composites Group Report, 1994. With permission.)

increased at a constant rate of 2–3°C/min until it reaches 110°C, and then it is held constant for approximately 1 h. During this stage, the polymer is in the liquid state. Next, the autoclave temperature is increased to and held at approximately 180°C for 2 h. During this stage, the polymer passes through gelation at a degree of cure of 0.46 and then approaches vitrification. Vitrification occurs when the polymer passes from the rubbery or gel state to the glassy state. In Figure 3.8, the vitrification point occurs prematurely at approximately 190 min into the cycle. Because the rubbery-to-glass transition occurs at vitrification, stresses developed as a result of the shrinkage of the polymer with cure progression may not relax during the remainder of the curing cycle. For the

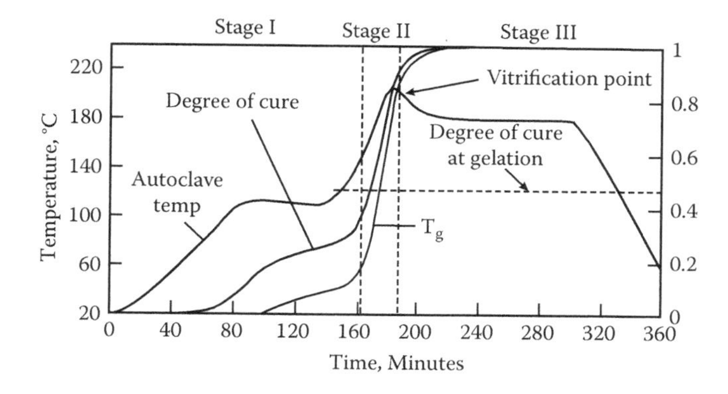

FIGURE 3.9
Cure cycle with premature vitrification. (From P. Hubert, University of British Columbia Composites Group Report, 1994. With permission.)

case in which vitrification is delayed until a point much later in the process close to cooling, much of this stress will be eliminated by completion of the cycle. Hence, the cure cycle can be tailored to the specific polymer to minimize residual stresses. Of course, thermal residual stresses will still develop in the laminate on cooling because of anisotropic thermal expansion, as discussed in Chapter 13.

3.3.1 Autoclave Molding

Figure 3.10 shows the vacuum bag layup sequence for a typical epoxy matrix prepreg composite. Different layup sequences can be used for other types of prepregs.

1. Thoroughly clean the aluminum plate (10) using acetone or a detergent. Then, apply mold release agent to the top surface of the aluminum plate twice.
2. Lay one sheet of Teflon film (1) and the peel-ply (2) (nonstick nylon cloth) on the aluminum plate. The Teflon film is used to release the layup from the aluminum plate, and the peel-ply is used to achieve the required surface finish on the laminate. Note: There should be no wrinkles or raised regions in the peel-ply, and its dimensions should be identical to those of the laminate.

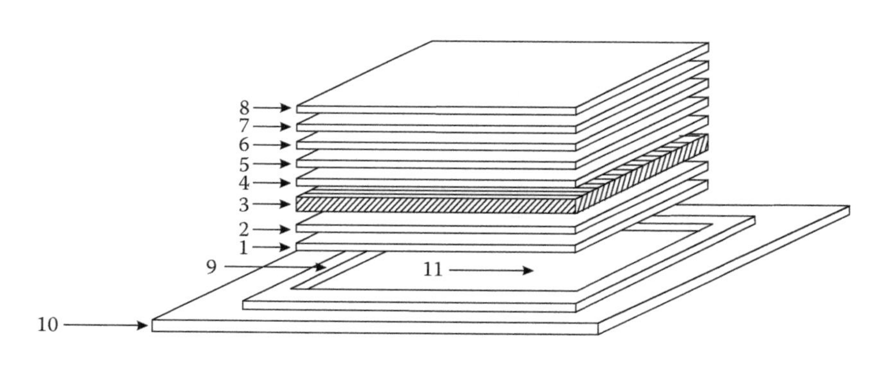

FIGURE 3.10
Vacuum bag preparation for autoclave cure of thermoset matrix composite.

3. Place the prepreg stack (3) on the plate, being sure to keep it at least 50 mm from each edge. Note: Do not cover up the vacuum connection in the plate.
4. Place a strip of the cork–rubber material (9) along each edge of the panel, making sure that no gaps exist and a complete dam is formed around the laminate. The dam around the layup prevents lateral motion of the panel and minimizes resin flow parallel to the aluminum plate and through the edges of the laminate (9).
5. Completely encircle the prepreg stack and dam with bagging adhesive, making sure that the adhesive material is adjacent to the dam. The purpose of the adhesive material is to form a vacuum seal.
6. Place a peel-ply (4) and a ply of Teflon-coated glass fabric (5) (with the same dimensions as the panel) on top of the prepreg stack. The purpose of the Teflon-coated glass fabric is to prevent the bleeder sheets (6) from sticking to the laminate.
7. Place the proper number of glass bleeder sheets (6) (e.g., style 181 glass cloth with the same dimensions as the prepreg stack) over the Teflon-coated fabric (5). The bleeder sheets absorb the excess resin from the laminate.
8. Place a sheet of perforated Teflon film (7) (0.025 mm thickness) over the bleeder material. The Teflon film, perforated on 50-mm centers, prevents excess resin from saturating the vent cloth (8).
9. Place a porous continuous vent cloth (8) (e.g., style 181 glass cloth) on top of the layup. Extend the cloth over the vacuum line attachment. Make sure that the vacuum line is completely covered by the vent cloth. The vent cloth provides a path for volatiles to escape when the vacuum is applied and promotes a uniform distribution of vacuum.
10. Place nylon bagging film over the entire plate, and seal it against the bagging adhesive. Allow enough material so that the film conforms to all contours without being punctured.
11. Place the plate in the autoclave and attach the vacuum line (Figure 3.11). An autoclave is generally a large pressure vessel equipped with a temperature and pressure control system. The elevated pressures and temperatures, required for processing of the laminate, are commonly achieved by electrically heating a pressurized inert gas (nitrogen). The use of an inert gas will reduce oxidizing reactions that otherwise may occur in the resin at elevated temperatures, and will prevent explosion of evolving volatiles.
12. Turn on the vacuum pump and check for leaks. Maintain a vacuum of 650 to 750 mm of mercury for 20 min and check again for leaks.

FIGURE 3.11
Vacuum bag sequence and tool plate placed in an autoclave.

13. After closing the autoclave door, apply the pressure and initiate the appropriate cure cycle (see example shown in Figure 3.12). As the temperature is increased, the resin viscosity decreases rapidly, and the chemical reaction of the resin begins. At the end of the temperature hold, at 127°C in Figure 3.12, the resin viscosity is at a minimum, and pressure is applied to squeeze out excess resin. The temperature hold controls the rate of the chemical reaction and prevents degradation of the material by the exotherm. The pressure is held constant throughout the cure cycle to consolidate the plies until the resin in the laminate reaches its glassy state at the end of the cooling phase.

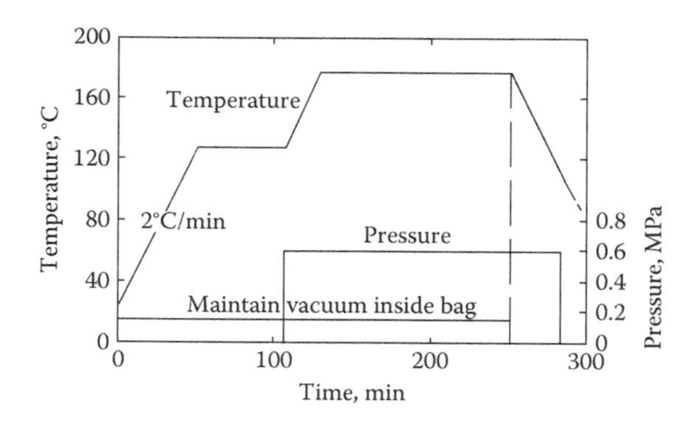

FIGURE 3.12
Typical cure cycle for a carbon/epoxy prepreg.

The vacuum should be checked throughout the cure cycle. The vacuum is applied to achieve a uniform pressure on the laminate and draw out volatiles created during the cure. Loss of vacuum will result in a poorly consolidated laminate.

14. After the power is turned off to the autoclave, maintain pressure until the inside temperature has dropped to about 100°C.
15. Carefully remove the laminate from the aluminum plate. Gently lift it in a direction parallel to the main principal direction of the laminate.
16. Clean the aluminum plate and store it for future use.

3.3.2 Resin Transfer Molding of Thermoset Composites

Resin transfer molding of composite laminates is a process wherein the dry-fiber preform is infiltrated with a liquid polymeric resin, and the polymer is advanced to its final cure after the impregnation process is complete. An extensive review of the RTM process can be found in the work of Advani (1994). The process consists of four steps: fiber preform manufacture, mold filling, cure, and part removal. In the first step, textile technology is typically utilized to assemble the preform. For example, woven textile fabrics are often assembled into multilayer laminates that conform to the geometry of the tool. Braiding and stitching provide mechanisms for the creation of three-dimensional preform architectures.

Typically, a thermosetting polymer of relatively low viscosity is used in the RTM process. There have been applications for thermoplastic polymers, but they are rare. Pressure is applied to the fluid polymer to inject it into a mold containing the fiber preform, and the mold may have been preheated. The flow of the fluid through the fiber preform is governed by Darcy's law (Advani 1994), wherein the velocity of the flow is equal to the product of the pressure gradient, the preform permeability, and the inverse of the polymer viscosity. Clearly, the lower the polymer viscosity, the greater the flow rate will be, and similarly, the greater the permeability, the greater the flow rate will be. Note also that because the fiber preforms typically exhibit different geometries in the three principal directions, permeability is a tensor and exhibits anisotropic characteristics. That is, for a given pressure gradient, the flow rates in three mutually orthogonal directions will differ. Flow through the thickness of a fiber preform that contains many layers of unidirectional fibers will be quite different from flow in the planar directions. In addition, the permeability of the preform depends on the fiber volume fraction of the preform. The greater the volume fraction, the lower the permeability is. It is important to vent the mold to the atmosphere to remove displaced gases from the fiber preform during the mold-filling process. Otherwise, trapped gases will lead to voids within the laminate.

After the polymer has fully impregnated the fiber preform, the third step occurs: cure. Cure begins immediately on injection of the polymer into the

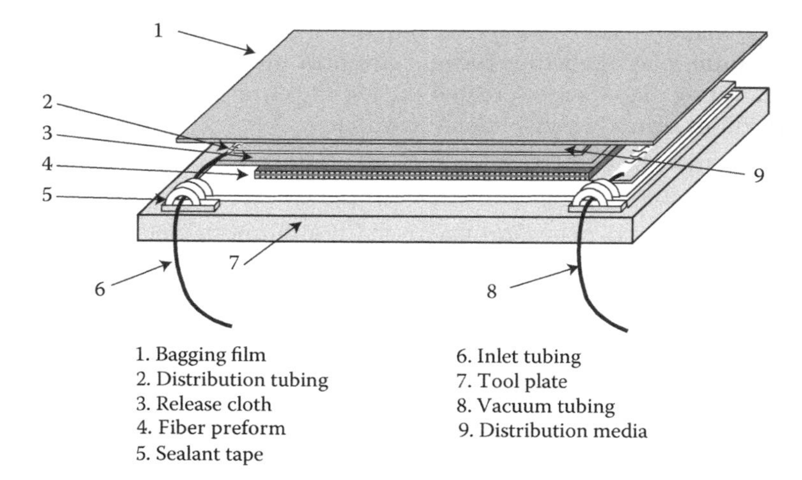

FIGURE 3.13
VARTM process. (Courtesy of B. Grimsley, NASA Langley Research Center, 2001.)

mold and will occur more rapidly if the mold is at an elevated temperature. As the cure of the polymer advances to the creation of a cross-link network, it passes through a gelation phase wherein the polymer viscosity increases and transforms the polymer into a viscoelastic substance, where it possesses both viscous and elastic properties. As this process proceeds and the cross-link network continues to grow, the instantaneous glass transition temperature of the polymer increases. Finally, vitrification of the polymer occurs when its glass transition temperature exceeds the laminate temperature. Should gelation or vitrification (or both) occur prior to completion of mold filling and preform impregnation, the resulting laminate will not be fully impregnated.

The viscosity of most polymers is highly dependent on temperature, and polymer cure kinetics are controlled by temperature as well. Therefore, heat transfer phenomena must be managed for successful RTM processes. Heat transfer between the polymer and the fiber preform, and between tool, preform, and polymer, as well as exothermic heat generation during the cure of the polymer, are three such phenomena that influence the process (Lee et al. 1982).

3.3.2.1 Vacuum-Assisted Resin Transfer Molding

Both open-mold approaches, where one surface is bagged with a flexible film, and closed-mold approaches to RTM are practiced. An example of open-mold RTM, VARTM, is a common method employed as an alternative to autoclave use. The VARTM process is illustrated in Figure 3.13. In VARTM, atmospheric pressure is utilized to achieve consolidation and impregnation by vacuum bagging the laminate. An inlet for the polymer is located at one or more points in the tool or bag, and vacuum outlets are located some distance away. The vacuum pump creates a pressure gradient of approximately 1 atm within the bag, which is sufficient for the impregnation of laminates large in size and complex in geometry. For processes in which final cure occurs after the mold is filled, completion of the cure may be carried out in an oven while atmospheric pressure is maintained on the impregnated laminate.

The VARTM procedure for a representative flat 61.0 × 30.5 × 0.64 cm panel (Figure 3.13) is described in the following steps:

1. *Tool surface.* The tool is a flat aluminum plate with planar dimensions sufficient to accommodate the composite panel. First, clean the metal tool surface using sandpaper and acetone. On the cleaned surface, create a 71 × 30.5 cm picture frame using masking tape. Apply several coats of release agent to the metal surface inside the masked frame. Remove the masking tape.
2. *Bagging tape.* In place of the masking tape, apply a 1.3 cm wide silicone bagging tape to the bare metal surface. The silicone tape should again form a 71 × 30.5 cm frame. Add a strip of the tape, 5 cm in length, to the outer edge of the length of the frame at either end. These two strips will provide an added adhesive surface for attachment of the inlet and outlet tubing. Leave the paper backing on the silicone tape to protect it during the remainder of the layup procedure.
3. *Preform.* Place the fiber preform stack on the coated tool, inside the tape frame. A 5.1-cm gap should exist between the silicone tape and both edges of the preform to allow room for tubing. No gap should exist between the silicon tape and fiber preform along the panel width to avoid providing a flow pathway outside the preform to the vacuum port.
4. *Release cloth.* Cut one layer of porous release film to 66 × 30.5 cm and place it on top of the preform. Place the cloth so that it completely covers the preform and allow 5.1 cm in length to overhang and contact the coated metal surface at the injection side of the layup. The release film will allow separation of the composite laminate from the distribution media. Cut a second piece of release cloth to 5.1 × 30.5 cm and place it on the tool surface at the vacuum side of the preform. This patch of cloth provides a clear path for the vacuum.
5. *Distribution media.* Cut one to six layers of highly permeable distribution media (e.g., biplanar nylon 6 mesh) to dimensions of 63.5 × 28.0 cm and stack them above the Armalon™ release cloth. Place the layers of media to leave a 2.5-cm gap on the top of the preform at the vacuum end. This gap will force the resin to fill through the thickness rather than be drawn directly into the vacuum port. The length of this gap will vary with the desired thickness of the composite panel. A 1.3-cm gap should exist between the media and the sides of the preform.

This will help prevent resin flow outside the preform. A 5-cm length of the media will overhang the preform at the resin inlet end of the layup.

6. *Distribution tubing.* Place a 28 cm length of distribution tubing across the width of the laminate at points 2.5 cm in front of the preform (inlet) and 2.5 cm away from the preform (vacuum). On the inlet side, place the tubing on top of the distribution media that overhangs the preform. At the vacuum side, place the tubing on the 5 × 30.5 cm piece of release cloth. Spiral wrap, 18-mm-diameter conduit is an ideal choice for the distribution tubing because it allows the resin to flow quickly into the distribution media and preform in a continuous line across the width. A plastic tube with holes at 2.5-cm intervals also works well. Attach a 13-mm portion of the spiral tubing to both the inlet–supply tubing and the vacuum tubing using Kapton™ tape (E.I. duPont de Nemours and Co.). Embed the free end of the spiral tubing in a 2.5-cm diameter roll of the silicone bagging tape and then affix it to the strip of bagging tape forming the frame of the laminate.
7. *Resin supply and vacuum tubing.* Use flexible plastic tubing (vinyl or Teflon, depending on temperature requirements) approximately 1.5 m in length to supply resin and draw vacuum on the laminate. Tape one end of the tube to the distribution tubing inside the bag. At a point just past this taped interface, wind one layer of silicone vacuum tape twice about the outer surface of the tubing. This 2.5-cm long sleeve of vacuum tape on the tube should match the tape frame and added strips that exist on the tool surface. Attach the taped tubes to the tool at these locations and place two more 7-cm long strips of tape on top of the tool and tape sealant to form a smooth, airtight joint when the bagging film is in place. Clamp the free end of the resin supply tubing to ensure a temporary airtight seal. Connect the free end of the vacuum tubing to a resin trap, which catches any resin that might be pulled into the tube on its way to the vacuum pump.
8. *Vacuum bag.* With the laminate complete and the tubing in place, the part can be bagged using an appropriate film. Take care to eliminate creases in the bag and ensure an airtight seal with the tool surface and silicone bagging tape. Once bagging is complete, the laminate should be fully evacuated to 762 mm Hg using the vacuum pump. Leaks can be detected by using either a listening device or by clamping the vacuum line and using a vacuum gage. Even a small leak in the system may result in voids and poor consolidation of the final composite part.
9. *Resin degassing.* Before infiltration can occur, the resin must be degassed to remove any air bubbles that were introduced during

mixing. Perform degassing separately in a vacuum chamber; degassing can typically require 1 to 4 h, depending on the resin viscosity. All air bubbles must be removed prior to infiltration. Contain the resin in a bucket.

10. *Resin infiltration.* With the bagged laminate under full vacuum, submerge the clamped end of the resin supply tubing in the degassed resin bucket. Remove the clamp while the tube end is submerged to prevent any air entering the tube and the part ahead of the resin. With the tube clamp removed, the resin flows through the supply tubing and into the distribution tubing. The spiral distribution tubing allows the resin to spread quickly across the width of the layup as it enters the distribution media. The distribution media provides the path for the resin to flow quickly down the length of the preform and then through the laminate thickness.
11. *Completion of infiltration.* The flow front of resin through the part can be viewed through the bagging film. Halt the flow of resin when the preform is fully infiltrated, as evidenced by resin beginning to enter the vacuum distribution tubing. Stop the resin flow by first clamping and severing the resin supply tubing and then clamping and severing the vacuum tubing. Again, these clamps must provide an airtight seal because any leaks during cure will result in poor consolidation of the part. It is recommended that a second envelope bag be used to pull vacuum on the part during cure. Finally, if elevated temperature cure is required, place the vacuum sealed part in an oven and heat it according to a cure cycle prescribed by the resin supplier.

3.4 Autoclave Processing of Thermoplastic Composites

Thermoplastic composites may be processed in a high-temperature autoclave. Figure 3.14 shows the autoclave layup sequence for a carbon/polyetheretherketone (PEEK) composite. Place Kapton film of slightly larger size than the panel, each side being coated with a release agent, on the tool plate. Place Kapton bagging film over the layup and seal the bag against the tool plate using A800 G3 (or equivalent) tacky tape. Place the tool plate and laminate in the autoclave and attach the vacuum line. The following processing cycle is recommended for a carbon/PEEK composite:

1. Maintain a vacuum of 650 to 750 mm Hg.
2. Apply a pressure of 0.5 MPa and simultaneously ramp the temperature as rapidly as possible to 390°C.

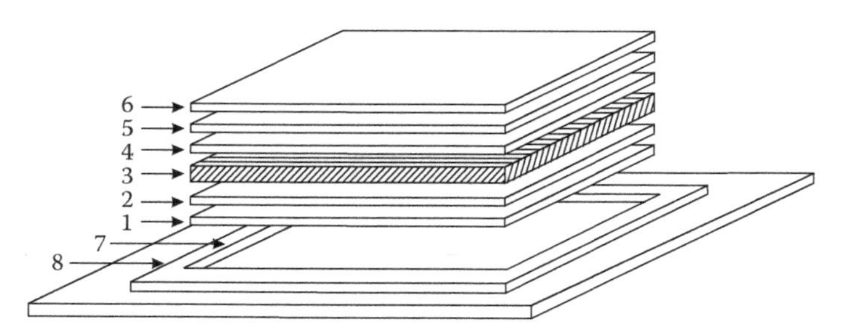

FIGURE 3.14
Vacuum bag preparation for autoclave processing of thermoplastic matrix composite.

3. Apply a consolidation pressure of 1.4 MPa.
4. Hold the pressure at a temperature of 390°C for 5 min per 8 plies, but not for more than 30 min.
5. Cool the laminate rapidly to room temperature. The degree of crystallinity for crystalline polymers is influenced by cooling rate. Pressure can be released as the laminate temperature falls below the glass transition temperature of the matrix (≈140°C).

3.5 Determination of Volume Fractions of Fibers, Resin, and Voids

As discussed in Chapter 2, the stiffness and strength properties of composites are strongly dependent on the fiber volume fraction, and this parameter thus constitutes an important quality measure of such materials. This section details measurement of fiber volume fraction for polymer matrix composites reinforced with glass, carbon, or aramid fibers.

The fiber volume fraction of a composite may be determined by chemical matrix digestion, burn-off, or photomicrographic techniques. The matrix digestion method is standardized, ASTM D3171 (2011), and consists of dissolving the (polymer) matrix in a hot digestion medium (concentrated nitric acid for epoxy matrix composites or sulfuric acid followed by hydrogen peroxide for polyimides and PEEK). Care must be taken to select a medium that attacks the matrix but does not attack the fibers. After the matrix is dissolved, the fibers are weighed. The volume fractions are calculated from the weights and densities of the constituents. The resin burn-off method (ASTM D2584 2011) is sometimes used for glass fiber composites because glass fibers (as opposed to carbon and Kevlar [E.I. du Pont de Nemours and Company] fibers) are resistant to oxidation at the temperatures required to burn off the matrix (500–600°C). Similar to the chemical matrix digestion method, the fibers are weighed after the matrix has been removed to enable calculations of fiber volume fractions.

The photomicrographic method is not an ASTM standard, but it provides an independent estimate of the fiber volume fraction. The method requires a photograph of a polished cross section of a composite and many samples to produce reliable results because the area viewed is only about a hundredth of a square millimeter. On the other hand, it provides an image of the distribution of fibers and voids.

3.5.1 Chemical Matrix Digestion Method

Equipment needed for this procedure (Figure 3.15) includes

1. Fume hood with a vacuum system
2. Large flask that can be attached to the vacuum system
3. Buchner funnel with filter
4. A 400-ml beaker
5. Nitric acid
6. Glass stirring rod

FIGURE 3.15
Acid digestion procedure.

7. Bunsen burner or electric heater
8. Desiccator
9. Precision balance
10. Rubber gloves and goggles

3.5.1.1 Procedure

1. Take a 50 × 50 mm composite sample and weigh it. Also weigh the dry Buchner funnel with the filter.
2. Put on rubber gloves and goggles and activate a hood vent fan. Place the sample in the 400-ml beaker and pour in 200 ml of the nitric acid (use the glass stirring rod for controlled, slow pouring of the acid). Heat the beaker with the Bunsen burner until the acid fumes, but avoid boiling; stir occasionally. Continue heating until the matrix is dissolved and the sample disintegrates, leaving bare fibers.
3. Insert the funnel into the large flask attached to the vacuum system and transfer the acid and the fibers into the funnel. Turn on the vacuum pump and wash the fibers three times with 20 ml of nitric acid and then follow with a water wash.
4. Remove the funnel and the fibers and dry them in an oven at 100°C for at least 90 min. Break up the fiber flocks occasionally with a glass rod to facilitate drying. Remove the funnel and the fibers and let them cool in a desiccator. Weigh the funnel containing the fibers.

3.5.1.2 Calculation of Fiber Volume Fraction

From the weights of the fibers and matrix (W_f and W_m, respectively) and their known densities (ρ_f and ρ_m, respectively), the volume fraction of fibers V_f is determined from

$$V_f = \frac{\rho_m W_f}{\rho_f W_m + \rho_m W_f} \tag{3.13}$$

where it is assumed that the void content of the composite is negligible.

As an example, consider the following data for a carbon/epoxy composite:

$W_f = 3.0671$ g; $W_m = 1.2071$ g (weight of composite minus W_f)

Table 3.6 gives densities for some current fibers and matrix resins. Using the densities in Table 3.6, $\rho_f = 1.73$ g/cm^3 and $\rho_m = 1.265$ g/cm^3, Equation (3.13) gives $V_f = 0.60$.

TABLE 3.6

Fiber and Resin Properties

Fiber Type	Carbon AS4	Carbon IM6	EGlass	Kevlar 49
Density	1.80	1.73	2.60	1.44
Matrix Type	**Epoxy N5208**	**Epoxy 3501-6**	**Polyimide**	**PEEK**
Density	1.20	1.265	1.37	1.30[a]

[a] 30% crystallinity.

Source: Daniel and Ishai, 2006. *Engineering Mechanics of Composite Materials*, 2nd ed., Oxford Unversity Press, New York.

3.5.1.3 Determination of Void Content

Voids may form in the composite as a result of gases and volatiles evolved during processing becoming trapped in the matrix. Voids are generally undesired. For example, for autoclave-produced composite parts, a void content of less than 1% is commonly desired. The procedure for measurement of void content is given in ASTM Standard D2734 (2009). Void content requires an accurate measurement of the density of the composite:

$$\rho_c = \frac{W}{V} \tag{3.14}$$

where W and V are the weight and volume of the composite, respectively. Methods for density measurements are presented in ASTM D1505 (2010) and D3800M (2011). To obtain the void content, consider the following condition for the various volume fractions:

$$V_f + V_m + V_v = 1 \tag{3.15}$$

where subscripts f, m, and v represent fiber, matrix, and voids, respectively. From Equation (3.15), an expression for the void content can be obtained:

$$V_v = 1 - \frac{(W_f/\rho_f + W_m/\rho_m)\rho_c}{W} \tag{3.16}$$

in which W_f, W_m, and W represent the weights of fiber, matrix, and composite, respectively ($W_f + W_m = W$). This method enables verification that an acceptable void content (e.g., <1%) has been achieved.

3.5.2 Photomicrographic Method

Equipment needed for this procedure includes

1. Polishing table
2. Specimen mounted (embedded) in a specimen holder
3. Metallographic optical microscope (×400) with a camera

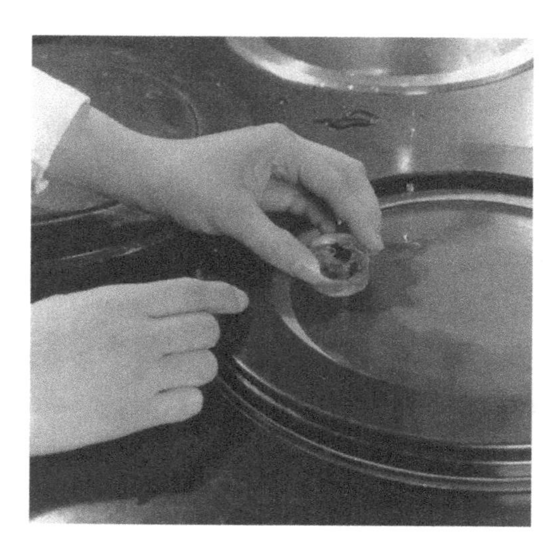

FIGURE 3.16
Polishing of the specimen embedded in mounting material.

3.5.2.1 Procedure

1. Cut the specimen perpendicular to the fiber direction to expose the desired cross section.
2. Place the specimen inside a mounting cup (e.g., Buehler SampleKup) and pour a potting material (epoxy) into the cup. After the mounting material is cured, the specimen is ready for grinding and subsequent polishing.
3. Grind the specimen by working through four sandpaper grades (180, 240, 320, and 400). Then, proceed to polish the specimen on a polishing table (Figure 3.16) using 5-, 1-, and if necessary, 0.3-μm particles. Polishing is the final step to obtain a flat surface with a mirror-like finish. Choose any direction to start the polishing and maintain that direction for that step. When changing to finer paper grades, alter the polishing angle by 90° each time to remove scratches from the previous step. Rinse the specimen after each step to remove grit.
4. When the specimen is polished, it is ready to be examined in the optical microscope. Take a photograph of a polished cross section like the one shown in Figure 3.17.

3.5.2.2 Determination of Fiber Volume Fraction

The fiber volume fraction can be determined from the photomicrograph in two ways, as illustrated in Figure 3.18. One way is to determine the total area of the fibers in a given area of the micrograph. This can be done directly with

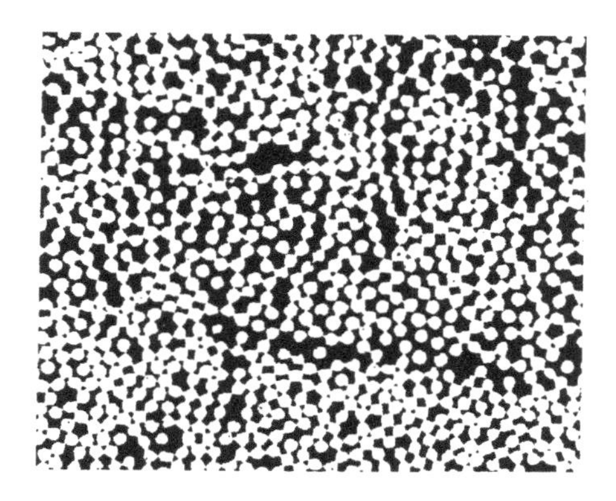

FIGURE 3.17
Photomicrograph of a polished cross section. (Courtesy of S. Nilsson, FOI, Stockholm, Sweden.)

a quantitative image analyzer or by counting the number of fibers in the area and calculating the total fiber area from their average diameter. The fiber volume fraction is determined as

$$V_f = A_f/A \tag{3.17}$$

where A_f and A are the total fiber area and the area of the selected region of the micrograph, respectively.

An alternative way, the line method (Figure 3.18), can also be used to determine the fiber volume fraction from the micrograph. In this method, a number of lines are randomly drawn on the micrograph. The fiber volume fraction is evaluated as the ratio of the cumulative length of fiber cross sections along the line to the length of the line. For a representative result, an average value should be determined from measurements along several lines.

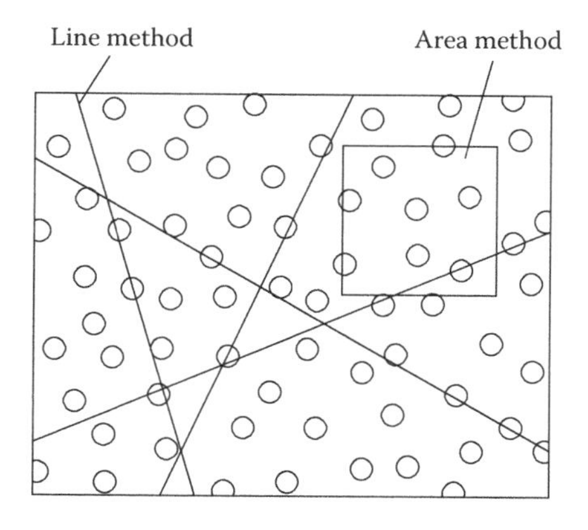

FIGURE 3.18
Illustration of area and line methods.

TABLE 3.7

Determination of Fiber Volume Fraction with the Line Method

Line	L_f (mm)[a]	V_f
1	44.5	0.58
2	40.8	0.54
3	54.9	0.72
4	45.3	0.60
5	48.3	0.63
6	48.1	0.63

[a] L_f = cumulative length of fiber cross sections. Total length of each line was 76.2 mm.

For a cross section of the carbon/epoxy composite discussed in the previous section, the results shown in Table 3.7 were obtained. From these data, an average fiber volume fraction was determined, $V_f = 0.62$, which can be compared with $V_f = 0.65$, determined with the acid digestion method. Differences between the two methods are likely due to the small region of the composite studied in the micrograph and to uncertainty in determining the length of the fiber cross sections. It is notable that the micrographic line method exhibits significant variability, as shown in Table 3.7, where the range is 0.54 to 0.72.

4

Test Specimen Preparation, Strain and Deformation Measurement Devices, and Testing Machines

4.1 Introduction

The quality of the experimental data ultimately obtained is a function of many factors. Composite processing was discussed in the previous chapter. Another critical step is specimen preparation. Care must be taken to avoid damaging the material when cutting individual specimens from a composite panel. Likewise, it is important that the testing equipment used is in good working order, and that measurement of forces, strains, and displacements are accurately conducted.

Many test methods require the use of tabs bonded to the test specimens. As described in relation to specific tests, tabs are used to reduce stress concentrations and protect the specimen from the aggressive action of the loading device (e.g., wedge grips). However, the tab-bonding procedure can introduce its own sources of error. The type of tabbing material can be inappropriate for the application or be of poor quality. The adhesive used can be inadequate or improperly applied.

4.2 Cutting the Composite Laminate

Care must be taken when cutting composite materials for use as test specimens (Abrate and Walton 1992, Srivatsan et al. 1994, Ramulu and Arola 1994). The material can be damaged in the process, resulting in reduced strength properties. One way this damage can be induced is by excessive heat buildup in the cutting zone. It may be necessary to alter the cutting tool speed, reduce the feed rate, use a different type of cutter, and use a cooling fluid. A suitable combination of these factors is often arrived at by experimentation for the particular composite being machined.

There might be concern when using a cooling fluid, which may be water based, that the composite properties would be altered by moisture absorption. All polymer–matrix composites absorb moisture, although to varying degrees and with varying consequences (Browning et al. 1977, Weitsman 2012). However, moisture diffusion is a relatively slow process (Shen and Springer 1976). Thus, brief exposure to water during the cutting process (typically measured in minutes) will usually not result in significant moisture absorption. However, it should be realized that once moisture enters the surface of the composite, it will in part continue to diffuse inward independent of the subsequent change in surface conditions. Even if the surface is immediately dried after cutting and the material is placed in a desiccator, some of the absorbed moisture will continue to diffuse inward (toward a lower-moisture concentration), while the remaining moisture diffuses back out. This is not usually a problem because the total absorbed amounts are small. Conversely, it is incorrect to assume that placing the composite in a desiccator or drying oven for even several times the water exposure time will totally dry the material again. Long-term storage in a desiccator will ultimately eliminate the moisture.

Glass-fiber composites are abrasive to cutting tools because of the inherent hardness of these fibers (Sakuma and Seto 1993, Komanduri 1993, Abrate and Walton 1992). In addition, the small glass particles produced in the cutting process can damage the wear surfaces (ways, bearings, lead screws) of the cutting device used. Boron, silicon carbide, and other ceramic fibers present similar problems. Using cutting fluids will help wash away these harmful abrasive particles. Carbon-fiber composites are much less of a problem because the particles produced act as a lubricant. However, carbon fibers are electrically conductive. Thus, the dust particles produced can cause shorts in electrical contacts such as switches and relays and create other electrical problems. The use of a cutting fluid to minimize airborne carbon particles can be advantageous.

Aramid (e.g., Kevlar®, E.I. du Pont de Nemours and Company, Wilmington, DE), polyethylene (e.g., Spectra®, Honeywell International, Inc., Colonial Heights, VA), and other organic fibers are particularly difficult to cut. Because these fibers are highly oriented (orthotropic), they tend to break up into subfibers (i.e., fray) under the cutting forces and thus do not cut cleanly. Special cutter designs have been developed to cut these fibers and their composites more effectively.

For general-purpose use, carbide milling cutters and drills and aluminum oxide abrasive cutoff blades and grinding wheels are an appropriate starting point. In general, medium-grit abrasive cutters perform better than fine-grit cutters because particles of the material being cut do not tend to become embedded in the cutter surface (i.e., the cutter does not load up as readily).

Diamond-particle-impregnated cutting tools are extensively used and are very durable, but they can also become clogged (loaded up) more readily with cutting debris. A clogged cutter is inefficient and can generate excessive amounts of frictional heat.

A surface grinder is a useful specimen preparation device. Its cutting depth, table translation speed, and traversing motion can all be controlled. It is desirable to employ a used (reconditioned) machine for this purpose because of the potential abrasive wear problems cited. Either a grinding wheel or a cutoff blade can be mounted in the grinder, depending on the operation to be performed. Although such devices frequently have magnetic tables, the composite being cut will probably not be magnetic. Thus, double-sided tape (e.g., carpet tape) is commonly used to hold the composite to a sacrificial plate. This sacrificial plate is preferably an easily cut material (for example, an inexpensive polymer such as polymethyl methacrylate [PMMA], commonly known as Plexiglas®, Röhm, GmbH & Co. KG, Darmstadt, Germany) because it will be scored as the cutter blade completes the cut through the composite panel thickness. This assembly can in turn be taped to the surface grinder table or to a steel plate held by the magnetic table.

A table saw with an abrasive blade, or a band saw with a suitable blade, can be useful for making rough initial cuts. A wire saw (which utilizes a thin wire impregnated with diamond particles) can be used for making cuts of very narrow kerf width (e.g., from 0.1 to 0.4 mm wide) to conserve material. It is also well suited for making cuts of complex shape because the wire can cut in any direction. However, the cutting rates tend to be slow. If curved cuts need to be made, it may be more practical to use a table router and routing jig. Commercial units are available. A computer numerically controlled (CNC) milling machine is also an option, although the abrasive cutting particles generated can be detrimental to the machine, as previously discussed.

Water-jet and abrasive water-jet cutting of composites is being used more often in composite structural component fabrication (Ramulu and Arola 1994). If properly controlled, these techniques can provide clean cuts with relatively little surface damage. They hold promise for laboratory specimen preparation as well and have been used as such, but not yet extensively. Such cutting devices tend to be expensive because of the high-pressure water pump required, and thus they are not readily available. Current-generation laser beam industrial cutters are used even less often. The heat developed at the surface of the cut can be detrimental to the properties of the test coupon material (Howarth and Strong 1990). The same is true of electrical discharge machining (EDM) (Lau and Lee 1991).

In Chapter 16, testing of laminates containing circular holes will be discussed. It is important to machine the hole, usually by drilling, without causing delamination of the hole edge, and to position the hole at the center of the specimen. One method of drilling holes without causing delamination is to back up the specimen with a strip of plastic such as PMMA. Position the drill at the center of the specimen within ±0.2 mm. A high-speed (about 2000 rpm), water-cooled, diamond-core bit works well to drill the holes, and a number of other suitable techniques exist (Abrate and Walton 1992). Inspect the quality of the hole and, in addition to measuring the hole diameter, measure the net section dimensions on both sides of each hole to check that the hole is centered.

Whatever method of cutting or drilling is used, the finished specimens should be carefully examined for indications of surface damage that could degrade the strength properties. Along with visual inspection, optical microscope, X-ray, ultrasonic scan, dye penetrant, and other techniques are usually suitable for inspection purposes.

4.3 Preparation of Neat Resin Panels and Test Specimens

Neat resin test specimens are typically flat. It is possible to mold the specimens in specially developed, open-face silicon rubber molds, as discussed for the single-fiber fragmentation specimen in Section 4.4. Such a procedure was used to mold dog-bone tensile specimens by Figliolini (2011). There are, however, several challenges with this method of specimen preparation; surface flaws and trapped gas bubbles are common. Such flaws are detrimental for the tensile strength of brittle materials. Surface concavity is another problem. Figure 4.1 shows a cross section of a tensile specimen from an open-face mold.

FIGURE 4.1
Surface concavity in a vinylester tensile specimen obtained from an open-face mold. After Figliolini (2011).

The concave shape of the top and edge surfaces is caused by shrinkage of the resin on cure (Chapter 5). The curved shape of the cross section makes gripping the specimen between flat metal gripping surfaces difficult. Furthermore, the application of strain gages and determination of the cross-sectional area is complicated by the concavity of the cross section.

As discussed by Figliolini (2011), an effective way for making flat resin specimens is to cast a flat sheet of resin and then machine the specimens to the desired shape and dimensions. The dimensions of the specimens become well defined, and bubbles and surface flaws are avoided. We describe here the manufacture of a rectangular sheet of dimensions 254 × 279 mm, thickness in the range from 4.57 to 6.35 mm, but these dimensions may be altered. In this specific case (Figliolini 2011), vinylester resin panels were manufactured.

To enable manufacture of a neat resin panel, prepare a mold consisting of two vertical square glass plates of 9.5 mm thickness and a side length of 305 mm (Figure 4.2). Create an outline of a rectangle defining the panel dimensions (254 mm by 279 mm) on the inner side of one glass panel using vacuum bag sealant tape. If needed, the tape may be doubled up to adjust the panel thickness (4.57–6.35 mm). The top side of the mold should be left open to allow for resin to be poured. Spacers should be placed on the three outer edges of the sealant tape to ensure a uniform thickness of the panel. The spacers should be slightly thinner than the tape thickness to allow the tape to seal when the panels are clamped together. Apply release agent (Freekote 700) to the bonding (inner) surfaces of the mold. Clamp the two glass panels together using C-clamps until both panels touch the spacers on the three

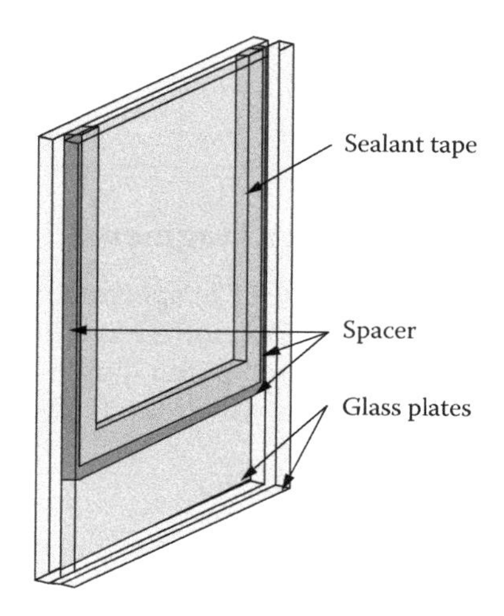

FIGURE 4.2
Mold for making a flat neat resin panel. The outside dimensions are 30.5 × 30.5 cm. After Figliolini (2011).

internal edges. Pour the resin into the rectangular opening between the glass panels on top of the mold. The vertical arrangement of the glass panels and a long gel time of the resin should allow air bubbles within the resin to travel to the top of the mold. This should leave the majority of the resin panel free of voids. Allow the resin to cure before separation of the mold.

To demold the resin panels after completion of cure, remove the clamps. Separation of the panels is often difficult without shattering the glass panels because some bonding of the resin to the glass plates tends to occur. A putty knife carefully pried between the glass plates and the resin panel may break the bonds.

Cylindrical test specimens, such as compression specimens, may be molded in 12.7-mm internal diameter PVC (polyvinyl chloride) pipes. Cut the pipes into segments of 254 mm long, rinse them with water, and let them dry. Coat the internal surface with Freekote® 700 NC release agent to ensure release of the cured resin from the pipe. Apply the release agent with a spray bottle and then distribute it on the internal surfaces of the pipe using a pipe brush. Close one end of the pipe by gluing a PVC slip cap to the end of the pipe using PVC plastic pipe cement. Place the pipe vertically in a stand and allow the cement to dry for about 20 min. Pour resin mixed with the hardener into the PVC pipe and allow the resin to cure.

To remove the molded resin cylinder from the pipe after cure, cut off the end cap. The separation between the resin and mold is assisted by the cure shrinkage of the resin. Cut the neat resin cylinder roughly to size using a band saw. Finishing end cuts are done using a lathe to ensure accurate dimensions and parallelism of the end planes. This is important to achieve a uniform distribution of load during the subsequent compression testing.

4.4 Preparation of Single-Fiber Fragmentation Test Specimen

The single-fiber fragmentation test (SFFT) specimen geometry used by Feih et al. (2004) is shown in Figure 4.3. The dotted line in the center of the specimen indicates the position of the embedded single fiber. Fabrication of the specimen requires a special mold in which the fiber is placed prior to pouring the resin.

4.4.1 Mold Fabrication

A "male" mold with multiple SFFT shapes matching the dimensions indicated in Figure 4.3 should be machined from a solid block of aluminum.

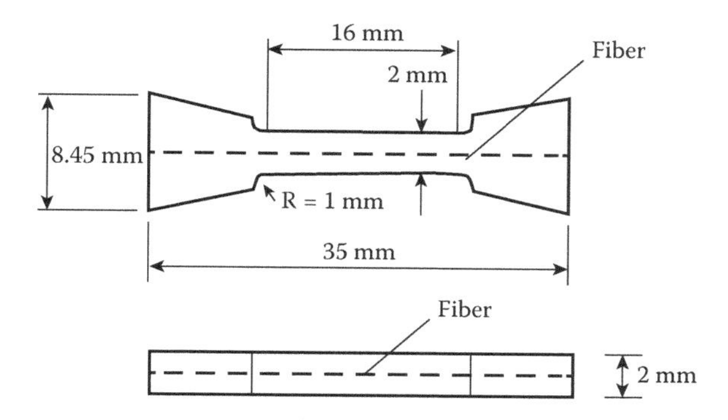

FIGURE 4.3
Geometry and dimensions of the SFFT specimen according to Feih et al. (2004).

Figure 4.4 shows an example of such a mold for the simultaneous preparation of eight SFFT specimens. The rectangular ribs at the ends of each specimen are to produce slots in the actual SFFT mold for positioning of the fiber across the mold cavity. The length of these ribs is about 20 mm, and the height above the mold bottom surface is half the specimen thickness to position the fiber axis at the centroid of the cross section of the specimen.

From the male mold, a "female" mold can be made from silicone rubber (e.g., RTV-644). Figure 4.5 shows a photo of a silicon rubber SFFT mold.

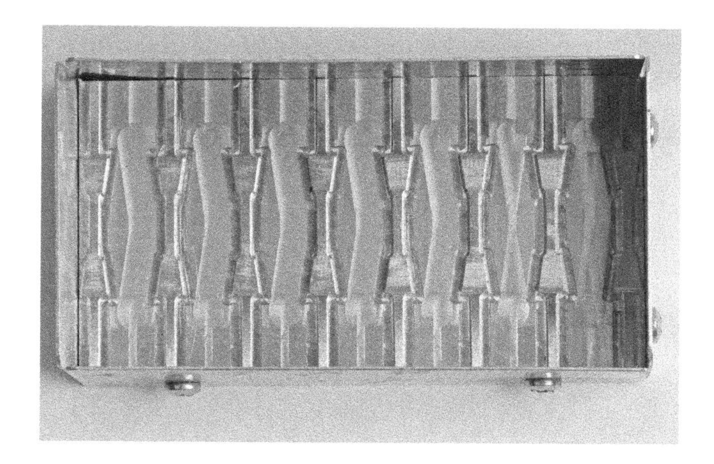

FIGURE 4.4
Aluminum "male" mold for preparation of eight SFFT specimens. (After Ramirez, F.A., 2008, Evaluation of water degradation of polymer matrix composites by micromechanical and macromechanical tests, master's thesis, Florida Atlantic University.)

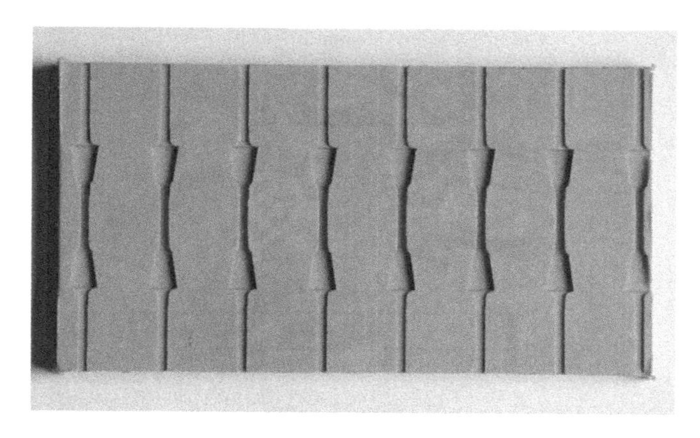

FIGURE 4.5
Silicone rubber SFFT specimen mold.

4.4.2 Molding of SFFT Specimen

A single fiber should be extracted from a fiber bundle with a dental pick. Avoid touching the fibers except at the ends of the bundle. Apply drops of rubber cement to the sprue regions at both ends of the mold cavity. Use a syringe and avoid getting glue in the mold cavity.

Carefully apply tension to the ends of the fiber to straighten it and then place the fiber in the mold. The tension will prevent fiber waviness due to cure shrinkage of the resin. Allow for cure of the rubber cement.

Resin can slowly be poured into the mold, using a syringe, to cover the fiber and fill the mold. Examine the specimen under a microscope at about ×20 magnification. Check for alignment and straightness of the fiber and embedded particles. Discard the specimen if necessary. Figure 4.6 shows an SFFT specimen after molding.

FIGURE 4.6
Single-fiber fragmentation test specimen.

4.5 Tabbing Materials

Some test methods, tensile and compressive tests in particular, require the use of tabs on the test specimen. As will be discussed in detail in relation to the individual test methods, tabs are used to transfer the applied loading into the test specimen from the loading device. Often, these loading devices are wedge grips, with roughened gripping surfaces. The tabs then also protect the surface of the composite test material from damage by the grips.

Currently, glass fabric/epoxy tabs are most commonly used (see Cunningham et al. 1985, Hart-Smith 1991, Xie and Adams 1995). Glass fabric/epoxy is approximately one-third as stiff as aluminum and only about one-tenth as stiff as steel. A more compliant tab material reduces the stress concentration induced because of the discontinuity at the tab end. Glass fabric/epoxy is also a tough and relatively strong material, able to absorb the surface damage induced when aggressive wedge grips are used in testing. The glass fabric/epoxy is usually used in a 0/90 orientation.

A common source of high-quality glass fabric/epoxy sheet for use as tabbing material is commercially available printed circuit board. This material is fabricated to tight thickness tolerances, which also is an important characteristic of a tabbing material. Printed circuit board is used in large quantities by the electronics industry and thus is readily available. It is also relatively low in cost. A commonly available sheet thickness of 1.6 mm is usually adequate. For use on relatively thick specimens of high-strength composites, it may be advantageous to increase the thickness, for example, to 3.2 mm (another readily available thickness) because the roughened grip faces tend to cut deeper into the tabs as the force required to fail the specimen increases.

4.6 Tab Bonding

Tabs can be applied to individual specimens, and years ago, before a better method was developed, they were, but this is a labor-intensive process, more prone to tab alignment errors. (A piece of tabbing material with uncured adhesive between it and the specimen can be very slippery). Even with a full composite plate, misalignment can be a problem if a suitable procedure is not used. Tabbing jigs have been developed and are also available commercially (Wyoming Test Fixtures). These consist of a base plate with projecting pins to index the composite panel and the tabbing strips against and a cover plate containing holes to receive the pins. After

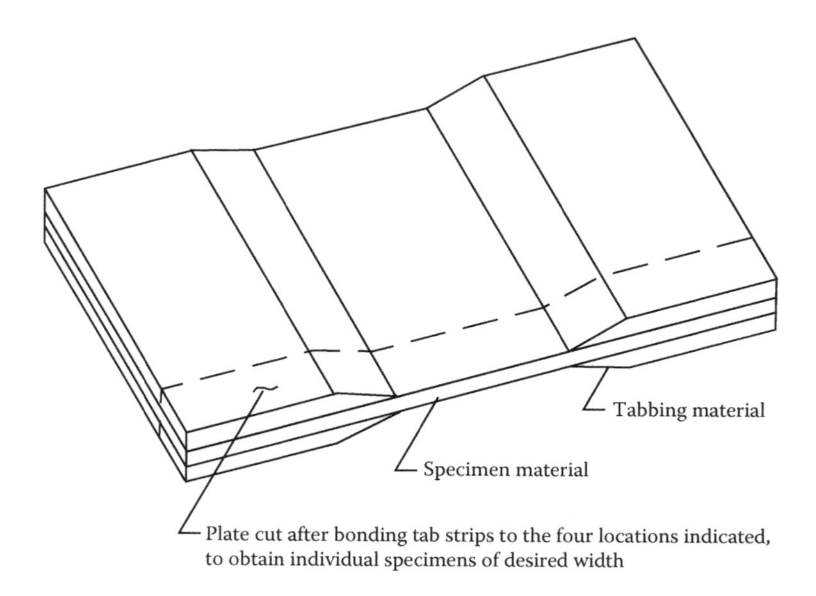

FIGURE 4.7
Fabrication of tabbed specimens.

the five components (the composite plate and four strips of tabbing material) are prepared (Section 4.6.1), they are positioned in the jig, the adhesive must be cured. Particularly for a room-temperature cure adhesive, a weight can be placed on the top plate to provide compaction during cure. However, it is usually more convenient to use a press. Even if only a relatively moderate compaction pressure of, for example, 70 kPa is desired, a force of 2.6 kN would be required for a 150 × 250 mm tab area. If atmospheric pressure is adequate, a vacuum bag can be used as an alternative to a press.

Sometimes, the particular adhesive used must be cured at an elevated temperature. The cure time can be reduced by subjecting it to a slightly elevated temperature. In these cases, it is convenient to cure the assembly of composite plate and tabs in a heated platen press. Because of this potential application, tabbing jigs are typically fabricated of an aluminum alloy because of its good heat conduction properties.

The plate is sliced into individual specimens, as indicated in Figure 4.7, perhaps using a thin abrasive disk in a surface grinder, as described in Section 4.2. If the tabs are to be tapered, as shown in Figure 4.7, they are tapered before being bonded to the composite plate, typically by grinding.

4.6.1 Suggested Tab-Bonding Procedure

Prior to bonding, it is necessary to carefully prepare the bonding surfaces of the specimen panel and the tabbing strips. A suggested procedure is as follows:

1. Lightly sand or grit blast the regions of the panel where the tabs are to be bonded and the bonding surfaces of the tab strips. Do not abrade the gage section surfaces of the specimen panel as this may weaken the material. Use a medium-fine sandpaper or emery cloth (about 180 grit). This step both cleans and slightly roughens the surfaces, enhancing adhesive bonding.
2. Use a wire brush to remove loose particles.
3. Clean the surfaces with a solvent such as acetone to remove any remaining loose particles. Do not touch the cleaned surfaces.
4. Prepare the adhesive. Cut film adhesives to the size required. Mix the components of a two-part adhesive. Desirable characteristics of the adhesive are low stiffness, high shear strength, and thick bond line. Low cure temperature is desirable but must be compatible with the subsequent testing temperature. Thus, a compromise must always be made since these characteristics are not mutually attainable. In the test environment, the shear strength of the adhesive must be adequate.
5. Apply the adhesive to the bonding surfaces of both the specimen panel and the tabbing strips and assemble as previously described. *Take care to keep the gage section regions of the specimen panel free of adhesive. Excessive adhesive is both difficult to remove after cure, and the removal process may damage the specimen.*
6. Cure the assembly as required for the adhesive being used.
7. Inspect the cured panel for proper positioning and alignment of the tabs, absence of excess adhesive, and bond lines of uniform thickness.

4.7 Compression Test Specimen Preparation

A significant source of potential problems in a compression test is in the test specimen. The specimen should be checked for proper fiber alignment, particularly for an axial test of a unidirectional composite lamina, because composite strength and modulus decrease rapidly with off-axis loading. Fiber alignment within ±1° is attainable if sufficient attention is given to mutual alignment of the individual plies of prepreg and the establishment of a visible reference axis during laminate fabrication (Adams and Odom 1991a). A few fiber threads of a contrasting color (e.g., glass fibers in a carbon–fiber composite) can be inserted during prepreg manufacture to assist in attaining alignment. As a final check after the unidirectional composite is cured, the panel could be fractured along the fiber direction. A clean break indicates that the individual plies are well aligned. This edge can also serve as a reference axis when individual specimens are cut from the composite. If a clean break is not obtained, the panel should be remade.

The ends of end-loaded and combined-loading specimens must be flat, mutually parallel, and perpendicular to the specimen axis. Specified parallel and perpendicular tolerances are typically on the order of 0.03 mm. Specimen end flatness is particularly important so that the loading is introduced uniformly to the specimen without end crushing, which is likely to occur if contact is nonuniform. Particularly for composites containing high-stiffness, relatively brittle fibers, such local crushing may then propagate along the specimen, negating the test results. Lower-stiffness, ductile fibers (e.g., Kevlar, Spectra, and other organic fibers) are much more tolerant of specimen end-flatness irregularities.

Uniform specimen thickness from end to end is also important. It is not uncommon for composite panels to have a taper from side to side or end to end, induced by poor alignment of the curing platens or caul plates during panel fabrication. This nonuniformity can result in loading misalignments when specimens are subsequently tested. Uniformity of specimen thickness from end to end within 0.06 mm is commonly specified.

Whenever specimen tabs are used, many additional sources of error exist. Just as for specimen thickness, variations in thickness of any of the tabbing strips can occur, along with variations in thickness of the adhesive bond lines. Not just end-to-end variations, but also differences in thickness of the tab and adhesive components on one side of the specimen relative to those on the other, are important. Symmetry must be maintained to achieve uniform axial loading.

4.8 Hinge Attachment for Double-Cantilever Beam and Mixed-Mode Bending Specimens

Double-cantilever beam (DCB) and mixed-mode bending (MMB) specimens for delamination fracture characterization (Chapter 17) require the attachment of loading tabs. These are typically in the form of hinges (so that no moment is introduced into the specimen). Commercially available metal hinges can be used, if desired. They can be adhesively bonded, bolted onto the surface, or both. Adhesive bonding is less time consuming

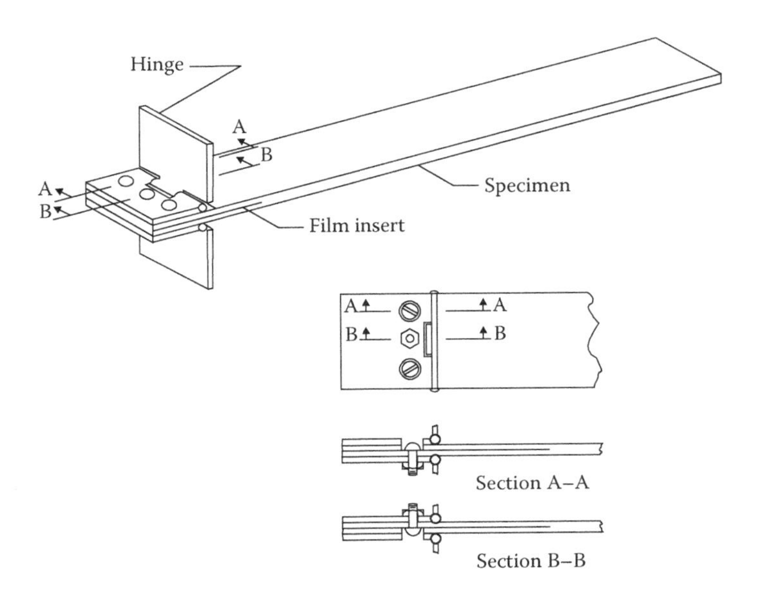

FIGURE 4.8
DCB specimen with hinges attached by bolting.

and is thus preferred, unless a sufficiently strong adhesive is not available. If the hinges are bonded onto the surface, the bonding length is typically on the order of 25 mm but can be shorter if sufficient strength can be attained. The bonding procedure can be the same as that suggested for tab bonding in Section 4.6. A typical specimen with hinges attached by bolting is shown in Figure 4.8. Note that each hinge is bolted to only half the specimen thickness by counterboring the fastener holes from opposite sides, as required.

The hinges must be aligned with the specimen axis and with each other. For this purpose, some type of hinge mounting jig is recommended, such as that shown in Figure 4.9. The hinges can be debonded and reused after completion of the specimen test if the hinges are undamaged.

Bolted hinges are used when the maximum test loading is greater than can be supported by adhesively bonded tabs. As indicated in Figure 4.8, the end of the specimen to which the hinges are to be attached is fabricated with a delamination at the thickness midplane. A hinge must be bolted to each half. This is typically achieved by drilling a hole through the total thickness of the laminate and then counterboring through the half-thickness to provide clearance for the bolt head, as schematically illustrated in Figure 4.8. These attachment holes are placed in a staggered pattern on opposite sides of the specimen, as required, and the hole patterns in the hinges are drilled accordingly. Proper alignment of the hinges must be achieved. The hinges can be adhesively bonded as well as bolted for additional strength if required. As

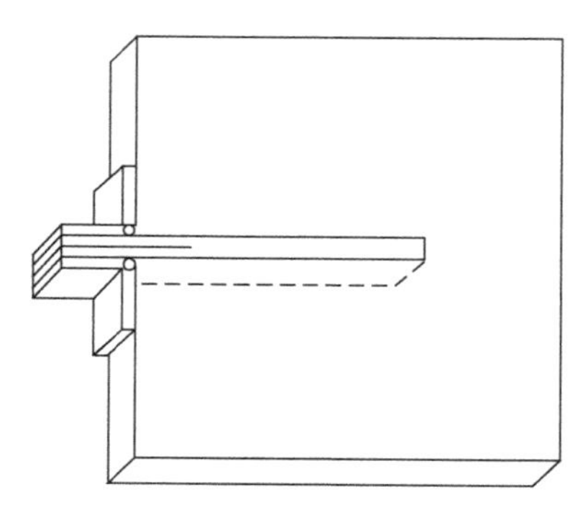

FIGURE 4.9
Jig for alignment and mounting of hinges for DCB and MMB specimens.

suggested, the use of bolted hinges is much more labor intensive and thus not normally used if avoidable. The hinges can be unbolted and reused if undamaged.

4.9 Sandwich Panel Manufacturing

Several of the sandwich beam test methods discussed in Chapter 19 utilize relatively narrow beam and column specimens cut from a larger panel. Sandwich panels may be fabricated using the VARTM (vacuum-assisted resin transfer molding) method described in Chapter 3, where the face sheets are impregnated while the resin forms an adhesive bond to the core. Alternatively, the face sheets may be adhesively bonded to the core as described by Aviles et al. (2009). The face sheets and core are cut to size (30.5 x 30.5 cm in Aviles et al. 2009). The bonding surfaces of the face sheets and core are sanded using 200-grit sandpaper and then cleaned thoroughly with acetone. The edges and edge regions on the nonbonding side of each face sheet should be covered with tape to prevent sticking of excess resin. Apply epoxy adhesive (Hysol® EA9039.3NA aerospace grade) over the core and face sheet bonding surfaces. The two faces and core should be lined up. Put the sandwich in a press and apply a pressure great enough to promote bonding without crushing the core. Aviles et al. (2009) used a pressure of 2.93 kPa. After application of the pressure, the edges should again be adjusted and lined up and excess adhesive removed. After cure, remove the border tape and sand the edges to allow inspection of the face/core interface.

Straight-sided column or beam test specimens may be cut from the sandwich panel using a fluid-cooled saw or water jet as discussed previously in this chapter and the ASTM (American Society for Testing and Materials) D5687 guide (2007).

4.10 Interlaminar Tension Specimens

Two specimen geometries are described in this section: flatwise tension and curved beam tension specimens. Both test geometries present interesting challenges in developing measures of the interlaminar strength and stiffness properties. Of the two, strength has dominated the development of these specimens and approaches.

4.10.1 Flatwise Tensile Specimens

A new ASTM standard, D7291 (2007), was released specifically for measuring solid laminate through-thickness properties. This new standard emphasizes the procedures for producing specimens consisting of a solid laminate adhesively bonded between two metal end tabs composed of either steel or titanium. The standard laminate specimen geometry is circular in cross section to eliminate stress concentrations at corners and is approximately 25 mm in diameter. It may be straight sided or necked down (through the thickness), as shown in Figure 4.10.

The stated minimum specimen thickness for strength determination is 2.5 mm. If through-thickness strain and modulus are to be determined as well as strength, the minimum specimen thickness is 6 mm, so that sufficient thickness is available for mounting strain gages. Increasing the thickness

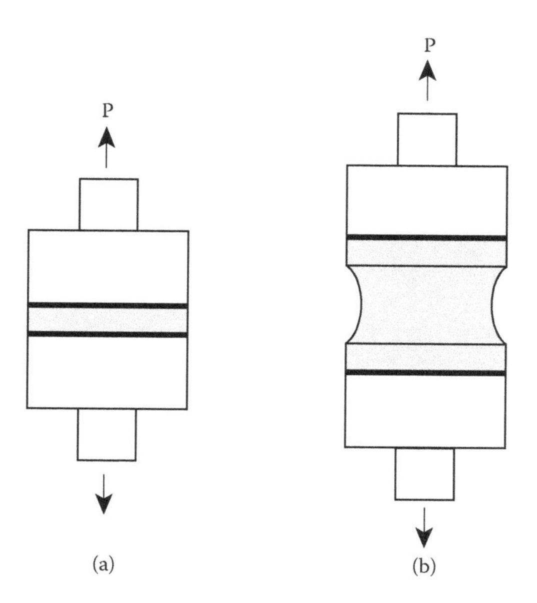

FIGURE 4.10
Typical through-thickness (flatwise tensile) specimens: (a) straight sided and (b) necked down.

also permits necking down (i.e., reduction in cross-sectional diameter of the specimen in the central region) (see Figure 4.10b). If the specimen is at least 25 mm in thickness, it is recommended that the cross section of the specimen be smoothly reduced to a minimum at its geometric center. This reduction in cross-sectional area promotes failure in the gage section as opposed to adhesive failure, particularly if the composite strength is of the same order as that of the adhesive. Note, however, that it may be necessary to reduce the cross section of even thin specimens to force failures into the test section. An efficient way of cutting straight-sided, round specimens from a solid laminate is with a core drill. It may be turned or ground to the desired final shape to achieve minimum cross section at the specimen half thickness.

The specimen preparation procedure outlined in ASTM D7291 (2007) is elaborate, involving the use of a special bonding jig with alignment bushings, as shown in Figure 4.11. It permits the bonding of up to 12 specimens simultaneously. Although not desirable, it may be necessary to machine the bonding faces of the circular specimen before bonding if the as-cured specimen does not meet the specified flatness and parallelism requirements of the standard. A very strong adhesive is recommended.

After the bonded specimens are cured in the alignment and bonding fixture, each individual specimen is subjected to a final machining of its

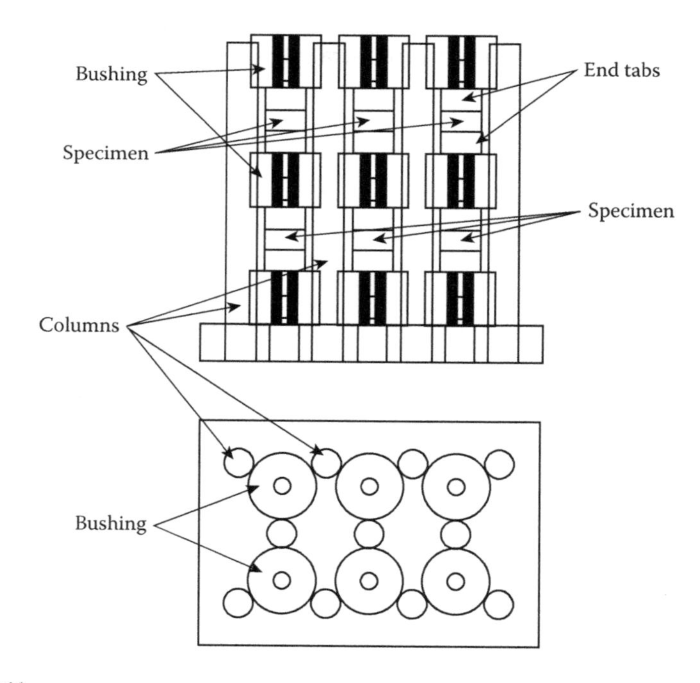

FIGURE 4.11
Drawing of specimen alignment and bonding fixture (showing 12 specimens) according to ASTM D7291.

diameter (and that of the end tabs) by either turning or grinding. This step can be labor intensive, as well as potentially inducing machining damage in the specimen material. Because the ASTM standard specimen preparation is quite involved, the user may consider developing a simpler and more cost-effective approach to specimen preparation after gaining some experience. Whatever approach to specimen alignment and bonding is taken, reducing the cross section of the specimen, even though relatively thin, is an attractive route to take to help control the location of the failure. Grinding or turning in a lathe are both viable approaches, either before or after bonding on the tabs.

4.10.2 Curved-Beam Interlaminar Tension Test Specimen

The preparation of the curved beam test specimen for determination of interlaminar tensile strength can require development of a composite specimen of constant curvature geometry or a specimen geometry consisting of two straight legs connected to a smaller segment of constant curvature. The latter is the recommended geometry. Manufacture of these specimens follows the general procedures outlined for flat laminates of a given composite material using the fabrication and processing approach consistent with the material system under investigation. For example, if the property to be measured is for an autoclave processed materials system, this approach should be followed for preparing the curved beam interlaminar test specimen. In the case of out-of-autoclave processes such as VARTM (discussed previously), those procedures should be followed for specimen preparation.

Preparation of curvilinear specimen geometries in composite materials requires tools with the appropriate curvilinear geometry. For the curved beam interlaminar tensile test specimen, two tool geometries are typically utilized. In the first case, the tool surface in the region of curvature is convex, while in the second approach the tool surface is concave. For the convex tool geometry, the tool surface is adjacent to the minimum radius or internal surface of the specimen, and for the concave tool geometry, the tool surface mates the maximum radius of the specimen (exterior surface). Since processing of thick laminates often encounters frictional resistance for prepreg plies to slip over one another, bridging of plies can result in incomplete consolidation in this region and lead to diminished interlaminar properties. For this reason, it is recommended that the tool surface geometry be convex.

Another consideration for the curved beam specimen is "spring-in," which may occur due to anisotropic shrinkage in the laminate due to cure and thermal expansion. Spring-in is a well-understood phenomenon in prepreg laminates. Essentially, the difference between the out-of-plane and in-plane shrinkages is accommodated by a small change in angle between the tangents at the extremities of the curved section. A simple expression that can be used to estimate the degree of spring-in for the specimen on manufacture is given in Equation (4.1) for thermal expansion, where the change in angle $\Delta\theta$ is proportional to the difference in thermal coefficients of expansion in the radial and in-plane directions β_R and β_L, respectively:

$$\Delta\theta \sim (\alpha_L - \alpha_R)\theta\Delta T \tag{4.1}$$

It should be noted that for the 0^0 laminate, the values of the coefficients of thermal expansion coincide with the minimum and maximum values for the material ($\alpha_L = \alpha_1$ and $\alpha_R = \alpha_3$).

Finally, machining of the test specimen to width is a challenge since the blades in most diamond wafering saws are of relatively small diameter. Water-jet machining can solve this problem, but care must be taken to ensure no edge damage results. Care must also be taken to support the specimen during the machining process so interlaminar damage is not induced in the specimen prior to test.

4.11 Specimen Conditioning

It is common practice to store and test specimens in the ambient laboratory environment. Often, this environment is relatively uncontrolled, although 23°C and 50% relative humidity (50% RH) is a commonly stated standard condition. It is important to note that the total amount of moisture a polymer absorbs from the surrounding air is directly proportional to the relative humidity (Shen and Springer 1976). Thus, given sufficient exposure time, a polymer–matrix composite at 50% RH will attain one-half of its maximum moisture absorption capability (one-half of its saturation level). If the influence of moisture on composite properties is a concern, and normally it should be, it is obviously important to control the conditioning environment.

Unlike thermal equilibrium, moisture equilibrium at any given percentage RH requires long exposure times, typically measured in months and even years, depending on the thickness of the composite and the exposure temperature. The rate of moisture diffusion increases with increasing temperature. Thus, it is common practice to moisture condition and dry polymer–matrix composites at elevated temperatures. It is important that the temperature not be too high, however, because the moisture gradient at the surface can produce damage. For example, consider a moisture-saturated polymer–matrix composite placed in a desiccator at a high temperature. Because of the high coefficient of moisture diffusion at that temperature, the surface will lose moisture rapidly, whereas the inner layers are still relatively unaffected. The surface material tends to contract (shrink) as it dries out but is restrained by the material below it. This induces tensile stresses in the surface material, which can become high enough to cause local cracking (termed microcracking) of the material (Mahishi and Adams 1984, Obst et al. 1996).

4.12 Strain and Displacement Measurements

In Chapter 2, the engineering constants were defined in terms of the stress–strain response of the composite, and the specimen compliance was defined in terms of the displacement at the point of load application. To measure strains and displacements of test specimens, the three most common transducers used are electrical resistance strain gages, extensometers, and linear variable differential transformers (LVDTs). Optical methods are also used, but to a lesser extent.

For strain measurement, electrical resistance strain gages are versatile, reliable, and accurate (Whitney et al. 1984, Dally and Riley 1991, Kobayashi 1993). Because strain gages consist of thin metal foils bonded to the surface of the specimen, they cannot be reused. Fortunately, their cost is moderate. A schematic of a typical strain gage is shown in Figure 4.12. If the composite is very inhomogeneous, such as a coarse-weave fabric composite, the gage must be large enough to cover a representative area of the microstructure, such as the weave structure (Tuttle and Brinson 1984, Masters and Ifju 1997).

The resistance of the gage used is important because of gage heating effects. The strain gage is normally wired into a Wheatstone bridge circuit and is activated by application of a constant voltage to the gage. When the specimen and gage are deformed, the gage resistance changes in proportion to the strain. This produces a calibrated voltage offset. Because the power dissipated by the strain gage is proportional to the square of the voltage divided by the resistance of the gage, higher-resistance gages cause less heating for a given voltage. Gage heating can be a problem with polymer–matrix composites because the polymer–matrix is a poor heat conductor. This allows heat to build up in the gage. The resulting temperature change causes a resistance change, which is falsely recorded as an apparent strain. Therefore, the use of 350-Ω strain gages is commonly recommended for composite testing, although 120-Ω gages are sometimes used. The gage excitation voltage should be less than 5 V. Sometimes, temperature compensation is necessary, as discussed further in Chapter 13.

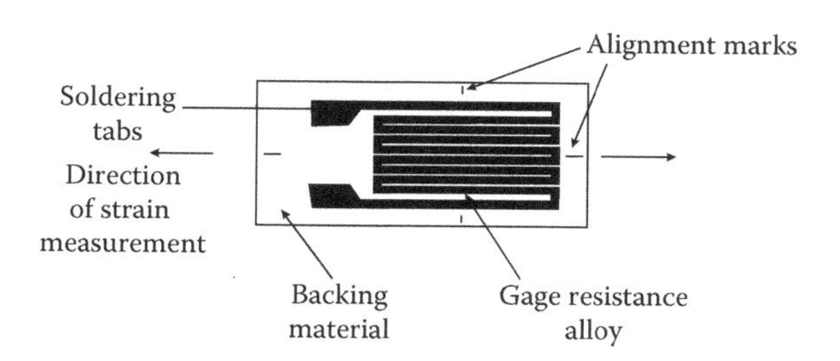

FIGURE 4.12
Schematic of an electrical resistance strain gage.

The accuracy of strain measurement using bonded-foil strain gages relies on a high-quality bond between the gage and the test specimen. The specimen surface should be carefully prepared (Perry 1988). The strain gage adhesive must be usable at the test temperature of interest.

Accuracy is also influenced by the transverse sensitivity of the strain gage. As indicated in Figure 4.12 the thin metal foil elements oriented in the direction of the desired strain measurement are connected by transverse foil elements to maintain continuity of the current flow. Because of Poisson effects for uniaxial loading, or deliberate biaxial loading, these transverse elements become strained and thus also change in resistance, adding to the total resistance change of the gage. If these transverse strains are not negligible, results can be modified to account for transverse sensitivity. Guidelines for such corrections are provided by the Measurements Group (1982).

Extensometers are also used for strain measurements. A typical single-axis extensometer is shown in Figure 4.13. Biaxial extensometers that measure both axial and transverse strains are also available (e.g., Instron Corporation, MTS Systems Corporation). Extensometers often contain strain gages bonded to the arms of the device. As the arms deflect, the strain gages are activated just as discussed. Thus, these units are sometimes also called strain gage extensometers to distinguish them from the now much less commonly used dial gage and optical (rotating mirror) extensometers.

The accuracy of strain gage extensometers can be equal to that of bonded strain gages, as might be expected because the basic technology is the same. For a strain gage, the gage length is the length of the foil grid (Figure 4.12). Commonly used gage lengths are in the 1.5- to 6-mm range. The gage length of an extensometer is the distance between contact points on the specimen. Typical extensometers have gage lengths in the range of 12 to 50 mm. A commonly used gage length is 25 mm. Thus, the gage length of an extensometer

FIGURE 4.13
Typical single-axis extensometer, mounted on a tensile specimen. (Courtesy of Epsilon Technology Corporation, 1997.)

tends to be much longer than that for a strain gage, which can be an advantage or a disadvantage depending on the application. For example, if there are steep strain gradients in the specimen, a long gage length is a disadvantage if a local (point) strain is desired. On the other hand, if it is desirable to average out local surface strain variations (such as for a coarse-weave fabric composite), a longer gage length is advantageous. Extensometers also accommodate local roughness of the specimen surface.

Strain gage extensometers are less commonly used in composites testing than they were some years ago. Extensometers have increased in price proportionally much more than strain gages since the 1990s. An extensometer is relatively expensive and thus must be used many times to amortize the purchase price. However, it is often desirable to leave the extensometer attached to the specimen until the specimen fails; this produces a complete stress–strain curve. Failures of composites can sometimes be rather violent because the large amount of stored elastic strain energy is suddenly released. This can damage or destroy the expensive extensometer. Extensometers can also be damaged or destroyed when they are accidentally dropped on the floor during installation (and perhaps then stepped on in the attempt to catch them). Thus, often the extensometer does not survive to reach its amortization life. As previously noted, strain gages are intended to be used only once, and the cost per gage is relatively low.

LVDTs, such as those shown in Figure 4.14, are mostly used for monitoring displacements. Examples include the center span displacement of a beam (as discussed in Chapter 11) or the load-point displacement of a fracture specimen (Chapter 17).

The LVDT produces an electrical output as its core moves. This output voltage, however, is directly proportional to the core displacement only over a specified range—the working range of that particular LVDT. The resolution of an LVDT depends on the electronic system used to convert the input signal. A displacement resolution on the order of 25 μm can be achieved.

Whatever type of device is used, it is important that it does not reinforce the surface of the composite and thus indicate erroneously low strains. This can be of particular concern when metal foil strain gages are used (Jenkins 1998). Membrane stiffness of the gage should not be greater than 10% of the substrate material. Optical (noncontacting) strain measurement devices have a distinct advantage in this regard because reinforcement of the surface is nonexistent. Although specialized, full-field techniques such as moiré, holography, and speckle interferometry exist (Whitney et al. 1984), these techniques are designed to measure strains throughout an area rather than at a point, and their use requires considerable training. In contrast, the use of a laser beam to track the relative movement of two points on the surface of the specimen being strained is much more straightforward, and commercial equipment is available (Instron Corporation, MTS Systems Corporation). However, the initial acquisition cost of a laser extensometer is typically considerably higher than that for a strain gage, extensometer, or LVDT measurement system, and

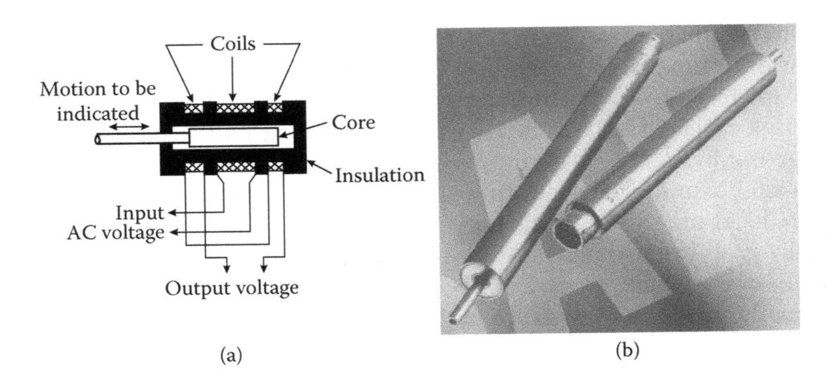

FIGURE 4.14
(a) Principle of a linear variable differential transformer (LVDT). (b) Photograph of typical LVDTs. (Courtesy of Lucas Shaevitz, Inc., 1990.)

the resolution of the currently available laser systems is generally not as high. The acquisition cost factor, in particular, has limited their use to date.

Recently much attention has been given to another type of optical technique that is able to provide quite accurate displacement and strain measurements over a finite area of the specimen. A very powerful method is the digital image correlation (DIC) method where an area of the specimen is illuminated by a laser (Sutton et al. 2011). The displacements of points on the surface of the body are expressed in pixels when the pictures are analyzed. The scale can vary from um to m depending on the accuracy required. A reference image is created for the undeformed object which is compared to an image of the deformed object. The locations of points on the surface of the deformed object are obtained by digital image correlation. By identifying the same points on the surface of the undeformed object it is possible to map the displacements and calculate the distribution of strains for the entire image. Although this method is most often applied to structural analysis, it has also found its place in materials testing, (see, for example, Kamaya et al. 2011 and Taher et al. 2012).

4.13 Testing Machines

Specialized test fixtures are described in relation to specific tests as they are discussed in the following chapters. In most cases, these specialized fixtures are designed to be mounted in a universal testing machine. As the name implies, these are general-purpose machines. They vary greatly in physical size, load capacity, versatility, and sophistication. The most basic machine should be capable of applying a uniaxial tensile loading (by controlling the motion of a moving crosshead) and indicating the corresponding force on the specimen. Most modern machines can apply both axial tensile and compressive loading. They can be operated in force or strain control as well as (crosshead) displacement control. An electronic load cell and multiple-channel strain–displacement signal conditioning electronics feed into a computerized controller, which processes these data and presents and stores the results in the forms desired. These can include stress–strain, stress–displacement, and strain–strain (for Poisson's ratio) plots; tabulations of axial stiffness; Poisson's ratio; ultimate strength; or whatever quantity is desired.

Each universal testing machine is designed to have a maximum load capacity. Small units may have a load capacity of only a few hundred newtons, sometimes even less. There is no limit on maximum load capacity. Machines of 10 MN capacity and larger exist and are used routinely. However, a common capacity for most composites testing is on the order of 100 kN; some test methods will require a larger machine. The size of the load cell may also limit how much force any machine can exert; these load cells are designed to be readily interchangeable. For example, a 100-kN machine can be used to test single fibers with breaking forces of less than 1 N if a sufficiently small load cell is used.

Most universal testing machines are of two basic types, electromechanical or servohydraulic. Electromechanical machines are typically screw driven, with electronic feedback. Servohydraulic machines, as the name implies, are hydraulically powered with electronically controlled servovalves. Typical examples of commercially available machines of each type are shown in Figures 4.15 and 4.16.

FIGURE 4.15
Typical electromechanical universal testing machine. (Courtesy of Instron Corporation, 1997.)

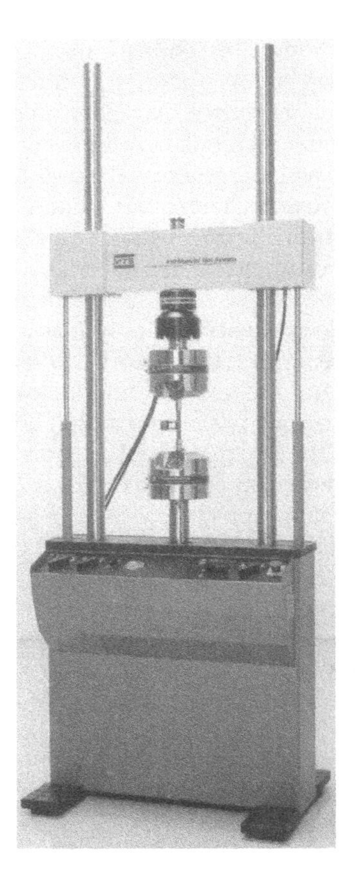

FIGURE 4.16
Typical servohydraulic universal testing machine. (Courtesy of MTS Corporation, 1999.)

In addition to universal testing machines, special devices are often developed for specific purposes. Examples include creep frames (for determining time-dependent deformations under constant applied force), resonant frequency fatigue machines, impact devices (e.g., drop weight and swinging pendulum), and biaxial loading machines.

5

Cure Shrinkage and Residual Stress Testing of Resins

5.1 Cure Shrinkage in Thermosets

Cure shrinkage is important because it can be responsible for changes in shape of the manufactured part geometry after cure is complete and because differential shrinkage can produce residual stresses in the material that can contribute to failure. Shrinkage in thermosetting polymer matrices is made up of two components: (a) shrinkage due to volumetric reduction resulting from cross link formation and (b) thermal contraction. While these phenomena occur together during the thermal curing process, it is the state of the polymer that determines the contribution of each in the development of residual stresses within the composite. Prior to gelation, the polymer offers little resistance to shrinkage, and in the rubbery state, the stiffness is significantly less than that after vitrification. Further, the viscous nature of the polymer prior to vitrification may also allow residual stresses to relax out over time. Thus, it is necessary to follow both the polymer shrinkage and its effective stiffness to determine the magnitude of residual stresses. To separate the chemical shrinkage from the thermal expansion behavior of the polymer, it is useful to examine a single thermal cycle of heating the initially uncured polymer above its glass transition temperature and cooling it to the initial temperature (see Figure 5.1).

The uncured polymer is heated from a reference temperature, T_o, along a–b to its cure temperature T_{cure} with a corresponding increase in volume due to the thermal expansion. It is held at this temperature until cure is complete, with a corresponding reduction in volume (b–c) due to cross link formation. On cooling, the polymer contraction exhibits one rate before (c–d) and a different rate after (d–e) it passes through the glass transition temperature T_g at d. The shrinkage in the region (c–d) occurs when the polymer is in the rubbery state, while it is in the glassy state in the region (d–e). Further, the T_g is a function of the degree of cure of the polymer; therefore, shrinkage of the polymer is related to the extent of cure that it has achieved. The model for cure kinetics

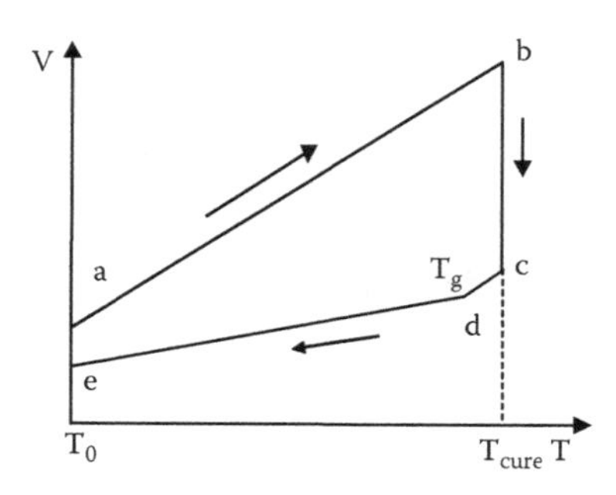

FIGURE 5.1
Volume change in the thermal cycle.

of the thermoset polymers in Chapter 3 can be utilized to determine degree of cure as a function of polymer characteristics and thermal history.

Measurement of shrinkage is complicated by the interaction of these phenomena. If the change in volume due to cure shrinkage is taken simply as the difference in volume before and after completion of the thermal cycle (a–e), as recommended in many test methods, the combined effects of volume change due to cure shrinkage, phase change from gel to glass, and thermal expansion cannot be separated. Rather, it is important to differentiate between these two phenomena if the thermal residual stresses are to be determined.

The test standard for measuring resin shrinkage, ASTM (American Society for Testing and Materials) D2566-86 (1986), *Standard Test Method for Linear Shrinkage of Cured Thermosetting Casting Resins during Cure*, was withdrawn in 1993 with no replacement. However, ASTM D2566-86 was not a successful method for separating chemical shrinkage from thermal expansion, as illustrated in Figure 5.1. Here, the true chemical shrinkage is clearly b–c, while the test standard result reports only a–e. Note that the true coefficient of thermal expansion of the polymer is d–e.

Further, the interaction between the thermoelastic deformation of elastic fibers and the shrinking polymer matrix is an important factor in the introduction of thermal residual stresses in the material. Taken as an orthotropic continuum, the laminae also exhibit expansion characteristics that are different in the fiber direction than perpendicular to the fiber direction. Thus, when laminae are assembled into a laminate, the lamina differential shrinkage is also responsible for thermal residual stresses, as discussed previously.

Measurement of shrinkage of the polymer matrix can be accomplished by several methods. The most direct is the use of the dilatometer to measure the change in length of a single dimension of the sample. Assuming isotropic deformation of the polymer, the volumetric changes follow. For the gravimetric approach, a polymer sample is suspended in a liquid, and its apparent weight is measured as the temperature is increased. With the increase in temperature of the suspending fluid and sample, the change in apparent weight reflects its change in density and thereby its volume according to:

$$\frac{\Delta V}{V} = \frac{\Delta W_{T_c}}{\rho_f V_{r,T_c}} \tag{5.1}$$

where ΔW_{T_c} is the change in the apparent weight of resin in the fluid bath at the cure temperature, ρ_f is the density of the suspending fluid, and V_{r,T_c} is the volume of the uncured resin system at the cure temperature (T_c).

Since the shrinkage of a given thermoset polymer is related to its cure state, several models have been developed for the prediction of polymer shrinkage during cure. Bogetti and Gillespie (1991) and Johnston et al. (2001) proposed a model wherein shrinkage initiates at a given degree of cure α_{c1} and progresses with degree of cure until it reaches a maximum at $\alpha = \alpha_{c2}$.

$$V_{rs} = \begin{cases} 0 & if \quad \alpha \leq \alpha_{c1} & (5.2a) \\ A\alpha_s + (V_{rs}^{\infty} - A)\alpha_s^2 & if \quad \alpha_{c2} \geq \alpha \geq \alpha_{c1} & (5.2b) \\ V_{rs}^{\infty} & if \quad \alpha \geq \alpha_{c2} & (5.2c) \end{cases}$$

$$\alpha_s = \frac{\alpha - \alpha_{c1}}{\alpha_{c2} - \alpha_{c1}} \tag{5.2d}$$

where V_{rs} is the polymer volume corresponding to a degree of cure α, and V_{rs}^{∞} is the polymer volume at $\alpha = 1.0$, while A is a constant. The second term in Equation (5.2b) involving α_s^2 is often neglected.

5.2 Modulus versus Cure State

Changes in polymer viscoelastic properties during the curing process are important to the understanding of the development of shrinkage stresses in thermosetting polymer matrices. However, the full viscoelastic characterization of these systems is quite complex and beyond the scope of this book. For this reason, a simplified approach is often taken wherein elastic properties are assumed to occur instantaneously during cure. One such approach

TABLE 5.1

Terms of the Modulus Development Relationship

Variable	Description	Units
a_{Tg}	Constant	°C or °F
T	Temperature	°C or °F
T_g^0	The uncured resin glass transition temperature	°C or °F
T^*	Difference between resin temperature and resin instantaneous T_g	°C or °F
T^*_{C1}	T^* when resin modulus begins to increase	K
T^*_{C2}	T^* when resin modulus has reached its full value	K
E_r	The temperature-dependent resin modulus	Pa or psi
E_r^∞	The resin elastic modulus at T_0 and $\alpha = 1.0$	Pa or psi

utilizes a linear model for modulus development with degree of cure α as shown in the following (Bogetti and Gillespie 1991):

$$T^* = \left(T_g^0 + a_{Tg}\alpha\right) - T \tag{5.3a}$$

$$E_r = \begin{cases} E_r^0 & if \quad T^* < T^*_{C1} \quad (5.3b) \\ E_r^0 + \dfrac{\left(T^* - T^*_{c1}\right)}{\left(T^*_{C2} - T^*_{C1}\right)}\left(E_r^\infty - E_r^0\right) & if \quad T^*_{C1} < T^* < T^*_{C1} \quad (5.3c) \\ E_r^\infty & if \quad T^* > T^*_{C2} \quad (5.3d) \end{cases}$$

The benefit of this approach is that the modulus development is a simple function of degree of cure and temperature. The symbols are defined in Table 5.1.

Note that the polymer modulus depends on both temperature and degree of cure. Further, the benefit of the relationship presented in Equation (5.3) is that the development of the polymer stiffness during cure can be tracked by its cure state α. Using the cure kinetics models presented in Chapter 3, it is possible to connect the stiffness of the polymer to the degree of cure and thereby model the complex interactions that produce residual stresses due to chemical shrinkage.

5.3 Coated Beam Cure Shrinkage and Residual Stress Test

The coated beam cure shrinkage and residual stress test was investigated in detail by Stone (1997), Flores (2000), and Farooq (2009). The approach relies on a method whereby a relatively thin coating of uncured polymer is applied

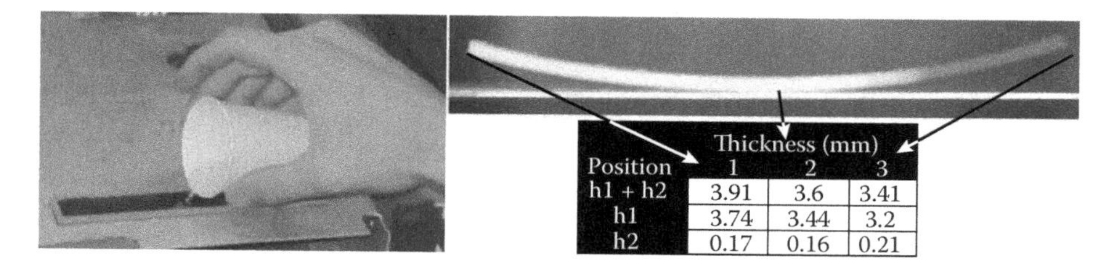

FIGURE 5.2
Liquid polymer poured onto elastic strip and layer thicknesses.

onto an elastic substrate (Figure 5.2). The coated substrate is then heated to produce a fully cured thermoset polymer coating. Shrinkage in the polymer produces a curvature in the elastic substrate so that shrinkage strains can be deduced from the observed bilayer beam curvature. Clearly, shrinkage in the liquid state occurs without inducing stress since the elastic modulus is zero. Further, shrinkage of the polymer in the liquid state can be expected to occur without dewetting the substrate surface. However, the transformation of the polymer from the liquid to the gel and then to the glassy state is accompanied by a change in the properties of the polymer [Equation (5.3)], and it is expected that the polymer bonds to the substrate in this step. Once the polymer is in the gel state, its stiffness is nonzero, and then it continues to increase with degree of cure of the thermoset polymer until it reaches the glass state. Therefore, as the polymer shrinkage occurs, it does so in a state of continuous change in modulus. The curvature of the polymer-coated elastic substrate therefore results from the combination of cure shrinkage and modulus development. In most applications, it is exactly this phenomenon that is of interest since it is one of the causes of residual stresses in composite materials and structures.

Another approach to model modulus change with cure state is to assume that the polymer modulus is zero until the state of cure is that of the gel point (approximately $\alpha = 0.5$) and to increase linearly to its maximum point at the final cure state of the thermoset polymer, as determined by the cure kinetic model in Equations (3.1) and (3.8) for an isothermal history. However, under isothermal conditions above room temperature (RT), the modulus of the curing polymer remains sensitive to temperature and does not achieve its value at RT. Rather, the polymer reaches the modulus appropriate to the isothermal temperature at which the cure process takes place as shown in Equation (5.3). Of course, the cure state develops in a nonlinear fashion with a sigmoidal shape with time, with the curing process beginning slowly, increasing rapidly, and then asymptotically approaching the final state of cure. Therefore, a linear increase in modulus with degree of cure will yield a nonlinear modulus increase with time.

When the uncured polymer-coated substrate is subjected to an isothermal thermal history sufficient to bring the polymer to its maximum degree of cure, the beam curvature can be related to the cure shrinkage, β^{ch}, according to

$$\kappa = \frac{1}{\rho} = \frac{6\left[1+\frac{h_1}{h_2}\right]^2 [\beta^{ch} + (\alpha_2 - \alpha_1)(T - T_0)]}{[(h_1 + h_2)]\left[3\left(1+\frac{h_1}{h_2}\right)^2 + \left(1+\frac{E_1 h_1}{E_2 h_2}\right)\left(\frac{h_1^2}{h_2^2} + \frac{E_2 h_2}{E_1 h_1}\right)\right]} \tag{5.4}$$

where the radius of curvature is ρ, while h_1, h_2, E_1, E_2, α_1, and α_2 are the thicknesses, moduli, and coefficients of thermal expansion of the resin (1) and substrate (2), respectively. This requires instantaneous use of the modified Timoshenko bimetallic strip Equation (5.4) because the modulus of the resin (E_1) changes with degree of cure. Equation (5.4) may used to calculate the change in curvature for time increments wherein the modulus and cure shrinkage are determined by the cure state. Note that Equation (5.4) includes both the thermal and chemical strains so that the curvature at the reference temperature contains both effects. However, if the reference state is the isothermal curing temperature, the thermal strain is not present. It is only when the substrate is cooled that the thermally induced curvature will be observed.

Figure 5.3 shows results from Stone (1997) for cure of vinylester resin (3.05 mm thickness) on an aluminum strip (0.41 mm thickness), where deflection

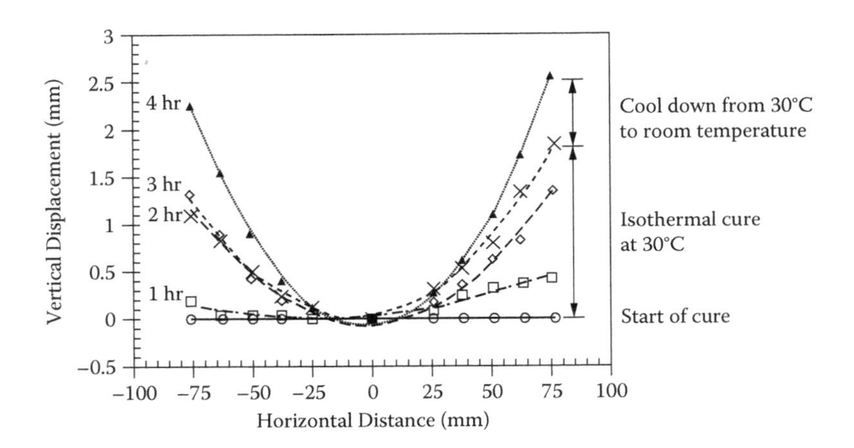

FIGURE 5.3
Bimaterial specimen deflection data due to cure shrinkage and thermal expansion.

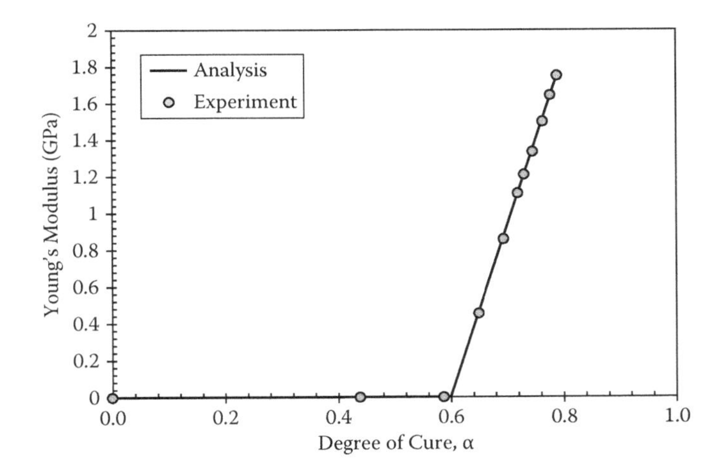

FIGURE 5.4
Young's modulus as a function of degree of cure and time for isothermal cure at 100°C for an epoxy resin.

of a beam is displayed at various times for a 30°C cure followed by a cool down to RT. A simple approach to determine radius of curvature from the displacements of the beam can be derived by arranging the coordinates of the beam centerline to be measured from the center of curvature, $x = 0$, $y = \rho$, where ρ is the radius of curvature and the curvature, $\kappa = 1/S$. κ may be expressed in terms of the deflection (y) of the beam at any given lateral position x.

$$\kappa = 2y/(x^2 + y^2) \tag{5.5}$$

After full isothermal cure and on cooling the polymer-coated substrate to RT, its curvature will increase solely due to thermal effects. Indeed, by increasing the temperature of the coated substrate between RT and the isothermal cure temperature, it is possible to determine the polymer coefficient of thermal expansion from Equation (5.4) (taking $\beta^{ch} = 0$ and E_1 = fully cured polymer modulus).

Clearly, proper interpretation of these test results requires knowledge of the change in Young's modulus of the polymer as a function of thermal history and cure state. Thus, a review of the discussion of thermoset cure kinetics in Chapter 3 is essential for the reader unfamiliar with the subject.

Consider example results for shrinkage strain measurements for an epoxy resin (Crosslink Technology, Inc., XRD 1014/CLH 6372) made with the bilayer beam: The substrate layer was a precured, single unidirectional ply of carbon-fiber composite (IM7/8552). Figure 5.4 shows the assumed

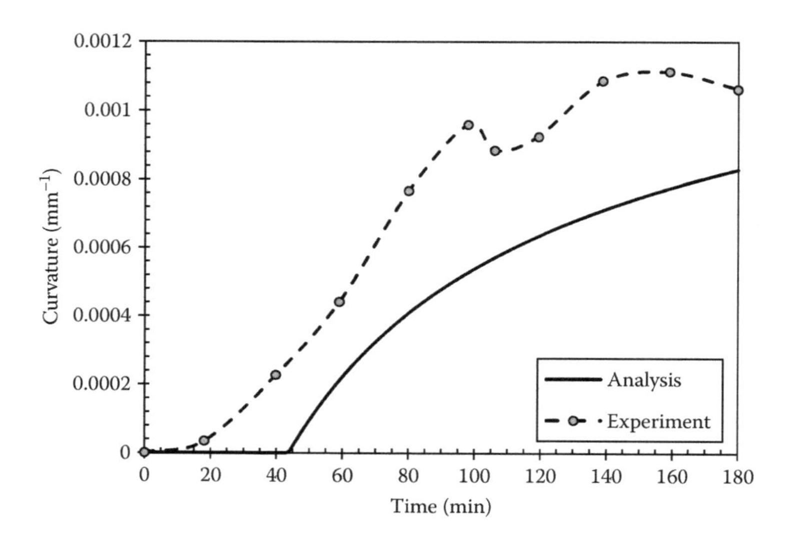

FIGURE 5.5
Experimental bilayer beam curvature (mm^{-1}) and predictions from Equations (5.3) and (5.4) as a function of time (minutes) for isothermal conditions (100°C).

linear relationship between resin modulus and degree of cure α. The initial point of increase of modulus is taken as $\alpha = 0.6$. Given that the modulus of the resin increases with cure according to Equation (5.3), Equation (5.4) is solved in an incremental manner. Following this approach for an isothermal cure at 100°C and using the results in Figure 5.4, the curvature versus time can be calculated. Figure 5.5 shows experimental results and prediction of curvature by Equation (5.4) using the incremental approach. The experimental data and predictions shown in Figure 5.5 deviate much, but it must be pointed out that successful implementation of the coated beam test is not without difficulty. Applying uncured, liquid polymer on a solid substrate to achieve a uniform thickness is problematic (Figure 5.2). Techniques to succeed with this test method are acquired after a considerable number of trials. The liquid resin on the substrate surface is typically constrained with elastomeric dams. While the polymer does not bond to the elastomer, it can exhibit some adhesion to the dam and thereby constrain deformation of the beam in the early phase of the experiment. Equation (5.4) must be solved in an incremental manner since the modulus and shrinkage change simultaneously.

Finally, the bilayer beam test would allow separation of the chemical shrinkage from thermal expansion when cure is either conducted under nonisothermal conditions or the test sample subjected to isothermal conditions is heated or cooled after it reaches the full extent of cure. In the latter case, the change in beam curvature with temperature is the thermal expansion measure since no further chemical shrinkage occurs after full cure is achieved.

6

Single-Fiber Tensile Testing and Matrix Characterization Tests

6.1 Introduction

Fiber tests are used extensively for quality control purposes and evaluation of the effect of potentially detrimental fiber surface treatments. They are also used to evaluate environmental degradation of fibers and to determine mechanical properties required for input in micromechanical analysis (Chapter 2). Most fiber test methods are designed to determine the modulus and strength of the fiber in uniaxial tension, although compression tests such as the recoil test have been approached (see Allen 1987 and Kozey et al. 1995).

6.2 Single-Filament Test

The single-filament test of ASTM (American Society for Testing and Materials) D3379 (1989) (withdrawn in 1998) provides methods to determine fiber tensile modulus and strength. For the test, a single filament is separated from a dry tow of fibers and mounted on a slotted cardboard tab as shown in Figure 6.1. The tabbed specimen is then inserted in a testing machine; the supporting tab carefully cut on both sides of the fiber or burned away using a hot wire, a soldering iron, or something similar; and the unsupported single filament tested to failure. The test requires a very low force load cell, such as a 5- to 20-N (0.5- to 2-kg) capacity load cell. Typically, there is large scatter in fiber modulus and strength data (Van der Zwaag 1989). Hence, a large number of specimens must be tested to obtain statistically significant results.

One source of variability in the test data is the fiber cross-sectional area used to determine modulus and strength. The fiber cross-sectional area is determined on the dry tow from which the fibers are selected for specimen mounting. This means that an average cross-sectional area is used, not the one specific for each fiber.

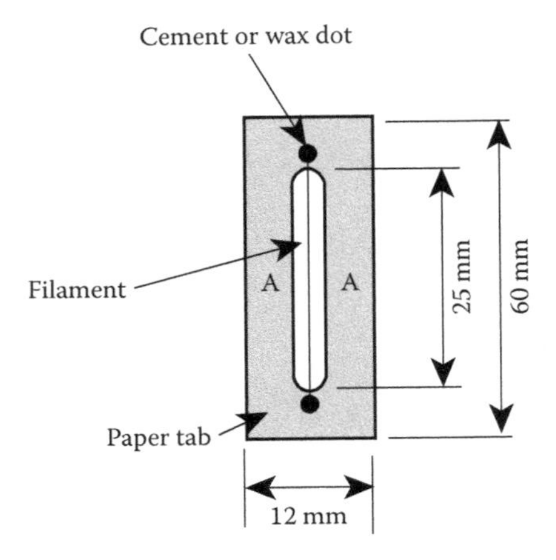

FIGURE 6.1
Single-filament test specimen.

6.2.1 Single-Filament Test Procedure

Choose a suitable single filament from a strand bundle. Care must be exercised to ensure that coalesced fibers are not chosen. Each fiber should be inspected under an optical microscope prior to specimen preparation. Center the filament over the tab opening with one end taped to the tab. Lightly stretch the filament and fix the free end to the tab using tape. Carefully place small dots of adhesive on the filament at the slot edges to bond the filament to the tab (Figure 6.1). The specimen gage length is defined by the length of the open slot in the end tab, nominally 25 mm as shown in Figure 6.1. Determine the gage length to within ±0.1 *mm*.

Mount the specimen in the grips of a properly calibrated testing machine. Visually check for alignment of the test specimen between the grips. Cut both sides of the mounting tab or burn the tab away at the midgage length. Use care not to damage the filament. Set the crosshead speed to achieve failure in about 1 min. Ramirez (2008) used a crosshead speed of 0.5 mm/min.

Apply load to the specimen while recording load and crosshead displacement until the filament fails. Fibers used as reinforcements of composites typically display linearly elastic behavior up to failure; examples of stress–strain curves for carbon and glass fibers are shown in Figure 6.2.

6.2.2 Data Reduction for Modulus and Strength

The fiber elastic modulus is defined as follows

$$E = \frac{\sigma}{\varepsilon} \tag{6.1}$$

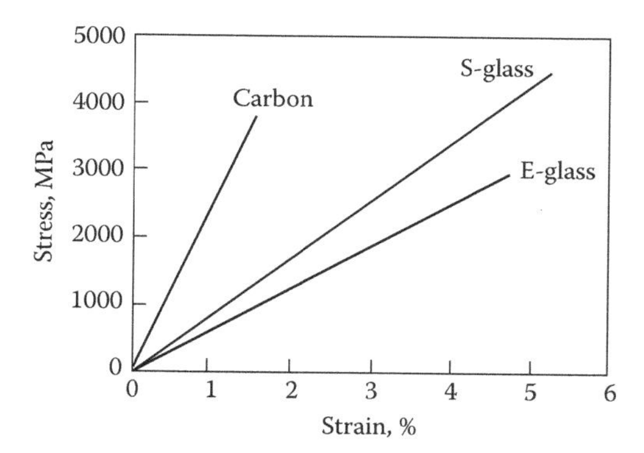

FIGURE 6.2
Typical stress–strain curves for carbon and glass fibers loaded in tension.

where σ is the fiber stress defined by

$$\sigma = \frac{P}{A_f} \tag{6.2}$$

where P is the load applied, and A_f is the fiber cross-sectional area. The strain is defined as the crosshead displacement (δ) divided by the gage length L:

$$\varepsilon = \frac{\delta}{L} \tag{6.3}$$

Notice that Hooke's law, Equation (6.1), demands strain in the linear region of material response, that is, $\varepsilon \ll 1$. For determining A_f, ASTM D3379 suggests planiometering a representative number (≥10) of fibers using photomicrographs of the fibers to determine the average cross-sectional area. Ramirez (2008) determined the cross-sectional area based on measurements of the fiber diameter with a scanning electron microscope. If the cross-sectional area of a tested fiber needs to be determined, photomicrographs of the broken ends may be prepared and examined.

The strain determined from the crosshead displacement, Equation (6.3), includes contributions from the load cell and gripping system. To accurately determine the strain of the specimen, it is necessary to subtract the system compliance from the apparent specimen compliance:

$$C = C_a - C_s \tag{6.4}$$

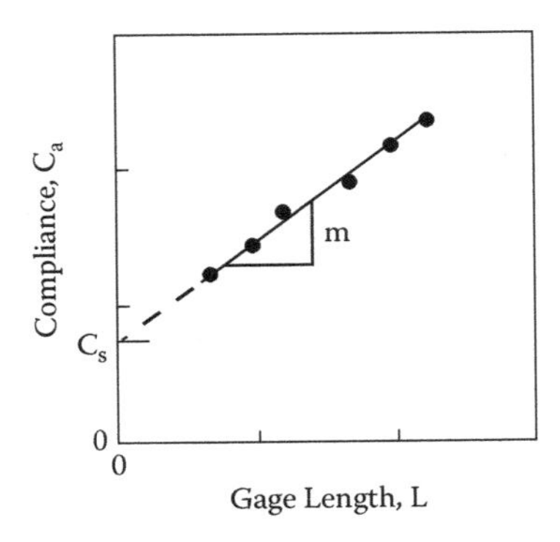

FIGURE 6.3
Typical graph of compliance versus gage length for a single filament.

where C, C_a, and C_s are the true specimen compliance, the apparent specimen compliance, and the system compliance, respectively. The true specimen compliance is given by

$$C = \frac{L}{EA_f} \tag{6.5}$$

where L is the gage length. ASTM D3379 suggests conducting tension testing (as described previously) on a material similar to the filament (or the actual filament) over a range of gage lengths at small loads to determine the apparent compliance C_a. Plotting C_a versus gage length should result in a graph similar to the sketch shown in Figure 6.3.

The system compliance C_s, defined as the intercept of the best-fit straight line at $L = 0$ is obtained by extrapolation. The fiber axial modulus E may be determined from the slope m of the line in Figure 6.3 as

$$E = \frac{1}{mA_f} \tag{6.6}$$

where it is assumed that the fiber cross-sectional area is known.

The ultimate tensile strength of the fiber σ_f is defined as the maximum stress before failure [Equation (6.2)].

6.3 Matrix Test Methods

Mechanical properties of matrix materials are mostly determined for performance evaluation and specification of matrix properties for product data sheets. Such properties can be used in micromechanics models that predict composite properties. It should always be remembered that neat resin properties are not necessarily equivalent to those that will be achieved by the resin in a composite. The fiber/matrix interface region in particular may differ from the bulk matrix.

The mechanical properties of the matrix material are normally measured in tension, compression, and shear. Since the materials are usually very brittle, extreme care should be exercised in preparation of specimens and in carrying out the test.

As described in Chapter 4, neat resin test specimens are often prepared by molding the resin. Some resins can be cast into void-free thick bars or dog-bone test specimens. Others are best cast into sheets. The presence of voids in casting will cause problems with premature failures of the test specimen. The quality of specimens is important. To the extent possible, specimen design should follow the standards or guidelines specified for the test procedure used.

Strain measurement should be accomplished through the use of extensometers or strain gages. If an extensometer is used, care must be taken so that the sharp knife edges of the extensometer do not promote failure. Typically, the specimen surfaces are smooth and glossy, and the knife edges of the extensometer may slip. A dot of White-Out correcting fluid has been found to enhance friction. If strain gages are used, they should be 350-Ω resistance (or greater) gages. Polymers do not dissipate heat very fast, and gage heating can be a problem. Excitation voltages should therefore be kept as low as possible. Strain gages may not perform properly for high-elongation polymers because the gage/gages tend to debond at moderate-to-large strains. This is, however, not a problem for most thermoset matrices used in advanced composites since they tend to be relatively brittle.

Polymers are often hydrophilic and sensitive to temperature changes. Specimen conditioning and the laboratory environment should be carefully controlled and monitored both prior to and during testing.

ASTM D638 (2010) specifies conditioning of the specimens at 23 ± 2°C and 50 ± 5% relative humidity for at least 40 h unless otherwise specified. Testing should be conducted under the same conditions. Because polymers are viscoelastic materials and strain rate sensitive, testing speeds must be well defined and controlled for valid comparison of data.

6.3.1 Matrix Tensile Testing

The standard tension test is described by ASTM D638 (2010). Several types of dog-bone specimen geometries are specified by this standard, depending on sheet thickness and polymer rigidity. Figure 6.4 shows a typical type I (rigid or semirigid polymer 7 mm or less thick) neat resin tensile specimen geometry and dimension. For modulus determination, a longitudinal strain gage or extensometer is required. Measurement of Poisson's ratio requires the use of two strain gages, a biaxial strain gage (0°/90° rosette), or a biaxial extensometer to measure longitudinal and transverse strains simultaneously.

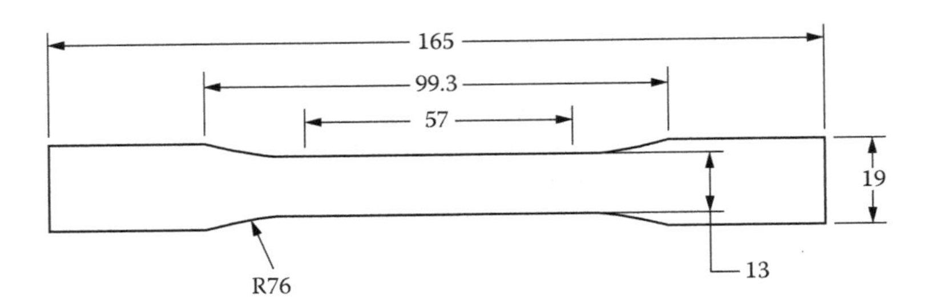

FIGURE 6.4
Tensile test specimen geometry (type I). All dimensions are in millimeters.

All specimens should be inspected to ensure that voids and surface flaws are not present. ASTM D638 specifies tolerance of thickness and gage section dimensions of ±0.4 and ±0.1 mm, respectively.

ASTM D638 stipulates a crosshead speed of 5 mm/min. To conform to most composite tests described in this text, however, a crosshead rate of 1–2 mm/min is recommended. Test at least five replicate specimens.

Figure 6.5 shows a typical stress–strain curve recorded for a vinylester matrix in tension. The specimen fails after displaying a short nonlinear

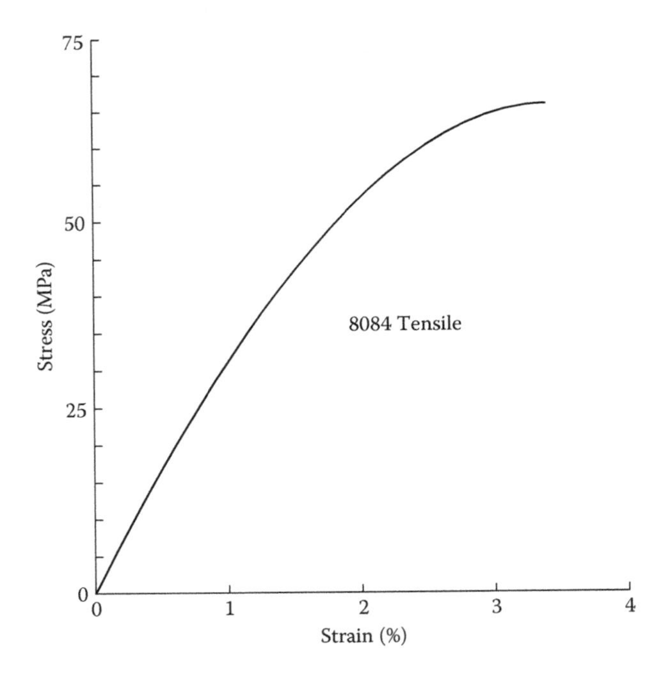

FIGURE 6.5
Stress–strain curve in tension for vinylester 8084 resin. (After Figliolini, A.M., 2011, *Degradation of Mechanical Properties of Vinylester and Carbon Fiber/Vinylester Composites Due to Environmental Exposure*, master thesis, Florida Atlantic University.)

response region at a strain of 3.78%. From this curve, it is possible to determine Young's modulus E, the yield stress (σ_{ys}), defined by a 0.2% offset strain (see Gere and Timoshenko 1997), and the ultimate tensile strength σ_t. For this particular specimen, $E = 3.01$ GPa, $\sigma_{ys} = 52.9$ MPa, and $\sigma_t = 66.1$ MPa.

6.3.2 Matrix Compressive Testing

Compression testing of the matrix material is conducted for purposes of research and development and quality control. Compressive properties include modulus of elasticity, yield stress, and compressive strength. The ASTM D695 (2010) standard describes test specimens of various forms, such as sheet and solid cylindrical rod. Although ASTM D695 includes testing of thin sheet-like materials of thickness under 3.2 mm, testing of such requires special support fixtures, and it is recommended here to test relatively short, solid, cylindrical rod specimens to avoid instability failures.

The standard specimen is 12.7 mm in diameter and 25.4 mm long. Such specimens may be prepared by molding longer cylinders (see Chapter 4) and machining the specimen in a lathe to achieve smooth and parallel end surfaces. At least five replicate specimens should be prepared and tested. Figure 6.6 shows a photo of an actual test specimen.

FIGURE 6.6
Compression test specimen. (After Figliolini, A.M., 2011, *Degradation of Mechanical Properties of Vinylester and Carbon Fiber/Vinylester Composites Due to Environmental Exposure*, master thesis, Florida Atlantic University.)

Prior to testing, the diameter and length of the specimen are to be measured to the nearest 0.01 mm at several points. Make sure the end surfaces are flat and parallel. Place the specimen between the parallel platens of the testing machine, centering the specimen to avoid introduction of loading eccentricities. Since the specimen is quite short, it may not be feasible to record strain using an extensometer or a strain gage. Hence, in many cases strain is defined as crosshead displacement divided by the original specimen length, which does not produce reliable results. As for the single-filament testing described in Section 6.2, improved modulus determination is obtained if the system (machine) compliance is subtracted from the apparent specimen compliance. The machine compliance for the compression test may be determined in a way similar to the procedure suggested for the single-filament test.

Before the actual test, the position of the crosshead of the test frame should be adjusted until it just contacts the specimen top surface. Set the crosshead speed at 1–2 mm/min and start the test while recording load and crosshead displacement. If the specimen does not fail catastrophically but continues to yield, the test may be interrupted.

Figure 6.7 shows a typical compressive stress–strain curve for a vinylester resin. Based on this curve, it is possible to determine an apparent elastic modulus E_a, yield stress σ_{ys}, and compressive strength σ_c.

The apparent elastic modulus E_a, is calculated based on an apparent strain measure (crosshead displacement/specimen length). From the stress–strain

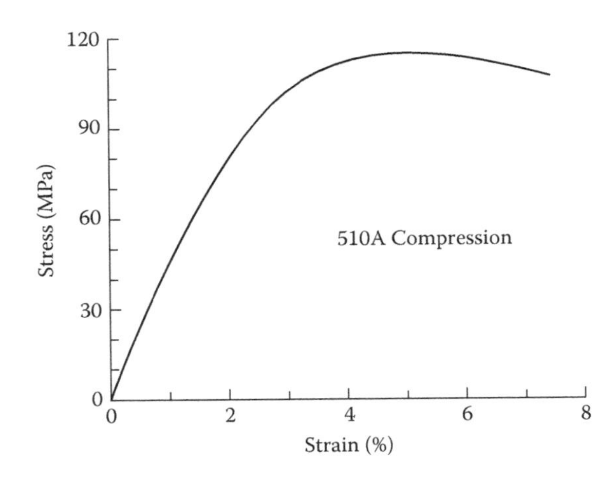

FIGURE 6.7
Compressive stress–strain curve for a 510 vinylester specimen. (After Figliolini, A.M., 2011, *Degradation of Mechanical Properties of Vinylester and Carbon Fiber/Vinylester Composites Due to Environmental Exposure*, master thesis, Florida Atlantic University.)

curve in Figure 6.7, the following data were reduced: $E_a = 4.00$ GPa, $\sigma_{ys} = 95.2$ MPa, $\sigma_c = 115$ MPa.

6.3.3 Matrix Shear Testing

One of the most important functions of the matrix in a composite is to transfer load into the stiff and strong reinforcing fibers. Such load transfer, discussed in more detail in Chapter 7, occurs through shear stresses at the fiber/matrix interface. Analyses of several loading situations of structural parts made from composite materials reveal transfer of load by shear. Consequently, the shear response of the matrix is of prime importance.

Although there is no ASTM standard for the determination of the shear stress–strain response of matrix materials, it is widely recognized (Adams 1990, Sullivan et al. 1984) that the most promising shear test of the matrix is the Iosipescu shear test method (ASTM D5379 2005). This standard, developed for composite materials (Chapter 10), is schematically illustrated in Figure 6.8.

As shown in Figure 6.8, the V-notched test specimen is held in the fixture, which is loaded in compression to produce a fairly uniform state of shear stress in the center region of the specimen, between the two notches.

A detail of the test specimen is shown in Figure 6.9. The specimen may be molded or machined from a flat plate. The top and bottom edges must be carefully machined to be flat and parallel to each other to avoid bending and twisting deformations when load is applied. The standard fixture allows specimens up to 12.7 mm thick, although neat resin plates typically are about 3 to 5 mm in thickness. Details of specimen preparation are provided in Chapter 4. The specimen is instrumented in the same way as the composite specimen (Chapter 10), typically with a two-element strain gage rosette configured to measure strain in the 45° and –45° directions (Figure 6.9).

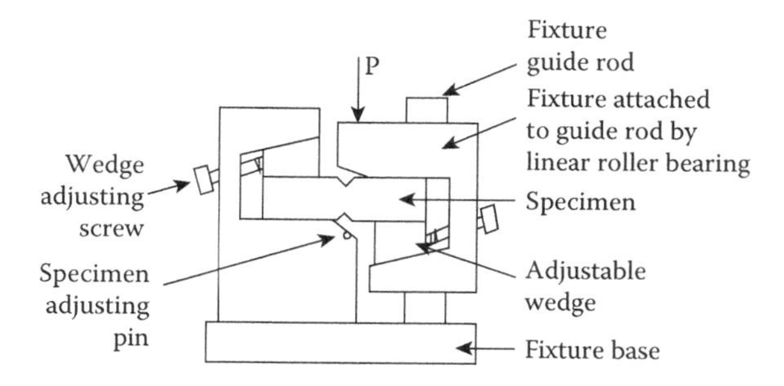

FIGURE 6.8
Schematic of Iosipescu shear test.

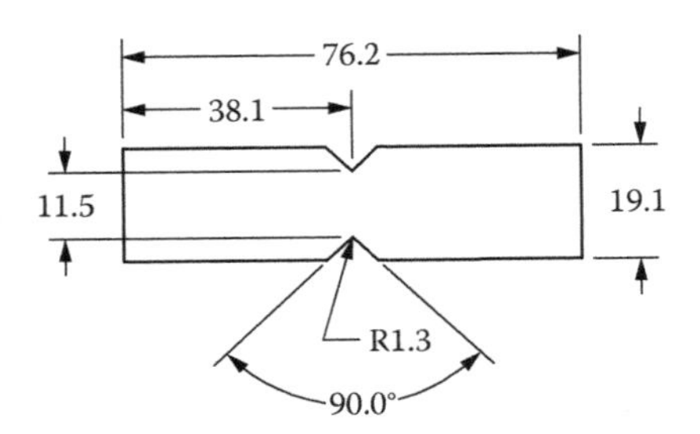

FIGURE 6.9
Details of matrix Iosipescu specimen. All dimensions are in millimeters.

Prior to testing, the specimen dimensions are measured at several locations to the nearest 0.01 mm. Insert the specimen in the test fixture. Apply load at a crosshead rate of 1–2 mm/min while recording load and strain until the specimen fails. Failure of brittle-matrix specimens tends to initiate at the notches and propagate in the 45° plane, where the principal stress is maximum.

For reduction of the load and strain gage readings, it is recognized that the shear stress is given by

$$\tau = \frac{P}{A} \tag{6.7}$$

where P is the load applied, and A is the cross-sectional area of the specimen between the notches. The shear strain is given by

$$\gamma = |\varepsilon(45°)| + |\varepsilon(-45°)| \tag{6.8}$$

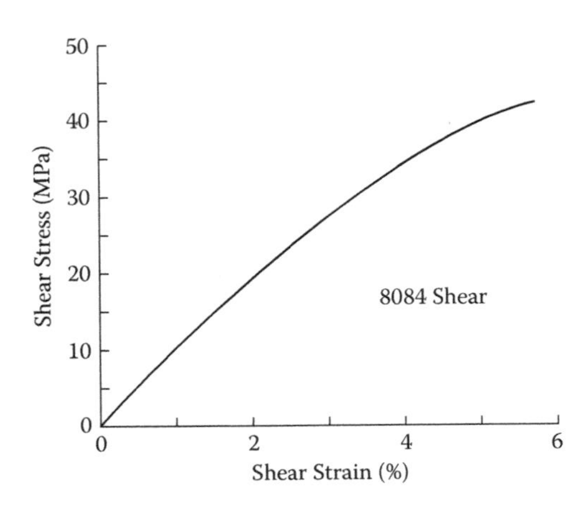

FIGURE 6.10
Shear stress–strain curve for vinylester 8084. (After Figliolini, A.M., 2011, *Degradation of Mechanical Properties of Vinylester and Carbon Fiber/Vinylester Composites Due to Environmental Exposure*, master thesis, Florida Atlantic University.)

This formula assumes that a ±45° strain gage rosette, or two 45° gages, is used. If a single gage oriented at 45° or −45° is used,

$$\gamma = 2\,|\varepsilon(45°)| \tag{6.9}$$

Figure 6.10 shows a representative shear stress–strain curve for a vinyl-ester Iosipescu specimen. Data reduction allows determination of the shear modulus G, shear yield stress τ_{ys}, and shear strength τ_u. Based on the curve shown in Figure 6.10, the following properties were reduced: $G = 0.93$ GPa, $\tau_{ys} = 33.3$ MPa, and $\tau_u = 42.2$ MPa.

7

Fiber/Matrix Interface Tests

7.1 Introduction

The fiber/matrix interface plays a vital role in the mechanical behavior of fiber-reinforced composites. To effectively utilize strong fibers as reinforcements of matrix materials, there must be strong adhesion between fiber and matrix to enable introduction of load into the fibers by shear stresses at the fiber/matrix interface.

Fiber composites consist of thousands of fibers, and the fiber strength displays substantial variability that properly must be described by statistical models, most commonly the two-parameter Weibull weakest-link model (Watson and Smith 1985, van der Zwaag 1989), discussed in more detail further in this chapter. If a dry bundle of fibers is loaded in tension, the weakest fibers will break early, followed by subsequent failures of the stronger fibers until the bundle breaks. When an individual fiber in the dry bundle breaks, the ligaments become fully unloaded and take no part of the load transfer of the bundle. If the fibers in the bundle are impregnated with a resin, however, thus forming a unidirectional composite, failure of an individual fiber will not end its role as a load-carrying member because the matrix will rapidly transfer load into the ligaments. As a result, an impregnated fiber bundle (composite) is substantially stronger than the dry counterpart. According to Peters (2002), the matrix acts to almost double the strength of a fiber bundle, and it also significantly reduces the dispersion of strength. The fiber/matrix interface is obviously also a critical link in the transfer of load in most other loading situations, such as transverse tension, longitudinal shear, and off-axis compression and tension.

Several fiber/matrix interface tests have emerged since the introduction of fiber composites. The oldest fiber/matrix interface test is the single-fiber pullout test (Broutman and McGarry 1963), illustrated in Figure 7.1.

The test consists of a single fiber partially embedded in a block of matrix. Part of the fiber extends out of the matrix to allow tensile load application. The embedded length must be short to promote shear failure of the fiber/matrix interface and subsequent pullout of the fiber. If the embedded length is too long, the fiber will fail in tension. Load and displacement at the point of load application are monitored during the test. Prior to failure, the

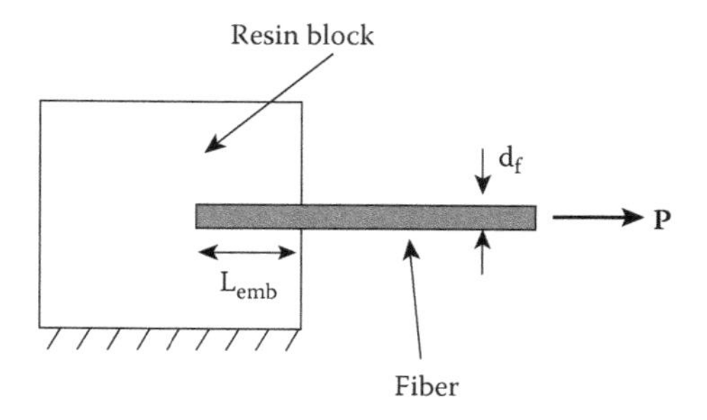

FIGURE 7.1
Schematic of single-fiber pullout test.

response is typically linear. At the instant of debonding, the load drops but is partially recovered by frictional effects as the debonded fiber slides against the matrix. Pigott and Chua (1987) and Greszczuk (1969) have studied the mechanics of the fiber pullout failure at great detail. Although Pigott (1987) suggests several failure mechanisms, the simplest one is to assume an ideal elastic-plastic process at failure, that is, that the interface shear stress is uniformly distributed along the embedded length at the instant of fiber debonding. This mechanism allows application of a simple force balance, yielding the interface shear strength as

$$\tau_0 = \frac{P_{max}}{\pi d_f L_e} \tag{7.1}$$

where P_{max} is the debonding force, d_f is the fiber diameter, and L_e is the embedded length.

As mentioned, this test requires fibers with short embedded lengths to fail the fiber/matrix interface. Failure of the fiber in tension requires a load

$$P_f = \pi d^2{}_f \sigma_f / 4 \tag{7.2}$$

where σ_f is the tensile strength of the fiber. Combination of Equations (7.1) and (7.2) (with $P_{max} < P_f$) yields a condition for the embedded length:

$$L_e \leq \frac{\sigma_f d_f}{4\tau_0} \tag{7.3}$$

For a typical T300 carbon–epoxy composite with an interface strength $\tau_y = 50MPa$, Equation (7.3) with the fiber data provided by Daniel and Ishai (2006) yields $L_e \leq 90\mu m$. Preparation of test specimens with controlled debond lengths of this small magnitude is at best challenging. The single-fiber pullout test is better suited to larger-diameter fibers with poor adhesion to the matrix.

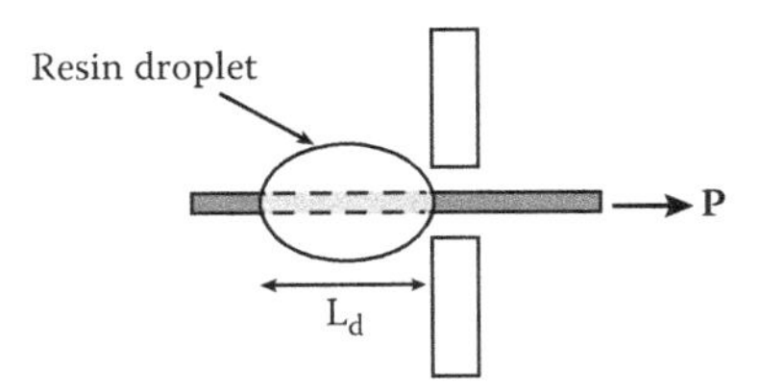

FIGURE 7.2
Schematic of the microdroplet test.

An alternative to the single-fiber pullout test is the resin droplet test proposed by Miller et al. (1987). In this test, a small droplet of resin is placed on a single fiber and pulled through a narrow slot formed by two blades (see Figure 7.2).

The axial load acting on the fiber is supported by shear stresses at the fiber/matrix interface. When the shear stress reaches its ultimate value (i.e., the fiber/matrix interface shear strength τ_0), separation of the fiber and matrix occurs, and the droplet will slide, supported only by frictional forces. The shear strength is calculated from Equation (7.1) with L_e replaced by the length of the droplet (L_d in Figure 7.2).

Similar challenges as discussed for the pullout test apply to the microdroplet test. Small-diameter fibers forming a strong interface with the matrix fail by tensile failure of the fibers unless the droplet is very small. Further, the droplet tends to form a meniscus on the fiber, obscuring the definition of droplet length and adding complexities to the state of shear stress inside the droplet (see Drzal et al. 2000). Furthermore, Rao et al. (1991) found that the small droplets formed on carbon fibers suffered from incomplete cure of the resin due to diffusion of the volatile curing agent, which results in low apparent values of the interface shear strength.

Several other fiber/matrix interface tests, such as the single-fiber compression tests proposed by Broutman (1969) and Outwater and Murphy (1969) and the microindentation test proposed by Mandell et al. (1980), have been used, but they have not reached widespread acceptance due to complexities associated with specimen preparation, testing, and data reduction.

Although all of these single-fiber tests continue to be used, the single-fiber fragmentation test (SFFT), presented in detail in the next section, is perhaps the most reliable and the most commonly used at present.

7.2 Single-Fiber Fragmentation Test

The SFFT test utilizes a small, dog-bone-shaped tension test specimen with an axially oriented single fiber embedded in the resin matrix at the center of the bulk of the matrix test specimen (see Figure 7.3).

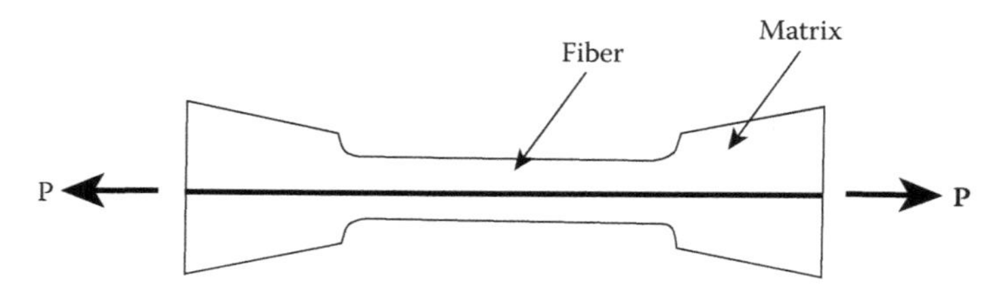

FIGURE 7.3
Schematic of single-fiber fragmentation test.

The SFFT emerged from the classical work of Kelly and Tyson (1965), who observed multiple fractures in the same fiber in a unidirectional composite consisting of low concentration of brittle tungsten fibers in a ductile copper matrix. The SFFT specimen was developed by Drzal and coworkers (1980). The basic test methodology and analysis are reviewed in this chapter. Key to this test is that the matrix strain to failure exceeds the fiber strain to failure. Further, when the fiber breaks, the matrix must be strong enough to support the applied load by itself. Figure 7.4 schematically illustrates the failure mechanism in a tension-loaded SFFT specimen at increasing strain. Initially, the fiber stress is uniform. At a certain load, the fiber will break at its weakest point, leading to zero axial stress at the location of the fiber break.

If loading is continued and the matrix is strong enough to carry the entire load after the fiber break, a shear transfer mechanism called *shear lag* will transfer load into the fiber on both sides of the fiber break. When the load applied to the SFFT with a broken fiber is increased, the fiber will break again at some weak point. The process will continue, and the number of

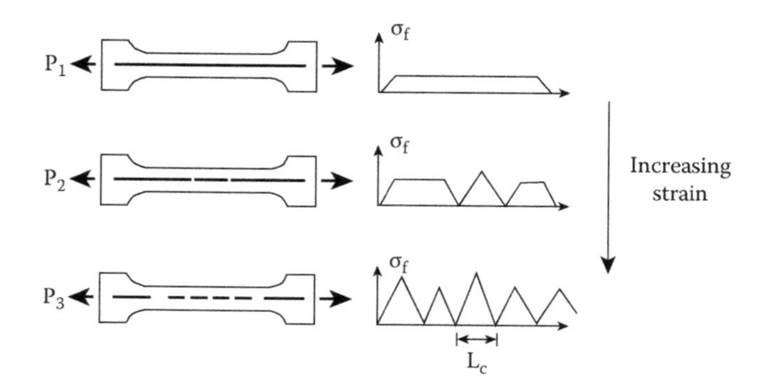

FIGURE 7.4
Failure mechanism (left) and fiber axial stress distribution (right) in an SFFT specimen loaded in tension. (After Rich, M.J., Drzal, L.T., Hunston, D., Holmes, G. and McDonough, W., 2002, *Proc. Am. Soc. for Compos., 17th Tech. Conf.*, Purdue University, West Lafayette, IN, October 21–23.)

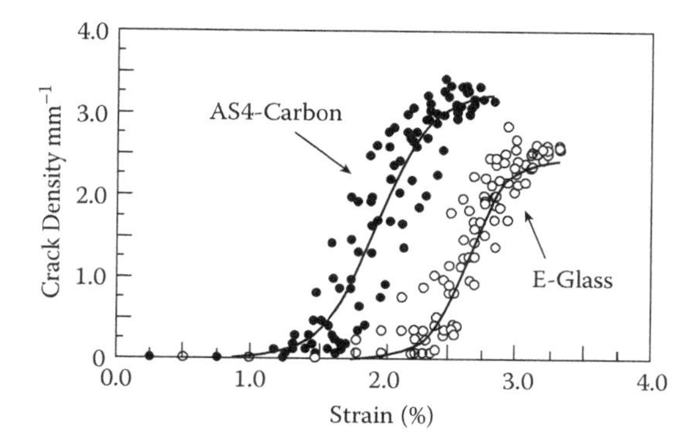

FIGURE 7.5
Number of fiber breaks versus applied strain in an SFFT specimen. (From Kim, B.W., and Nairn, J.A., 2002, *J. Compos. Mater.*, 36, 1825–1858.)

fiber fragments will increase until their lengths become insufficient to cause enough of load introduction into the fiber fragment and any further breaks. This state is called *fiber break saturation,* and the final fragment length is referred to as the *critical length.* Actual test results for SFFT specimens consisting of AS4 carbon and E-glass fibers in epoxy resin are illustrated in Figure 7.5, in which the number of fiber breaks per millimeter is plotted versus applied strain.

The number of fiber breaks was normalized with the length of the gage section. The number of breaks at a given strain depends on the time after the application of strain. For the carbon fibers, the number stabilized after about 5 min, while with the glass fiber the number stabilized after about 20 min. The time dependency has been attributed to the viscoelastic nature of the polymer matrix (see Kim and Nairn 2002 and Moon and McDonough 1998). Figure 7.5 shows that the first break of the carbon fiber occurs at a strain of about 1%, while corresponding strain for glass fiber is about 2%. After the first break, the number of breaks increases rapidly with increasing strain until saturation occurs. The saturation crack densities are about 3.3 and 2.7 mm^{-1} for the carbon and glass fibers, respectively. This corresponds to fragment lengths $\ell = 0.30$ and 0.37 *mm.*

The critical length can be used to determine the shear strength of the fiber/matrix interface. Consider the free-body diagram of a fiber segment inside the matrix shown in Figure 7.6.

The axial fiber stress $\sigma_f(x)$, increases by an amount $d\sigma_f$ over the length dx of the segment. The shear stress at the fiber/matrix interface is denoted by τ. Equilibrium of the element requires

$$-\frac{\pi}{4}d_f^2\sigma_f(x) - \pi d_f \tau dx + \frac{\pi}{4}d_f^2(\sigma_f(x) + d\sigma_f) = 0 \tag{7.4}$$

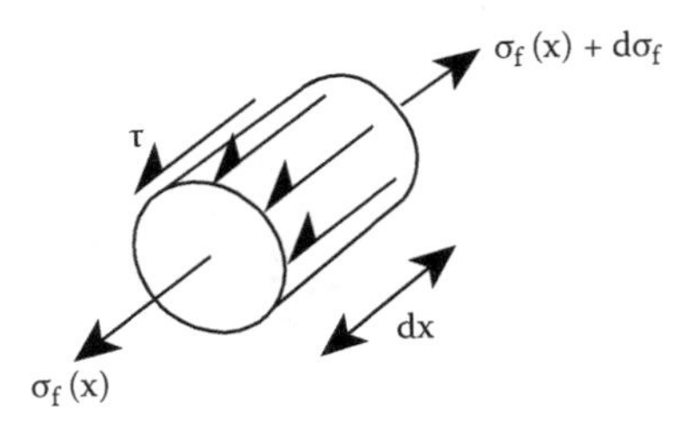

FIGURE 7.6
Free-body diagram of a fiber segment in an SFFT specimen.

Solution for $\sigma_f(x)$ yields

$$\sigma_f(x) = \frac{4}{d_f}\int_0^x \tau(x)dx \tag{7.5}$$

where it has been implicitly assumed that the fiber stress is zero at the broken end ($x = 0$), consistent with the physical situation. As for the previously discussed pullout test, it is assumed here that the shear stress at saturation is uniform, $\tau = \tau_0$, where τ_0 is the fiber/matrix interface shear strength. Substitution of a constant shear stress (τ_0) into Equation (7.5) yields the fiber stress distribution:

$$\sigma_f(x) = \frac{4\tau_0 x}{d_f} \tag{7.6}$$

This expression is valid only for the left half of the fiber fragment (i.e., $\{0 \le x \le L_C/2\}$) because of the shear stress reversal dictated by symmetry and overall equilibrium of the fiber fragment. The resulting fiber axial stress distribution is shown in Figure 7.7.

Notice that the maximum fiber stress (Figure 7.7) occurs at $x = L_c/2$ and is given by

$$\sigma_f(\max) = \frac{2\tau_0 L_C}{d_f} \tag{7.7}$$

From the failure condition that $\sigma_f(\max) = \sigma_f$, where σ_f is the fiber strength (at length L_C), it is possible to rearrange Equation (7.7) to obtain an expression for the shear strength of the fiber/matrix interface:

$$\tau_0 = \frac{\sigma_f d_f}{2L_C} \tag{7.8}$$

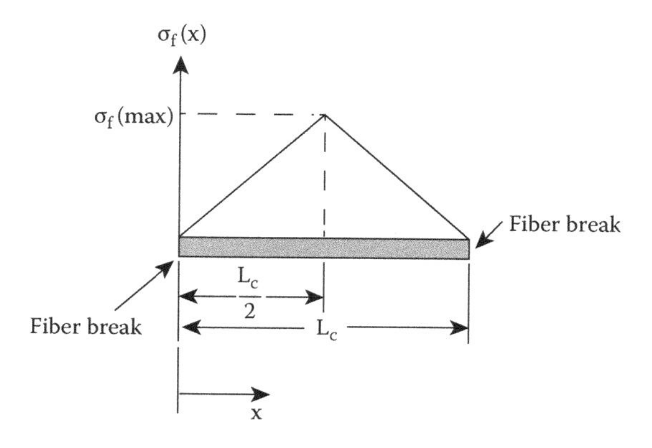

FIGURE 7.7
Distribution of axial fiber stress in a short fiber fragment.

By this test, it thus becomes possible to determine the fiber/matrix interface shear strength based on knowledge of the fiber strength and diameter and the measured value of the critical length L_C.

As discussed previously in this chapter, the fiber failure is controlled by flaws that are randomly distributed along the fiber length. A long fiber has a higher probability of containing a severe flaw than a shorter fiber. As a result, the longer fiber is weaker. Measurements of fiber strength at different gage lengths support this observation. Figure 7.8 shows an example of the fiber tensile strength versus gage length for boron fibers (Rosen and Hashin 1987). As the gage length decreases, the mean fiber strength increases, as well as the dispersion of strength.

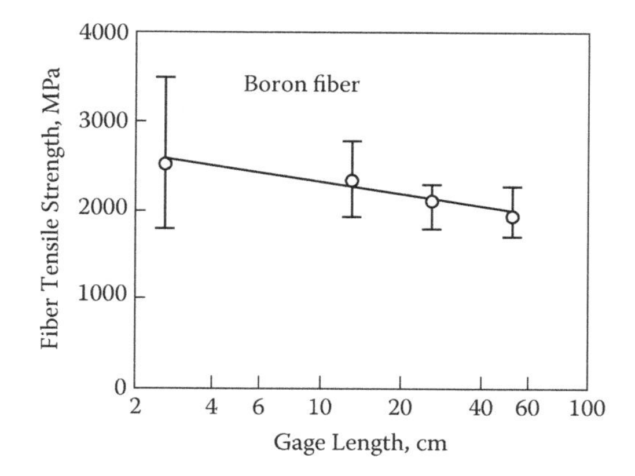

FIGURE 7.8
Fiber tensile strength versus gage length for boron fibers. (From Rosen, B.W., and Hashin, Z., 1987, in *Engineered Materials Handbook, Vol. 1, Composites*, T.J. Reinhart, tech. chairman, ASM International, Metals Park, OH, pp. 185–205.)

The critical length is typically of the order of 1 mm, making it impossible to directly measure the fiber strength at the appropriate gage length L_C.

For the purpose of establishing the fiber strength at the appropriate length L_C, the Weibull weakest-link theory (Weibull 1951, Pickering and Murray 1999) is typically used:

$$\frac{\sigma_f(L)}{\sigma_f(L_0)} = \left(\frac{L_0}{L}\right)^{1/\alpha} \tag{7.9}$$

where $\sigma_f(L)$ is the mean fiber strength measured at gage length L, while $\sigma_f(L_0)$ is the mean fiber strength at any arbitrary reference gage length L_0.

The Weibull parameter α may be estimated from fiber strengths measured at different gage lengths, such as the data shown for boron fibers in Figure 7.8. Equation (7.9) may be linearized by taking the logarithm of both sides:

$$\log \sigma_f(L) = \log \sigma_f(L_c) + \frac{1}{\alpha}\log L_0 - \frac{1}{\alpha}\log L \tag{7.10}$$

Hence, by plotting measured fiber strength data versus gage length in a log-log plot, $\log \sigma_f$ vs. $\log L$, the Weibull shape parameter may be determined from the slope of the line (slope = $-1/\alpha$). Figure 7.9 shows such a plot constructed based on the boron fiber data in Figure 7.8. For convenience, the data are tabulated in Table 7.1.

A least-squares linear fit to the data in Figure 7.9 yielded a slope of -0.096, corresponding to a Weibull parameter α = 10.4. The intercept at $\log L_0 = 0$ ($L_0 = 1m$) is 3.26, which corresponds to a "reference strength," $\sigma_0 = 1,820$ MPa at 1 m gage length. Equation (7.9) becomes

$$\sigma_f(L) = 1,820\left(\frac{1}{L}\right)^{0.096} \tag{7.11}$$

where the units of length and stress are m and MPa, respectively.

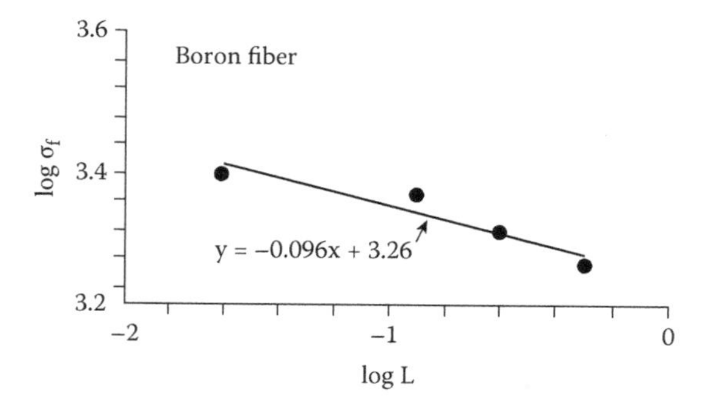

FIGURE 7.9
Log-log plot of fiber strength versus gage length (based on Figure 7.8). Units of strength are MPa and length units are m.

TABLE 7.1

Mean Fiber Strength at Various Gage Lengths

Gage Length (cm)	Fiber Strength (MPa)
2.54	2510
12.7	2340
25.4	2085
50.8	1872

Source: From Rosen, B.W., and Hashin, Z., 1987, in *Engineered Materials Handbook, Vol. 1, Composites*, T.J. Reinhart, tech. chairman, ASM International, Metals Park, OH, pp. 185–205.

This equation may be used to determine the fiber strength at the critical fiber length L_c (by extrapolation). For example, if the critical fiber length $L_c = 100\mu m$, Equation (7.9) yields $\sigma_f = 4{,}406$ MPa.

When conducting the SFFT, Feih et al. (2004) recognized that any fragment with a length slightly more than L_c will break in two, resulting in a range of fragment lengths from $L_c/2$ to L_c. Thus, the average fragment length at saturation should be $\overline{\ell} = 3L_c/4$. Hence, the critical fiber length becomes

$$L_c = \frac{4\overline{\ell}}{3} \tag{7.12}$$

This length is used in Equation (7.8) to determine the fiber/matrix interface shear strength, with the fiber strength determined from Equation (7.9) as explained previously. As an example, Ramirez (2008) conducted SFFT testing of E-glass/epoxy and found the following results for the fragment length in the SFFT: $\overline{\ell} = 834 \pm 213\mu m$. The corresponding critical length [Equation (7.12)] is $L_c = 1{,}112 \pm 284\mu m$. The fiber strength at this length is $\sigma_f = 3{,}280$ MPa, and the fiber diameter is $d_f = 14\mu m$. Equation (7.8) yields the following fiber/matrix interface shear strength: $\tau_0 = 20.7 \pm 7.1$ MPa.

The large scatter range is here solely attributed to the dispersion in the critical length. Additional scatter could originate from scatter in fiber strength and fiber diameter. The scatter in fiber strength is given by Weibull statistics as

$$SD(\sigma_f) = \sigma_f \sqrt{\Gamma\left(1+\frac{2}{\alpha}\right) - \Gamma^2\left(1+\frac{1}{\alpha}\right)} \tag{7.13}$$

where $SD(\sigma_f)$ represents the standard deviation around the mean fiber strength σ_f, and Γ is the gamma function (see Pickering and Murray 1999).

Another method, proposed by Drzal et al. (1982), consists of fitting the measured distribution of fragment lengths ℓ to a two-parameter Weibull distribution. This results in modified expressions for the fiber/matrix interface shear strength and its variability. Wimolkiatisak and Bell (1989) found that the dispersion in fragment length could also be well represented by Gaussian statistics.

Even though the SFFT does not account for failure processes other than fiber failures, such as fiber/matrix debonding and matrix cracking, it has been experimentally demonstrated that the critical fiber length is a sensitive measure of the level of fiber/matrix adhesion (see, e.g., Drzal et al. 1982, 1983).

7.3 Single-Fiber Fragmentation Test Setup

The small size of the SFFT specimen (Figure 7.1) and the need to accurately determine lengths of the fiber fragments along the gage length of the SFFT specimen require special test fixtures and instrumentation. Many commercial matrix materials are transparent (e.g., epoxy, vinylester), and the locations of fiber breaks can be determined using photoelasticity or, although not as sensitive, just a microscope and light reflecting off the free surfaces. Figure 7.10 shows a test setup used by Ramirez (2008) in an investigation involving glass and carbon fibers in vinylester matrices. Similar test setups were developed and used by Drzal and coworkers (1980) and Kim and Nairn (2002).

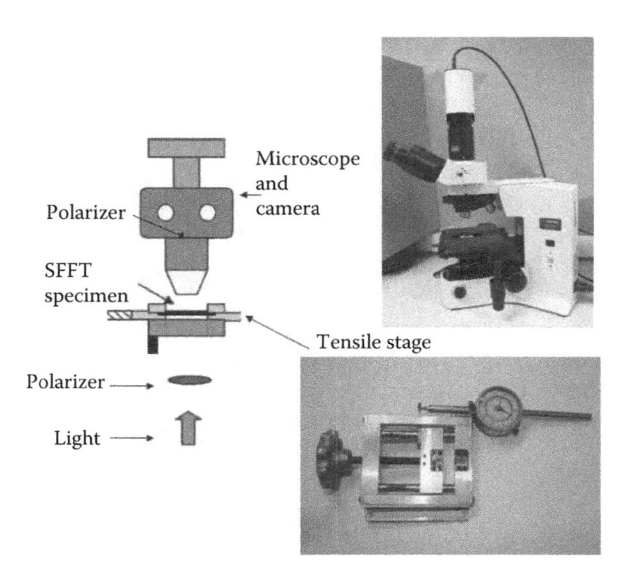

FIGURE 7.10
SFFT setup. (From Ramirez, F.A., 2008, Evaluation of water degradation of polymer matrix composites by micromechanical and macromechanical tests, master thesis, Florida Atlantic University.)

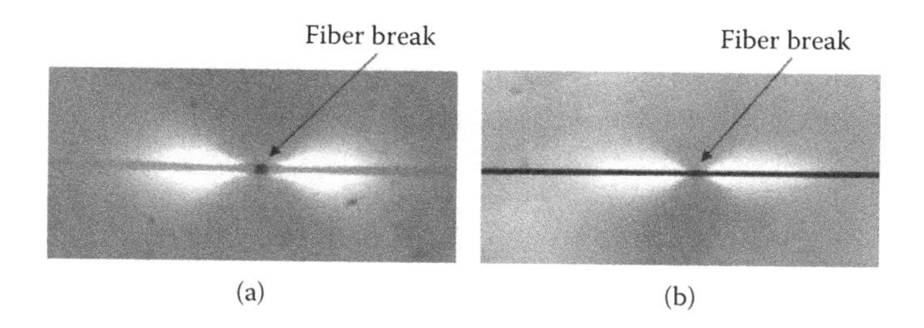

FIGURE 7.11
Birefringence patterns around fiber breaks in SFFT specimens: (a) glass/epoxy; (b) carbon/epoxy.

The central components of the test setup are an optical transmission microscope equipped with a camera, a hand-operated tensile stage with a dial indicator for strain measurement, and a light source. By placing the tensile stage (and SFFT specimen) in the microscope between crossed polarizers, application of load to the specimen will reveal a photoelastic pattern that allows analysis of the in situ distribution of stresses around the fiber break. When the specimen is under load, the region around a fiber break exhibits a colored pattern called birefringence, caused by the presence of shear stresses in the matrix associated with the perturbation of the nominally uniform distribution of stress in the matrix due to the fiber break (Timoshenko and Goodier 1970).

The SFFT specimen is loaded in small incremental steps by imposing a displacement. Inspection of the specimen in the microscope will reveal breaks occurring in the fiber. Figure 7.11 shows examples of birefringence patterns around fiber breaks in SFFT specimens made from carbon and glass fibers in an epoxy matrix. Since glass fibers are transparent and have a different refractive index than the epoxy matrix, the fiber break is easily detected as a black gap (even without photoelasticity) (see Figure 7.11a). On the other hand, detection of fiber breaks in carbon-fiber SFFT specimens is more difficult and requires photoelasticity.

An alternative technique to measure the fragment lengths was developed by Waterbury and Drzal (1991) using a special software called Fibertrack. Such a method is claimed to substantially reduce the level of effort in the evaluation of fiber fragment lengths and fiber/matrix shear strength.

7.4 SFFT Test Procedure

A specially developed small tensile loading apparatus (Figure 7.10) is mounted on an x-y stage fitted to an optical microscope. A transmission microscope is suitable for the observation of the fiber breaks. One polarizer is placed above the specimen and one below the specimen. The planes of polarization of the polarizers should be at 90°. To observe the fiber failures, the microscope should be equipped with a long-distance lens with a magnification of at least ×20. Ramirez (2008) used an Olympus Bx41 microscope equipped with a QICAM-FAST 1394 digital camera.

Prior to the actual test, a 20 mm long transparent coverslip is placed over the SFFT specimen and is adhered to the specimen surface using mineral oil. Load the specimen slowly using the loading screw in increments of about 1 mm. After each load increment, observe the gage section of the SFFT specimen. Detect the position of each fiber by translating the *x-y* stage along the fiber axis. Apply more load increments until the number of fiber fragments does not increase or some other, unanticipated failure, such as matrix cracking or extensive fiber/matrix debonding, occurs. At the end of a successful test, a set of discrete fiber fragment lengths is achieved that allows reduction of the fiber/matrix interface shear strength as outlined in Section 7.3.

8

Lamina Tensile Response

8.1 The Need for Lamina Testing

For most composites in use today, the individual lamina (i.e., the individual layer or ply) is the basic unit or building block, whether it is in the design, the analysis, or the fabrication process stage. This lamina may be a unidirectional prepreg, a fabric, a chopped strand mat, or another fiber form, with or without the matrix present prior to laminate fabrication. However, some composites are not fabricated of individual layers. As discussed in Chapter 3, such alternate processes include resin transfer molding (RTM) and vacuum-assisted resin transfer molding (VARTM). However, the resulting composites are typically still nonhomogeneous and anisotropic in terms of strength, stiffness, and physical properties. Thus, even in these cases the designed composite component still consists of individual layers (or regions) of differing material properties.

In summary, whatever the material form or fabrication process, the properties of the individual laminae (regions, layers, or whatever form the composite takes) must be known for design and analysis purposes. Determination of the tensile stiffness and strength properties of these individual laminae is the topic of the present chapter.

8.2 Introduction to Tensile Testing

Tensile properties are often the first to be thought of when a composite or any other type of material is considered for a design application. Although tensile properties do not always ultimately determine the design, as shown in subsequent chapters, they are nevertheless among those of primary importance.

The difficulty of performing an acceptable tensile test typically increases as the orthotropy of the material increases, that is, as the ratio of the axial stiffness (or strength) to the transverse stiffness (or strength) and the longitudinal shear stiffness (or strength) increases. Thus, a unidirectional composite lamina is often the most difficult material form to test. For this reason, particular emphasis here and in the following three chapters is on suitable methods for testing unidirectional composites. Often, if a unidirectional composite can be successfully tested using a particular procedure, almost any other laminate form can be tested as well. Conversely, it may be possible to successfully test a material of less orthotropy using alternative procedures that are more efficient in terms of time and cost.

For example, it may be possible to successfully test a $[90/0]_{ns}$ cross-ply laminate and then use classical lamination theory (Chapter 2) to "back out" the unidirectional ply properties. This technique has been examined extensively for compression testing (as discussed in Chapter 9) but has not yet been shown to be acceptable for tensile testing. Other examples are discussed in the remainder of this chapter.

8.3 Load Introduction

Proper introduction of the applied force into the test specimen is one primary concern. For materials of low orthotropy, this may be as simple as pin loading through a hole in each end of a dog-bone-shaped specimen (or dumbbell-shaped, as termed in the 2010 ASTM [American Society for Testing and Standards] Standard D638). However, the high bearing stresses induced by the pins would cause local failures around the holes for most composites. Thus, it is more common to use a clamping grip at each end of the dog-bone-shaped specimen. A sketch of a dog-bone-shaped specimen is shown in Figure 8.1a. Several specific dog-bone specimen shapes are suggested in ASTM D638.

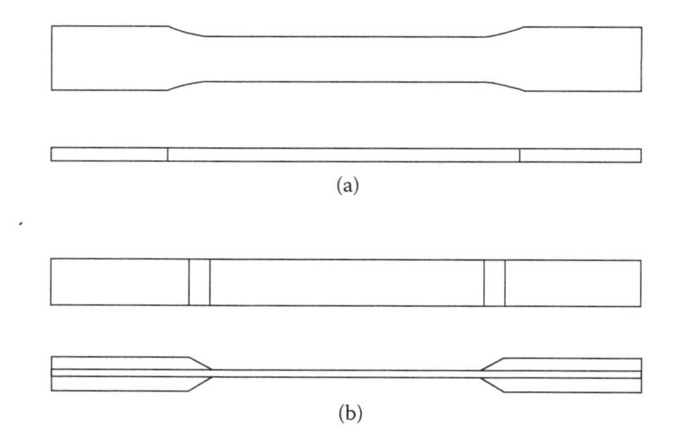

FIGURE 8.1
Typical tensile test specimen geometries. (a) dog-bone, (b) straight-sided end tabbed.

When clamping grips are used, the tensile loading in the specimen is induced via shear at the clamp–specimen interface. This shear force is equal to the clamping force times the effective coefficient of friction at the interface. Usually, little can be done to increase the coefficient of friction of the specimen surface without potentially degrading the properties of the material being tested. However, the faces of the clamp can be altered to increase friction. The most common technique is to roughen the grip surfaces by machining or by coating (Coguill and Adams 1999a, Wyoming Test Fixtures). Machined patterns of increasing aggressiveness (e.g., from swirls to cross-hatches to straight grooves) can be used, with deeper and sharper profiles more effectively penetrating the surface of the specimen and thus producing a higher effective coefficient of friction. An alternative is to coat the clamps with a friction-enhancing material. Rubber coatings can be used when testing low-strength materials. Emery cloth placed between the clamps and the specimen surfaces has long been used to increase the gripping force. This has led to the development of grip surfaces thermal sprayed with tungsten carbide particles (Coguill and Adams 1999a). The roughness of the clamp faces is dictated by the size of particles used. Typically, the roughness is on the order of 80 to 150 grit.

The more severely the surface is penetrated, the greater the danger of premature specimen failure whatever type of clamp face is used. The alternative is to increase the clamping force. However, the through-thickness compressive strength of most composites is low relative to the axial tensile strength. Thus, there is a limit to how much clamping force can be applied without crushing or otherwise degrading the specimen material.

For low-strength composites, simple mechanical clamps or pneumatically actuated grips and relatively smooth grip faces may be adequate. However, wedge grips are more commonly used. Two types of wedge grips are in common use: mechanical and hydraulic (Instron Corporation, MTS Corporation). Examples of each are shown in Figures 8.2 and 8.3, respectively. The wedging action of mechanical wedge grips is in direct proportion to the magnitude of the tensile force applied and inversely proportional to the angle to which the wedges are machined, which is typically about 10°. If the wedge angle is too low, the clamping force may crush the specimen being gripped. The wedges of hydraulic grips are loaded by hydraulic pressure prior to the start of the tensile test. The magnitude of the hydraulic pressure must be predetermined. Suppliers of commercially available hydraulic wedge grips provide guidelines for the pressure required as a function of the anticipated total axial force to be applied by the grips.

Although hydraulic wedge grips perform well, they are much heavier and bulkier than mechanical wedge grips and much more expensive. However, for repeated loading (e.g., fatigue), they perform better than mechanical wedge grips. Mechanical wedge grips tend to progressively tighten with successive cycles and may eventually crush the specimen ends (Odom and Adams 1983).

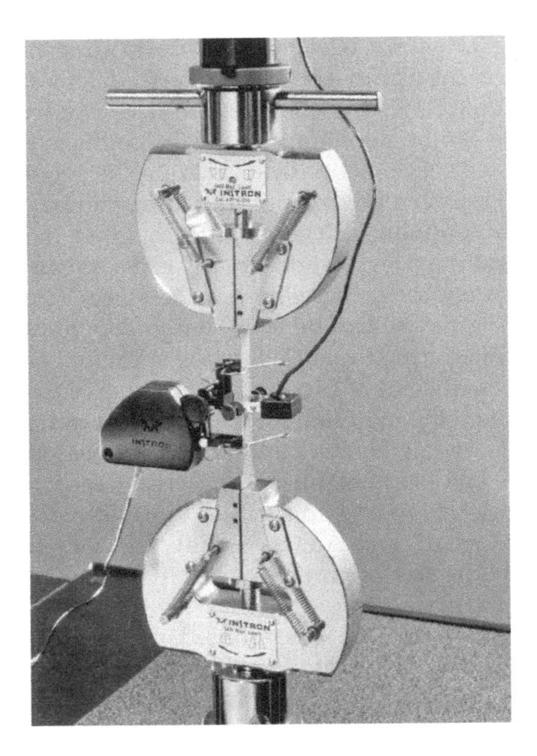

FIGURE 8.2
Typical mechanical wedge grips, with axial and transverse extensometers attached to a tensile test specimen. (Photograph courtesy of Instron Corporation.)

The foregoing discussion is particularly directed toward composite materials of low orthotropy, emphasizing the use of untabbed, dog-bone-shaped specimens. For highly orthotropic composite materials, dog-bone-shaped specimens become less suitable. The longitudinal shear strength of highly orthotropic composites becomes progressively less relative to the axial tensile strength. Thus, at some point the wide ends of the dog-bone-shaped specimen begin to fail in longitudinal shear near the gripping regions, effectively converting the dog-bone-shaped specimen into a straight-sided specimen (a specimen of constant width).

Thus, if this type of shear failure is likely to occur during the test, it is logical simply to start with a straight-sided specimen. This configuration is the basis for ASTM Standard D3039 (2001), see Figure 8.1b. If one assumes that the material has a relatively high axial tensile strength, it has to be gripped firmly to prevent slipping or more aggressive grip faces must be used. In either case, stress concentrations are induced in the unprotected specimen ends, promoting premature (grip) failures. Thus, it becomes necessary to

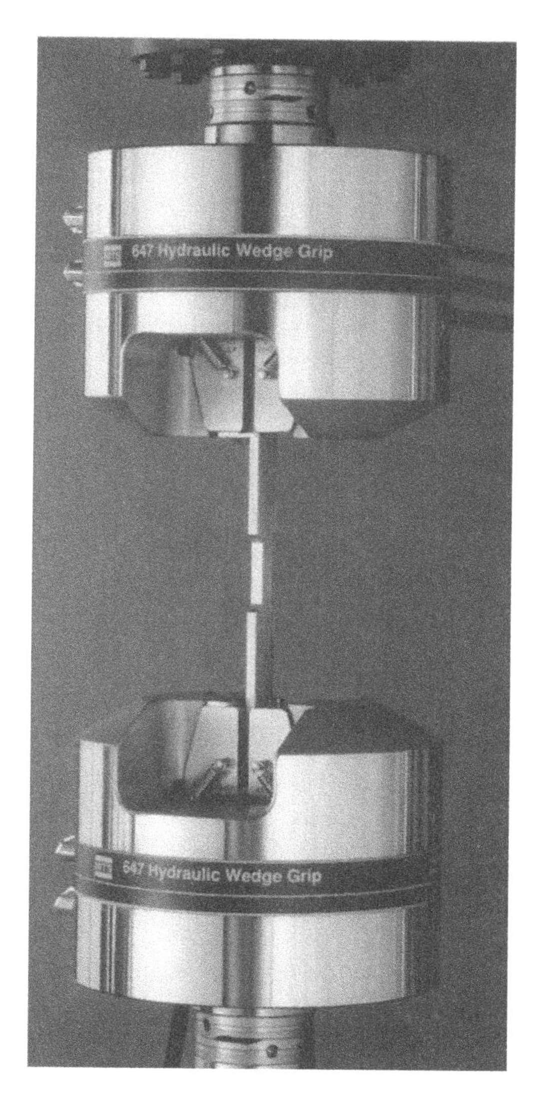

FIGURE 8.3
Typical hydraulic wedge grips. (Photograph courtesy of MTS Corporation.)

use tabs, as discussed in Chapter 4, in an attempt to reduce these stress concentrations and to protect the specimen ends from grip damage, see Figure 8.1b. Unfortunately, the presence of the tabs induces new stress concentrations because of the relatively abrupt change in specimen thickness at the tab ends. Thus, the tabs are typically tapered, as shown in Figure 8.1b, to reduce the abruptness of the thickness change. Stress analyses of tabbed specimens by Chatterjee et al. (1993a), Oplinger et al. (1982), and Kural and Flaggs (1983) indicate that significant out-of-plane (through-thickness) peel (i.e., transverse tensile) and interlaminar shear stresses are induced in the test material at the tab ends, as well as axial tensile stress concentrations, particularly for tapered tabs. It appears that the geometric discontinuity reduction benefits of tab tapering are offset by the through-thickness (peel) stresses induced in the specimen. ASTM D3039 does not suggest a specific taper angle and indicates only that the angle should be between 5° and 90° (a rather broad guideline). (Note that an untapered tab is defined as having a taper angle of 90°.) Results of a round-robin experimental study reported by Hojo et al. (1994) suggest that there is no significant difference in tensile strength of unidirectional carbon–epoxy composites over a range of tab taper angles between 10° and 90°. On the basis of these results, the International Organization for Standardization (ISO) has adopted the untapered tab specimen as their tensile specimen configuration.

Consistent with this discussion, ASTM D3039 recognizes the possibility of successfully testing untabbed, straight-sided specimens of lower-strength materials, such as fabric-reinforced composites and $[\pm 45]_{ns}$ laminates.

8.4 Specimen Configurations and Test Procedures

Recommended dog-bone-shaped and straight-sided specimen configurations and dimensions are presented in ASTM Standards D638 and D3039, respectively. Although considerable detail is presented, these standards recognize and accept that certain composite materials and test conditions may require modifications. Prior experience must then be relied on whenever possible.

For unidirectional composites of 0° fiber orientation, a specimen width of 12.7 mm and a specimen thickness of 6 to 8 plies are common, assuming a typical ply thickness on the order of 0.127 mm. Unidirectional 90° specimens are typically 25 mm wide and 16 to 24 plies thick; the number of plies depends on the actual ply thickness. Loading eccentricity may arise because of variations in tab and specimen thicknesses. Tolerances for tab and specimen thicknesses are ±1% and ±4%, respectively (Chatterjee et al. 1993a). The tab length (see Figure 8.1b or 8.4) should be at least 38 mm, and the tab

FIGURE 8.4
Photograph of tensile specimen with axial and transverse strain gages attached.

material should be 1.6 to 3.2 mm thick, see also Chapter 4. The gage length (distance between tabs) is commonly 125 to 155 mm. Variations of the specimen width should not exceed 1%. If Poisson's ratio is desired, a 0/90 strain gage rosette should be bonded in the center gage section region of the specimen (see Figure 8.4), or a biaxial extensometer should be used. If only axial stiffness and strength are desired, a longitudinal strain gage or a uniaxial extensometer is sufficient.

Accurate measurement of the specimen cross-sectional area in the gage section is particularly important, along with careful alignment of the specimen in the grips of the testing machine. Observers should wear adequate eye protection during the test procedure. Composite materials, particularly axially loaded high-strength unidirectional composites, can splinter and fragment violently on failure.

Measure the cross-sectional dimensions at several points on the specimen. Insert the specimen in the grips of a properly aligned and calibrated test frame. Set the crosshead rate at 2 mm/min. Avoid unprotected eyes in the test area, especially for the 0° tensile test. The strain readings may be recorded continuously or at discrete load intervals. If discrete data are taken, a sufficient number of data points must be recorded to reproduce the stress–strain behavior. At least 25 data points are needed in the linear response region. A total of 40 to 50 points is desirable to establish the total stress–strain response. Monitor all specimens to failure.

8.5 Data Reduction

The elastic moduli E_1 and E_2 and the Poisson ratios v_{12} and v_{21} are defined as follows:

- E_1: The initial slope of the stress–strain curve ($\Delta\sigma_1/\Delta\varepsilon_1$) for the 0° tensile test
- v_{12}: The negative ratio of the transverse to longitudinal strains ($-\varepsilon_2/\varepsilon_1$) for the 0° tensile test
- E_2: The initial slope of the stress–strain curve ($\Delta\sigma_2/\Delta\varepsilon_2$) for the 90° tensile test
- ν_{21}: The negative ratio of the transverse to longitudinal strains ($-\varepsilon_1/\varepsilon_2$) for the 90° tensile test.

The longitudinal and transverse tensile strengths X_1^T and X_2^T are defined as the ultimate values of σ_1 and σ_2 for the 0 and 90° tensile tests, respectively. The ultimate strains e_1^T and e_2^T are the strains corresponding to X_1^T and X_2^T, respectively.

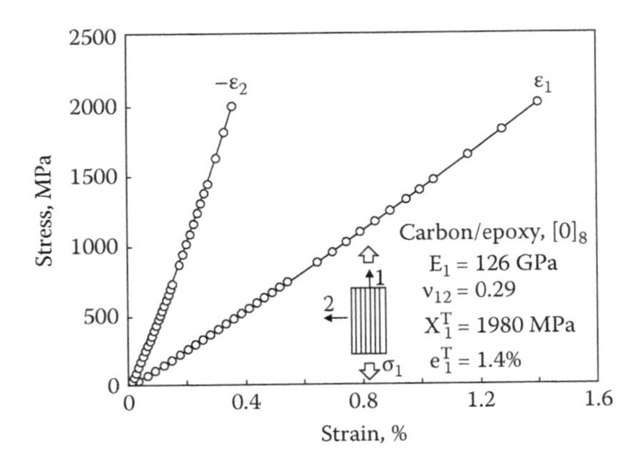

FIGURE 8.5
Tensile stress–strain response of a $[0]_8$ carbon–epoxy specimen.

A representative example of stress σ_1 versus strains ε_1 and $-\varepsilon_2$ curves for a $[0]_6$ unidirectional carbon–epoxy composite is shown in Figure 8.5. The stress σ_1 is defined as the force applied to the specimen divided by test section cross-sectional area. The modulus E_1 was obtained using a least-squares linear fit (Miller et al. 1990) to the linear initial portion of the curve σ_1 versus ε_1; Poisson's ratio ν_{12} was determined from the ratio of initial slopes of σ_1 versus ε_1 and σ_1 versus $-\varepsilon_2$. Reduced values of E_1, ν_{12}, and X_1^T are listed in Figure 8.5. Figures 8.6–8.9 show examples of stress–strain curves for other unidirectional ([0] and [90]) composites. In addition, a sample laboratory report for tensile test is given in Appendix C.

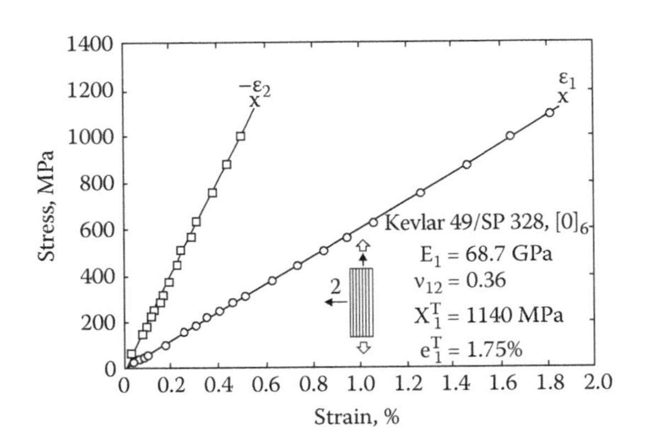

FIGURE 8.6
Tensile stress–strain response of a $[0]_6$ Kevlar–epoxy specimen.

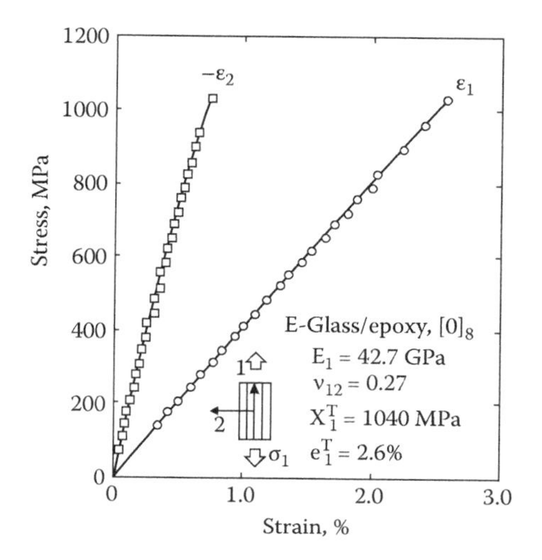

FIGURE 8.7
Tensile stress–strain response of a $[0]_8$ E-glass–epoxy specimen.

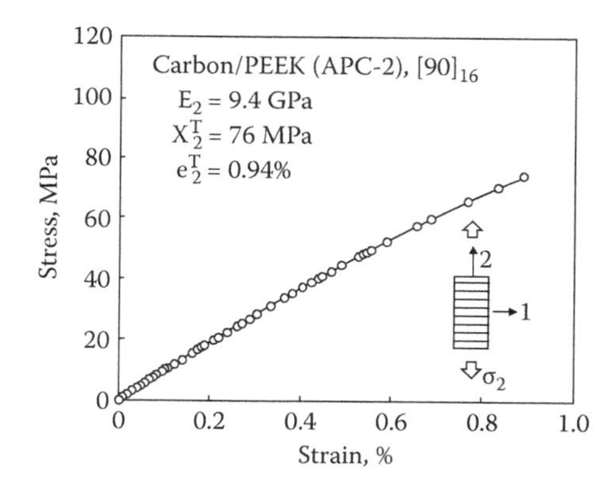

FIGURE 8.8
Tensile stress–strain response of a $[90]_{16}$ carbon–polyetheretherketone (PEEK) specimen.

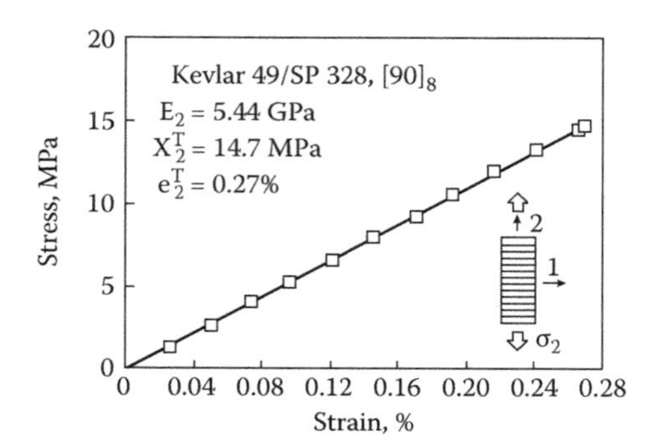

FIGURE 8.9
Tensile stress–strain response of a $[90]_8$ Kevlar–epoxy specimen.

9

Lamina Compressive Response

9.1 Introduction

When fiber-reinforced composites containing unidirectional plies oriented in, or at small angles to, the loading direction are loaded in compression, the fibers may buckle in small regions of the test section (Rosen 1965). This is followed by the formation of kink zones (Evans and Adler 1978) and failure of the fibers at the boundaries of the kink zones because of locally large bending stresses (Figure 9.1).

Compression loading of a composite perpendicular to the fibers involves failure of the matrix and fiber–matrix interface, often exhibited as a shearing type (inclined failure planes), such as illustrated in Figure 9.2. A detailed discussion of the various compressive failure modes observed in fiber-reinforced composites has been presented by Fleck (1977) and Odom and Adams (1990).

Many compression tests are available (Wilson and Carlsson 1991, Chatterjee et al. 1993b). Each method should be judged by its ability to produce compression failure without introducing loading eccentricity and severe stress concentrations at the loaded ends, while avoiding global buckling instability (Euler buckling) of the specimen. The fulfillment of those criteria makes the determination of the true compressive strength difficult. In fact, measurement of the true compressive strength of the composite has only rarely been achieved and perhaps is of minimal practical interest because it is seldom achieved in practical applications (Welsh and Adams 1997). Indeed, compressive strength is determined by test specimen geometry and loading constraints.

The evaluation of test methods for determining lamina compressive response has received exceptional attention in recent years. Prior to 1975, there was no standard for compression testing of composite laminae, although a number of methods had been proposed during the prior 20 years, and some were being used within the composites testing community. The first ASTM standard focused on the Celanese compression test fixture, so called because it had been developed by the Celanese Corporation (Park 1971). Its several deficiencies were soon established, leading to the development of the Illinois

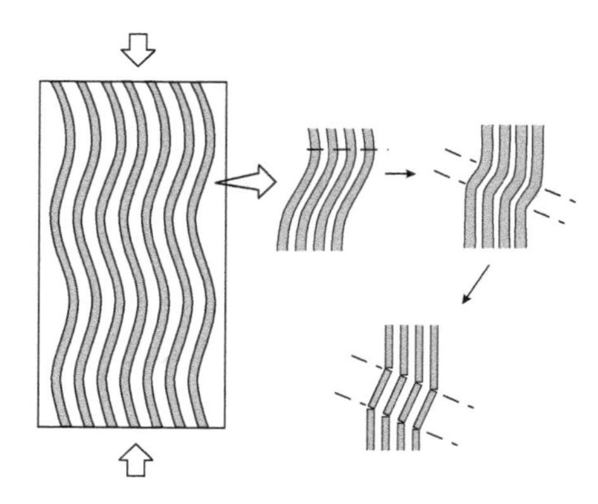

FIGURE 9.1
Mechanism of compressive failure through microfiber buckling and kink band formation.

Institute of Technology Research Institute (IITRI) test fixture (Hofer and Rao 1977), which became ASTM Standard D3410 (2008). Both of these are shear-loaded specimen test methods; that is, the load is introduced into the specimen via the shearing action of wedge grips.

Another form of load introduction that has received minor attention over the years is that achieved by the flexural testing of a sandwich beam consisting of thin laminate face sheets bonded to a core material. The core of the sandwich is typically a honeycomb material. The face sheet on the compressive side of the beam is the test material, whereas the face sheet on the tensile side, designed to have an axial stiffness equivalent to that of the test laminate, can be of any material sufficiently strong that it will not fail before the compressive face sheet. This test method is designated as ASTM D5467 (2010), although this test has not been commonly used for several reasons. The specimen is relatively large (560 mm long, 25 mm wide, and 40 mm thick) and thus consumes considerable test material. The specimen is also relatively expensive to fabricate, and proper sandwich beam fabrication requires special skills not always available to the test laboratory. Improper design or fabrication may lead to core shear failure, core crushing and composite face wrinkling, tensile face sheet failure, or failure of an adhesive

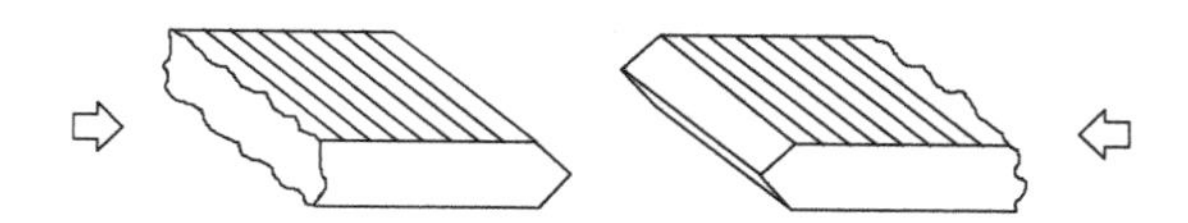

FIGURE 9.2
Shear type of compression failure in a composite loaded in the transverse direction.

bond line prior to compressive face sheet failure, thus invalidating the test. Some of these issues are discussed in Chapter 19.

A more obvious method of applying compression is direct end loading of the specimen. However, if relatively thin laminae are to be tested, lateral supports must be provided to prevent gross buckling. Such a procedure is already standardized for the compression testing of plastics in ASTM D695 (2010), first published in 1942. In the early 1980s, a major modification of this general testing concept was developed for high-performance composites (Boeing Specification Support Standard BSS 7260 1988). This test is commonly termed the Modified ASTM D695 compression test method, although ASTM has not adopted it. It is, however, a recommended test method (SRM 1-94) by the Suppliers of Advanced Composite Materials Association (SACMA) (1994).

There are advantages and disadvantages of both shear loading and end loading. Thus, a combined loading compression (CLC) test method has been developed, ASTM D6641 (2009), which combines the favorable features of both loading types.

9.2 Shear-Loading Test Methods

The IITRI compression test method persists as the shear-loading test methods of choice. A schematic of a typical IITRI fixture is shown in Figure 9.3, and a photograph of an actual fixture is shown in Figure 9.4. Its principal features are the use of flat wedge grips and a pair of alignment rods and bearings. Either untabbed or tabbed specimens are permitted. However, tabbed specimens are most commonly used to protect the test material from the high clamping forces of the wedge grips and surface damage if aggressive grip faces are used. The flat wedge grips permit variation of specimen thickness within the range of movement of the wedges (Adsit 1983). With the use of interchangeable inserts of different thicknesses in the upper and lower blocks of the fixture, a broad range of specimen thicknesses can be accommodated. Some commercially available fixtures allow specimens ranging in thickness from 4 to 15 mm. The standard specimen length is 140 mm, with a 12.7-mm gage length (unsupported length). A specimen width of 12.7 mm is common, although some fixtures can accommodate a width up to 38 mm.

The alignment rods and linear ball bearings of the IITRI fixture will not bind and provide minimal frictional resistance. One disadvantage of the IITRI fixture is its relatively large size. The result is a heavy (>40 kg) and relatively expensive fixture.

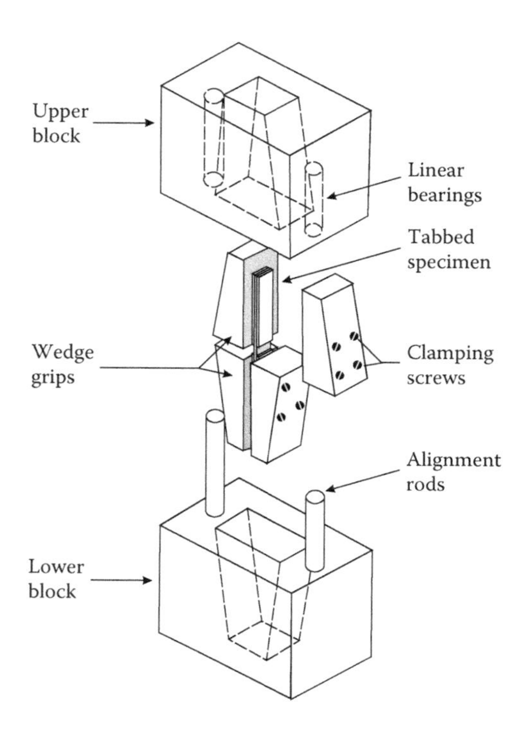

FIGURE 9.3
Sketch of the basic IITRI compression test fixture.

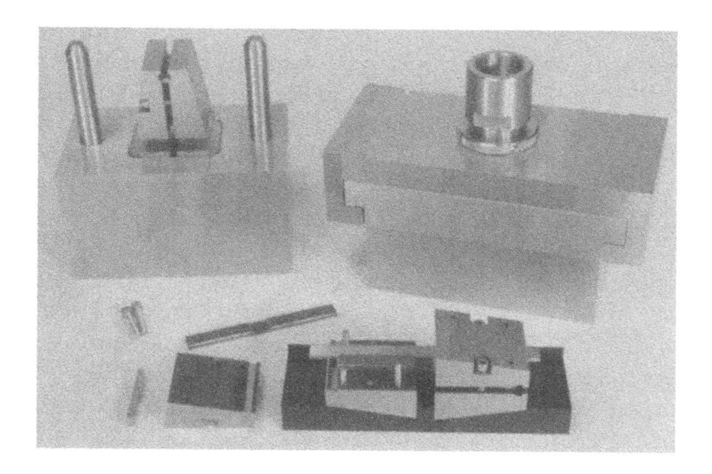

FIGURE 9.4
Photograph of a typical IITRI compression test fixture, with an extra set of wedge grips mounted in a specimen installation jig. (Photo courtesy of Wyoming Test Fixtures, Inc.)

9.3 End-Loading Test Methods

The Modified ASTM D695 method is the most common purely end-loaded test method at present. The fixture is compact, weighs only about 2 kg, and is easily handled. Its only feature in common with the actual standard ASTM D695 (2001) test method is the shape of the lateral supports used to prevent specimen buckling. The standard ASTM D695 specimen is dog bone shaped and untabbed (see Figure 9.5a), whereas the modified specimen is straight sided and tabbed (Figure 9.5b). A photograph of the Modified ASTM D695 fixture is shown in Figure 9.6. This version of the fixture was adopted by the Boeing Company as Boeing Specification Support Standard BSS 7260 (1988).

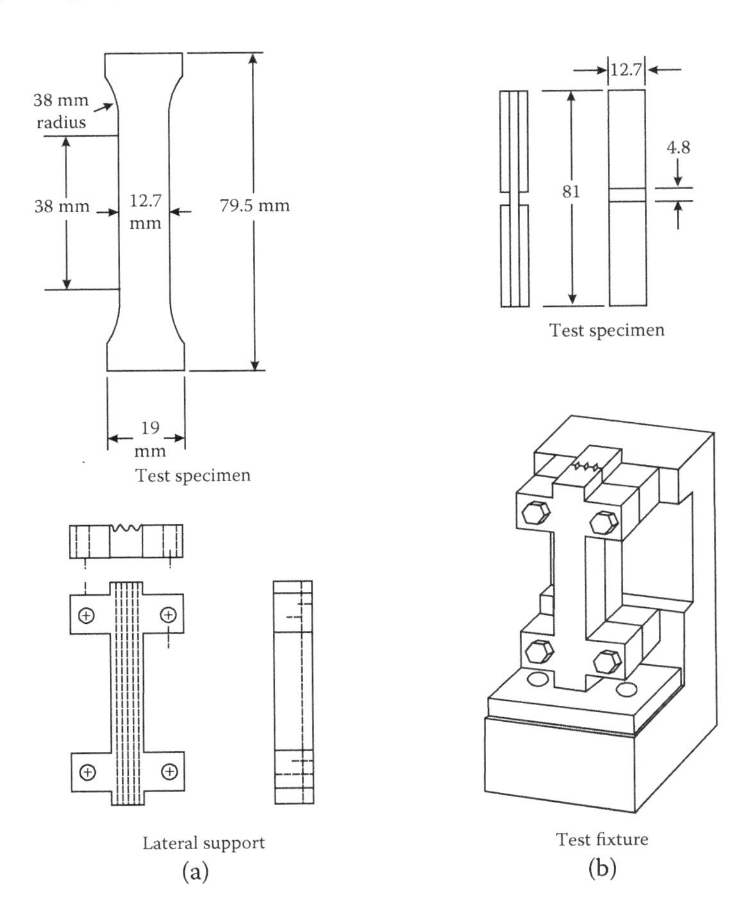

FIGURE 9.5
Sketches of ASTM D695 and Modified D695 compression test fixtures. (a) ASTM D695 dog-bone-shaped compression test specimen and lateral supports; (b) Modified D695 straight-sided tabbed compression strength test specimen and loading fixture with integral lateral supports.

FIGURE 9.6
Photograph of a Modified D695 compression test fixture with tabbed strength specimen and both strength and stiffness (cutout) lateral supports. (Photo courtesy of Wyoming Test Fixtures, Inc.)

This test is not an ASTM standard. It was, in fact, evaluated during a round-robin testing for ASTM standardization but was not recommended (Adsit 1983). The straight specimen geometry acknowledges the fact that highly orthotropic (e.g., unidirectional composites) will simply split longitudinally under load, thereby converting the dog-bone-shaped specimen into a straight-sided specimen. Because the purpose of the dog-bone shape of the specimen is to increase the load bearing area at the ends and thereby eliminate end crushing, tabs were added to the straight-sided specimen to achieve the same result.

The specimen is nominally 80 mm long, Figure 9.5. The standard dog-bone-shaped specimen is 79.5 mm long, while the length of the tabbed, straight-sided specimen is 80.8 mm. There is no technical reason for the modified specimen to be slightly longer, and the same test fixture can be used for both specimens.

Because the modified specimen was developed to test relatively thin laminae (typically an eight-ply unidirectional composite about 1 mm thick), it was necessary to keep the gage length (the unsupported central section between the tabs) short to prevent gross (Euler) buckling. A gage length of 4.8 mm was adopted, Figure 9.5b. There was concern among many potential users that the strength results obtained would be influenced by the close proximity of the tabs (Smoot 1982). Because this gage length is also too short to permit the practical attachment of strain instrumentation (strain gages or extensometers), it is necessary to test a second, untabbed, specimen if compressive modulus is to be determined. An untabbed specimen can be used because only a fraction of the ultimate force need be applied to obtain sufficient data to establish a modulus (the slope of the initial portion of the stress–strain curve). A complete stress–strain curve to failure is thus not available when using this test method because the untabbed specimen will end crush if the loading becomes too high. In addition, because two tests rather than one must be performed, the total cost of testing is increased significantly.

Lateral supports without cutouts are used for the strength test, as shown in Figures 9.5 and 9.6. If a strain gage is used for the modulus test, a lateral support with a shallow cutout (such as that shown at the bottom of the photograph of Figure 9.6) is substituted. The cutout provides clearance for the strain gage. If back-to-back gages are used, two lateral supports with cutouts are needed. If an extensometer is used, it can be clipped onto either edge of the specimen, because the standard 12.7-mm-wide specimen is slightly wider than the central portion of the lateral support. In this case, a lateral support with a cutout is not needed. The recess in the back of the base provides additional clearance for the body of the extensometer.

Another problem with the Modified ASTM D695 fixture (Figure 9.5) is the potential introduction of a redundant load path. This occurs when the fixture screws are tightened to press the lateral supports against the specimen. Friction between the specimen and lateral supports provides the redundant load path, which results in both apparent strength and modulus values that are higher than the actual values. The error introduced is proportional to the degree of tightening of the screws. The Boeing procedure calls for the screws to be torqued from 0.68 to 1.13 N·m, which is not much more than finger tight. However, it is not uncommon for considerably higher torques to be used, with corresponding increases in the induced error (Welsh and Adams 1997). Many users of the fixture apparently do not realize the negative consequence of increasing the torque. It perhaps is intuitive that because the purpose of the lateral supports is to prevent buckling, the tighter they are clamped against the specimen the better.

Despite these various disadvantages and limitations, the Modified ASTM D695 test method was popular during the 1990s. The fixture is compact, lightweight, relatively inexpensive, and comparatively easy to use. However, its use is now waning, because of both the limitations discussed and the introduction of improved fixtures, such as the Wyoming CLC test method, ASTM D6641 (2001), described in Section 9.4.

9.4 Combined Loading Compression Test Methods

The early concept of combined load introduction apparently evolved from the desire to eliminate end crushing of end-loaded compression specimens. The Royal Aircraft Establishment (RAE) test method developed in the early 1970s utilized a straight-sided specimen, adhesively bonded into a slot in an aluminum block at each end (Ewins 1971). The adhesive transmitted a portion (perhaps on the order of 20%) of the applied loading via shear through the adhesive into the specimen faces. (To further minimize end crushing, the specimen was also thickness tapered in the gage section to reduce the cross-sectional area.) Although acceptable results were provided, the concept did not become popular and was never standardized. The requirement to clean up the aluminum blocks for reuse and the specification of thickness tapering were definite drawbacks of the test method.

More recently, a CLC test method was standardized as ASTM D6641 (2009). It was developed at the University of Wyoming (Adams and Welsh 1997, Wegner and Adams 2000) to combine the best features of shear loading and end loading and minimize the deficiencies of each. In addition, acknowledging the difficulties of successfully testing highly orthotropic, strong composite materials, it promotes the testing of cross-ply laminates and then backing out the unidirectional lamina strength using classical lamination theory, as discussed in Section 9.8.

A sketch of the test fixture is presented in Figure 9.7, and a photograph is shown in Figure 9.8. The fixture is compact and relatively lightweight (5.5 kg). The shear-loading component is achieved by clamping pairs of lateral support blocks, which have high-friction contact surfaces, to each end of

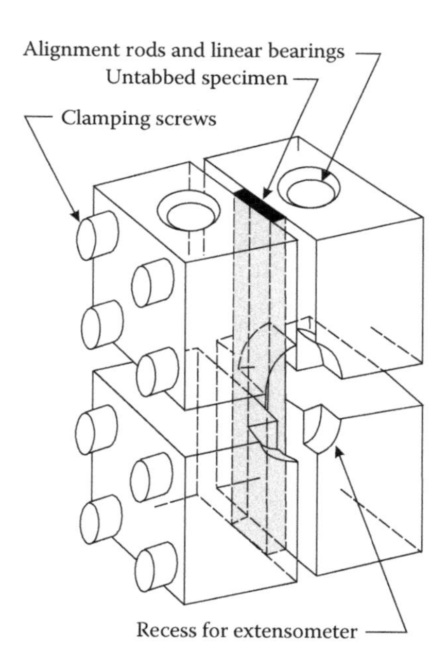

FIGURE 9.7
Sketch of Wyoming CLC test fixture.

FIGURE 9.8
Photograph of a partially assembled Wyoming CLC test fixture, with specimen-centering strips. (Photo courtesy of Wyoming Test Fixtures, Inc.)

the specimen. The end-loading component is induced directly because each end of the specimen is flush with the outer surfaces of the support blocks. The ratio of shear loading to end loading can be controlled by varying the torque in the clamping screws. The goal is to achieve just enough shear loading to avoid end crushing of the specimen. Because the fixture grip surfaces contain relatively small tungsten carbide particles, they do not significantly penetrate and damage the specimen surfaces. Combined with the fact that the through-the-thickness clamping forces on the specimen do not have to be very high, it is typically possible to test an untabbed, straight-sided specimen. This simplifies specimen preparation considerably.

Compared to the IITRI (shear-loaded) test method, the clamping forces on the CLC specimen are much less, thus inducing lower transverse stresses. Because the specimen is not tabbed, the stress concentrations associated with geometric discontinuities are minimized. In addition, there is no redundant load path in the fixture because there is a gap between the upper and lower pairs of support blocks. Alignment of the two halves of the test fixture is maintained by posts and linear bearings (similar to the IITRI fixture).

Cross-ply, angle-ply, quasi-isotropic layup, and similar laminate forms can all be readily tested using an untabbed specimen. But data obtained by testing such laminates, and then indirectly determining unidirectional lamina properties, are sometimes not accepted by the potential user. However, testing untabbed, straight-sided specimens of highly orthotropic, high-compressive-strength, unidirectional composites using the CLC fixture does present some problems. Relatively high clamping forces may be required to prevent end crushing (Coguill and Adams 1999b). This induces stress concentrations, which are undesirable. The solution is to use tabs to increase the end-loading

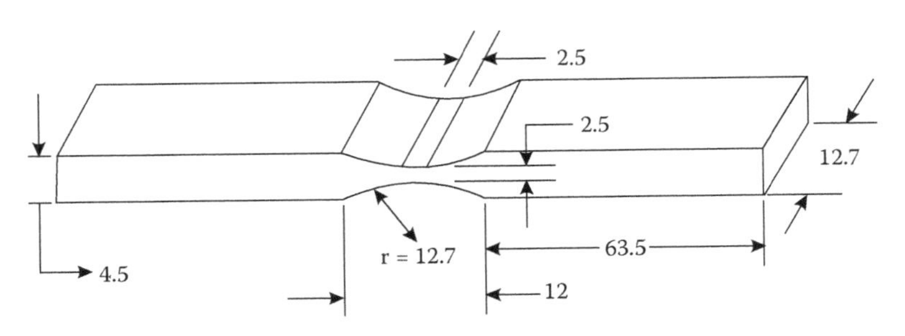

FIGURE 9.9
Sketch of a typical thickness-tapered compression test specimen (dimensions in millimeters).

area (as for the Modified ASTM D695 test method). This permits reducing the clamping force (shear-loading component) required, which reduces the induced stress concentrations. As a result, ASTM D6641 was expanded in 2009 to include a procedure B, which permits the testing of tabbed unidirectional composites, in addition to procedure A for untabbed specimens.

An alternate approach is to use an untabbed, straight-sided, thickness-tapered specimen, such as that shown in Figure 9.9. This thickness tapering has not been approved by ASTM D6641, but greatly improved machining techniques and a composites community, which is now more willing to permit machining of the as-cured surfaces of composites, make thickness tapering more acceptable today (see Adams and Finley 1996).

9.5 Compression Test Procedures

The three compression test methods emphasized in this chapter require certain test procedures that are common to all three, as outlined in this section. Test specimen preparation is discussed in Chapter 4, Section 4.6.

Proper alignment of the testing machine, and of the associated test fixture relative to the machine, is particularly important for compression loading, considering the stability (buckling) issues involved. Any misalignment can induce bending of the specimen and promote buckling. The use of a spherical seat platen or other alignment device in the load train is sometimes encouraged to accommodate misalignments. However, a more positive approach is to achieve proper alignment by careful checking and adjusting of the test setup prior to beginning testing. Specimen bending and buckling cannot usually be detected visually during the test or by microscopic examination of the failed specimen. As discussed in the various standards, the use of two axially oriented strain gages, mounted on opposite faces in the gage section, is the only reliable method of detecting bending and buckling. Although some bending can be tolerated, buckling cannot. The governing ASTM standards specify an acceptable bending limit of 10%, although Wegner and Adams (2000) found that greater amounts of bending, as much as even 20% or 30%, may not be detrimental.

In all cases, the specimen should fail in the gage section. Unfortunately, failure often occurs at the very end of the gage section. This is due to the unavoidable stress concentrations induced in these regions by the tabs (if used), grips, or lateral supports. Thus, although undesirable, such failures are commonly accepted.

The general test procedure is summarized as follows:

1. Examine each individual specimen for fabrication and material defects, and either note the nature and severity of the defect or discard the specimen if deemed necessary.
2. Carefully measure and record the critical dimensions of each specimen (overall length, gage length, width, thickness) and verify that the specified parallelism, perpendicularity, and flatness requirements are met. Reject all out-of-tolerance specimens.
3. Attach electrical resistance strain gages, if they are required, and verify that the quality of the gage installation is acceptable.
4. Align the specimen in the test fixture, verifying that the fixture itself is clean, properly lubricated, and undamaged.
5. Mount the fixture in the testing machine.
6. Set and record the required load range and specified loading rate of the testing machine.
7. Initiate the test and manually record data or confirm that the automated data acquisition system is functioning properly.
8. During the test, using suitable eye protection and a transparent shield surrounding the specimen if deemed necessary, carefully observe the specimen for indications of end crushing, tab debonding, and other anomalies. Note any anomalies and also the observed failure mode.
9. Verify that the test results and failure mode are reasonable before proceeding to test the next specimen. Suspend testing and identify and correct any suspected problems.

9.5.1 IITRI Test Procedure

Many of the existing IITRI fixtures have serrated or knurled grip faces that are relatively aggressive. In such cases, the use of tabs is essential to protect the specimen surfaces. ASTM D3410 does not require a specific type of grip surface. Some newer fixtures have thermal-sprayed surfaces (tungsten carbide particles embedded in the grip surfaces), typically producing a roughness equivalent to only about 100-grit emery cloth or less. In these cases, it may be possible to use untabbed specimens. However, because of the high clamping forces, some type of cushion, such as bonded (or unbonded) tabs or softer pads, is commonly used.

Because of the massiveness of the IITRI test fixture, its main components normally remain in the testing machine. During specimen installation, just the grips are removed. The specimen is mounted in these grips, the assembly is placed back in the cavity in the lower block of the fixture (Figures 9.3 and 9.4), and then the upper block (which is attached to the crosshead) is lowered to engage the upper grips (and the alignment rods).

Because the sloping surfaces of the wedge grips make their handling awkward during specimen installation, it is convenient to use some type of installation jig when mounting the specimen in the grips. This aids in properly aligning the specimen in the grips. Two methods of maintaining alignment are discussed in ASTM D3410.

Back-to-back strain gages are typically used on at least the first few of a batch of specimens being tested, to detect buckling, if present.

Testing is conducted at a crosshead rate of 1 to 2 mm/min.

9.5.2 Modified ASTM D695 Test Procedure

The Modified ASTM D695 test fixture is shown in Figure 9.6. During use, its base rests on a flat platen in the base of the testing machine. A second flat platen mounted in the crosshead is used to apply load to the upper end of the specimen, which extends approximately 3 mm above the lateral supports. Because the fixture is not directly attached to the testing machine, it can be readily moved to a bench for specimen installation.

It is important that the specimen be axially aligned in the fixture. This is typically achieved visually, using the edges of the lateral supports as a guide. Special alignment jigs have also been developed (Wyoming Test Fixtures) but are not commonly used.

As previously discussed in Section 9.3, it is important that the clamping screws be torqued lightly to avoid the development of a significant friction effect through the redundant load path.

During the strength test (for which tabs are used), it is important to watch for specimen end crushing, possibly induced by debonding of one or more of the tabs, which would invalidate the test. This can occur if the wrong combination of tab material, tab thickness, and adhesive is used for the specific composite material being tested. If the tabs are too stiff or too thick relative to the test specimen, they will carry too high a percentage of the applied force and will either end crush or debond. If the tabs are too compliant or too thin, the specimen will end crush. The proper ratios are typically determined experimentally.

9.5.3 CLC Test Procedure

The CLC test fixture, shown in Figures 9.7 and 9.8, is not attached to the testing machine. It rests on the base and is loaded by a second flat platen mounted in the crosshead of the testing machine. The entire fixture can be moved to a bench for specimen installation.

To install a specimen, the two halves of the fixture are separated, and the eight socket-head cap screws (Figures 9.7 and 9.8) are loosened sufficiently to permit insertion of the (typically untabbed, straight-sided) specimen in the lower half. Two uniform-width metal strips for centering the specimen are provided with the standard fixture: The wider strip is used with 12.7-mm-wide specimens, and the narrower strip is used with 25.4-mm-wide specimens. Additional strips can be fabricated as required for specimens of other widths. Indexing pins are provided in the lower half of the fixture, and the appropriate strip is inserted between these pins and the specimen to center the specimen while the four fixture screws are being tightened sufficiently to hold it in position. The outer end of the specimen should be flush with the end of the fixture. This can be achieved easily by performing the installation on a flat surface. The centering strip is then removed, and the upper half of the fixture is mated with the lower half, which is aligned automatically by the fixture posts and bearings. With the upper end of the specimen flush with the end of the upper fixture half, the four screws are tightened sufficiently to hold the assembly in position.

The assembly can then be placed on its side, with the screws facing up, and all eight screws tightened to the desired final torque level in several increments using a controlled tightening pattern, so that the clamping forces on the specimen are distributed as uniformly as possible on each end. The desired torque level is that just sufficient to prevent end crushing of the specimen. Excessive torque increases the detrimental clamping stresses and provides no positive benefit. It has been found that, for most composite materials, torques on the order of 2.3 to 2.8 N·m are adequate. This is a low torque, resulting in low clamping stress concentrations relative to, for example, the IITRI fixture, in which the clamping forces cannot be controlled.

Back-to-back strain gages are typically used on at least the first few of a batch of specimens being tested to detect buckling should it occur.

Testing is conducted at a crosshead rate of 1 to 2 mm/min.

9.6 Failure Modes

Some typical compressive failure modes are sketched in Figure 9.10. As already noted, gross buckling (Figure 9.10d) is unacceptable. However, the other failure modes are typically accepted. A description of acceptable failure modes is presented in both ASTM D3410 and ASTMD 6641.

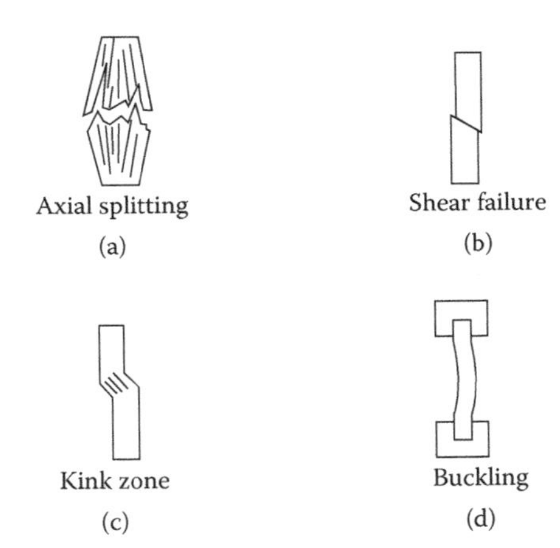

FIGURE 9.10
Typical failure modes for compression specimens: (a) axial splitting, (b) shear failure, (c) kink zone, and (d) buckling.

As previously discussed, the only reliable way of detecting gross buckling is by the use of back-to-back strain gages. This is illustrated in Figure 9.11. At the applied stress level at which the specimen buckles, the strains suddenly diverge. Divergence of the strain readings prior to this level (also shown in Figure 9.11) is an indication of specimen bending.

The kink band formation in the 0° composite indicated in Figure 9.10c is due to a local instability at the fiber level when the lamina is axially loaded. It progresses from elastic deflections of the fibers to actual fiber fractures, as indicated in Figure 9.1. In contrast, when a 90° lamina is loaded, a shear failure is typical, such as indicated in Figure 9.10b. Figure 9.2 indicates an

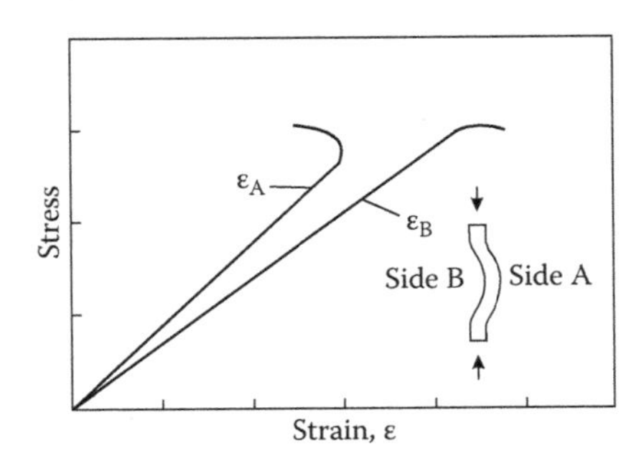

FIGURE 9.11
Schematic of compressive stress–strain responses for back-to-back strain-gaged specimen that failed by Euler buckling.

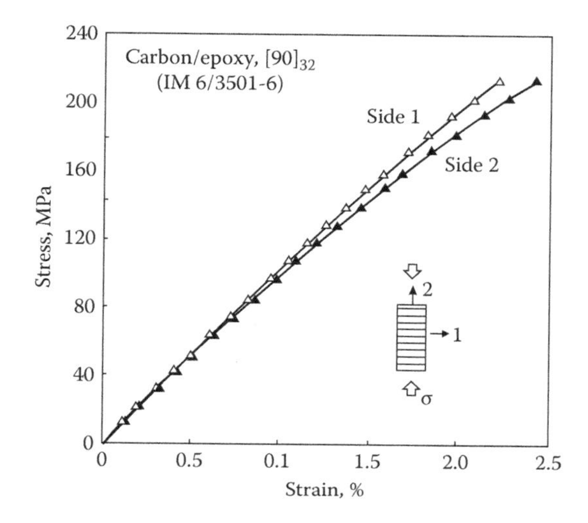

FIGURE 9.12
Compressive stress–strain response for a back-to-back strain-gaged $[90]_{32}$ carbon–epoxy specimen.

alternate form of this shear failure mode. An example of stress–strain curves for a uniformly loaded specimen is shown in Figure 9.12.

9.7 General Data Reduction

One can plot the load versus strain data for each gage if back-to-back gages are used, and examine the plots for global buckling or excessive bending (Figure 9.11). To determine the compressive moduli, recall the definitions of Chapter 2:

- E_1: The initial slope of the stress–strain curve $(\Delta\sigma_1/\Delta\varepsilon_1)$ for the 0° compressive test
- E_2: The initial slope of the stress–strain curve $(\Delta\sigma_2/\Delta\varepsilon_2)$ for the 90° compressive test

The modulus is thus evaluated from the initial slope of the average response curve, which can be obtained, for example, by using a linear least-squares regression fit of the data. If plots from two gages are available, the reported modulus is the average of the two values obtained. The various standards outline specific procedures in detail. The compressive strengths X_1^c and X_2^c are obtained directly from the plots, whereas the ultimate strains e_1^c and e_2^c are taken as the average of the two values obtained if two plots are available. Figures 9.13–9.16 show examples of compressive responses for various composites.

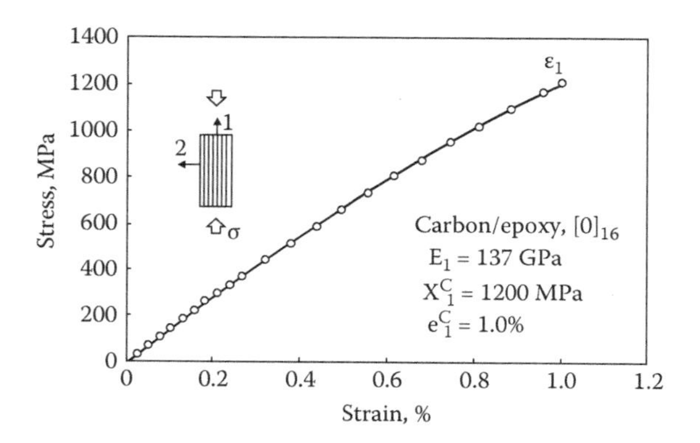

FIGURE 9.13
Compressive stress–strain response for a $[0]_{16}$ carbon–epoxy specimen.

9.8 Indirect Determination of Unidirectional Lamina Strength from a Test of a Cross-Ply Laminate

In the early 1990s, considerable interest developed in the possibility of compression testing a general laminate of arbitrary layup orientation but containing some 0° plies and then indirectly determining the strength of these 0° plies using a suitable analysis (Hart-Smith 2000). That is, the compressive strength of a unidirectional composite X_1^C is estimated by multiplying the

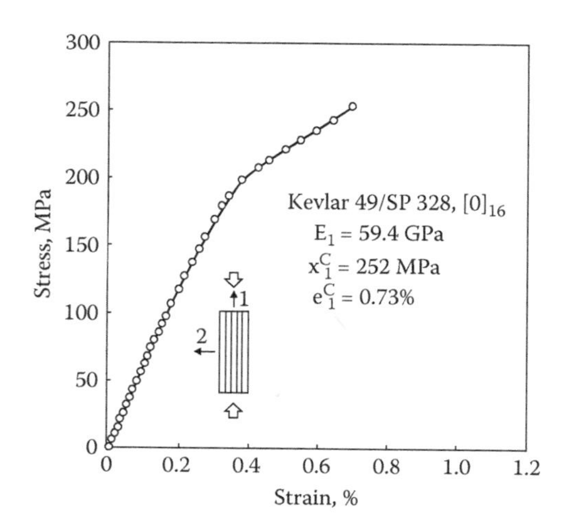

FIGURE 9.14
Compressive stress–strain response for a $[0]_{16}$ Kevlar–epoxy specimen.

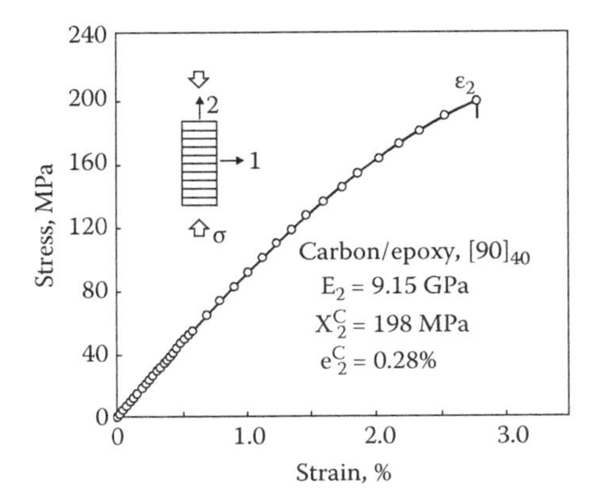

FIGURE 9.15
Compressive stress–strain response of a $[90]_{40}$ carbon–epoxy specimen.

measured compressive strength of the chosen laminate X^C by the back-out factor (BF):

$$X_1^C = BFX^C \tag{9.1}$$

The BF is calculated from the ply stiffnesses and layup of the laminate as discussed further in this section.

Although a wide range of both cross-ply and angle-ply laminates was initially considered, it was soon realized that using a simple $[90/0]_{ns}$ cross-ply laminate has significant advantages. In particular, there is no shear

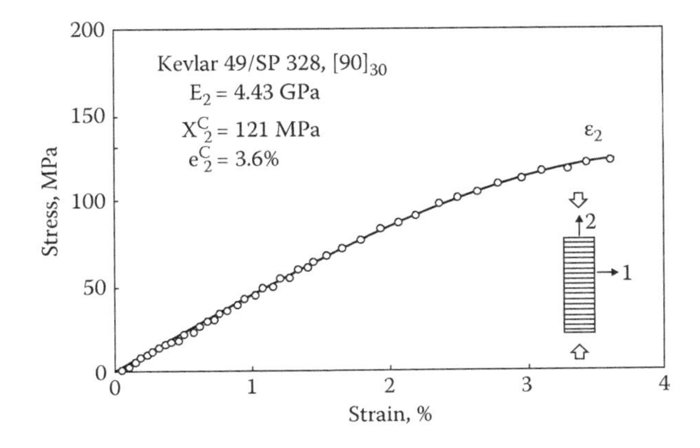

FIGURE 9.16
Compressive stress–strain response for a $[90]_{30}$ Kevlar–epoxy specimen.

stress in the plies (see Chapter 2), eliminating the need to consider the lamina shear modulus in calculating the BF. The use of a $[90/0]_{ns}$ rather than a $[0/90]_{ns}$ laminate is deliberate; the 90° plies on the surfaces protect the immediately adjacent 0° plies from fixture- and tabbing-induced stress concentrations. Of course, as n becomes larger, the importance of protecting the 0° plies next to the surface becomes less important because they become a smaller percentage of the total number of primary load-bearing plies (i.e., 0° plies) present.

The BF for $[90/0]_{ns}$ and $[0/90]_{ns}$ cross-ply laminates is (Welsh and Adams 1996)

$$BF = \frac{\frac{1}{2}E_1(E_1 + E_2) - (\nu_{12}E_2)^2}{\frac{1}{4}(E_1 + E_2)^2 - (\nu_{12}E_2)^2} \tag{9.2}$$

where E_1 and E_2 are the axial and transverse moduli of the unidirectional lamina, respectively, and υ_{12} is the major Poisson's ratio of the unidirectional lamina. These three stiffness properties must be known from axial (E_1, υ_{12}) and transverse (E_2) tests of a unidirectional composite. This is not a difficulty, however, because the composite does not have to be loaded to high-stress levels to determine these stiffness properties.

Note that the BF in Equation (9.2) will typically not be very sensitive to υ_{12} because its value (on the order of 0.25) is small, and $E_2 < E_1$. That is, the second terms in the numerator and denominator will be small relative to the first terms. This yields

$$BF \approx \frac{2}{1 + E_2/E_1} \tag{9.3}$$

In addition, if $E_2 \ll E_1$, $BF \approx 2$ (implying that the 90° plies of the laminate carry a minor portion of the applied force).

Equation (9.1) is thus relatively easy to implement in backing out the 0° lamina compressive strength from the results of a test of a $[90/0]_{ns}$ cross-ply laminate. Again using classical lamination theory, corresponding equations can be derived for other types of laminates as well.

10

Lamina Shear Response

10.1 Introduction

A shear test of a composite material is typically performed with the desire to determine the shear modulus and shear strength. Ideally, both properties can be determined from the same test, but this is not always the case. In addition, the shear response of a composite material is commonly nonlinear, and full characterization thus requires generating the entire shear stress–strain curve to failure. However, many shear test methods are not capable of generating the entire curve, sometimes not even a portion of it.

Figure 10.1 defines the in-plane shear stress τ_{12} (and τ_{21}) and shear strain γ_{12} (and γ_{21}). The in-plane shear modulus is denoted by G_{12} and the shear strength by S_6. Additional definitions and notation are presented in Chapter 2.

The major deficiency of all existing shear test methods for composite materials is the lack of a pure and uniform state of shear stress in the test section. Thus, compromises have to be made. Although many shear test methods are described in the literature (Chatterjee et al. 1993c, Adams 2000a), only a relatively few are in common use.

In particular, the torsional loading of a thin-walled, hoop-wound tube is not detailed here because it is seldom used since the specimen is difficult to prepare and test properly. Most discussions of shear testing start with the statement: "The torsional loading of a thin-walled tube produces a uniform state of shear stress, but … ." Then, some or all of the following negative aspects of tube testing are enumerated: Fabrication of the tube, which is typically hoop-wound using the filament winding process, requires special equipment and expertise. Fabrication of a tube with fibers oriented along the axis of the tube is even more difficult. In both cases, the resulting tube is relatively fragile. Equally important, a tube is usually not representative of the material form used in the eventual structural design. For example, because of the radical differences in the processes used to fabricate tubes versus flat (or curved) structural panels, the material may not have the same strength properties. One example of the fabrication of unidirectional tubes is provided by Whitney et al. (1971). A torsional loading machine of sufficiently low torque capacity is required and often not

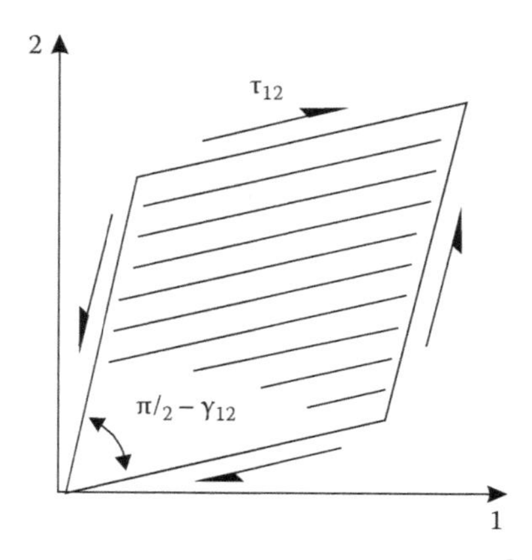

FIGURE 10.1
Definition of in-plane shear stress τ_{12} and shear strain γ_{12}.

available. The tube specimen must be reinforced at each end in some manner so that it can be gripped within the testing machine without damaging it. A hoop-wound tube in particular is very susceptible to inadvertent bending loads induced during testing. Any induced bending stresses combine with the shear stress to induce premature failure and thus low apparent shear strength. However, even for the more robust test geometries with longitudinal fiber orientation, the gripping of the specimen is the central problem. Typical tabbing materials adhesively bonded for load introduction restrict the radial deformation of the tube when under load, and this lack of freedom to contract or expand in the radial direction induces local bending in the thin-walled tube, thereby introducing stress concentrations in the gripping region. When premature failure occurs in this region, the data are inappropriate measures of the shear strength of the material.

The six most popular current shear tests all employ flat specimens. They include the Iosipescu shear test (American Society for Testing and Materials [ASTM] D5379 2005); the two- and three-rail shear tests (ASTM D4255 2007); the V-notched rail shear test (ASTM D7078 2005); the $[\pm45]_{ns}$ tension shear test (ASTM D3518 2007); and the short-beam shear test (ASTM D2344 2006).

10.2 Iosipescu Shear Test Method

The Iosipescu shear test method and specimen configuration shown in Figure 10.2 are based on the original work with metals by Nicolai Iosipescu of Romania (1967), from which the test method derives its name. The Composite Materials Research Group (CMRG) at the University of Wyoming

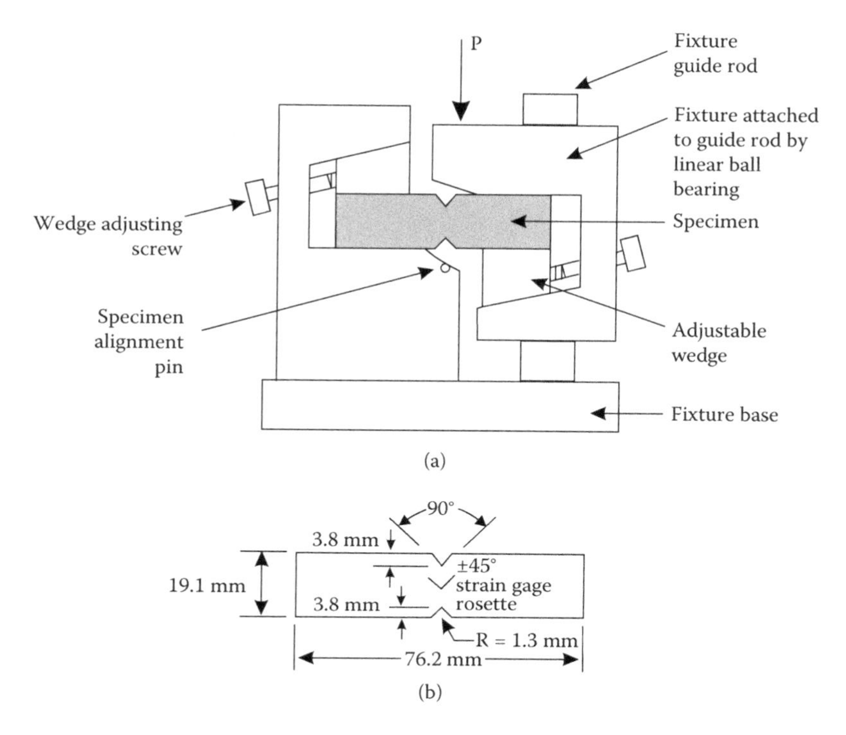

FIGURE 10.2
Sketches of (a) Iosipescu shear test fixture and (b) test specimen.

led its application to composite materials (Adams and Walrath 1982, Walrath and Adams 1984). The test method became an ASTM standard for composite materials in 1993. Analysis of the specimen under load reveals that a state of uniform shear stress exists in the center of the notched specimen on the cross section through the notches, although not in the immediate vicinity of the notch roots (Bergner et al. 1977, Walrath and Adams 1984, Ho et al. 1994). In addition, the normal stresses (the nonshear stresses) are low everywhere on this cross section. By orienting the specimen axis along any one of the three axes of material orthotropy, any one of the six shear stress components, representing the three independent shear stress components (see Chapter 2), can be developed.

For example, Figure 10.3 shows the required specimen orientations for achieving the two (nonindependent) in-plane shear stress components τ_{12} and τ_{21} for a unidirectional composite. However, note that a 0° orientation (fibers parallel to the long axis of the specimen) forms a much more robust specimen and is strongly preferred over a 90° orientation. A $[0/90]_{ns}$ (cross-ply) specimen is even more robust. Because there is no shear coupling between the plies of a $[0/90]_{ns}$ laminate (see Chapter 2), this orientation will theoretically produce the same shear properties as those of a unidirectional composite. In

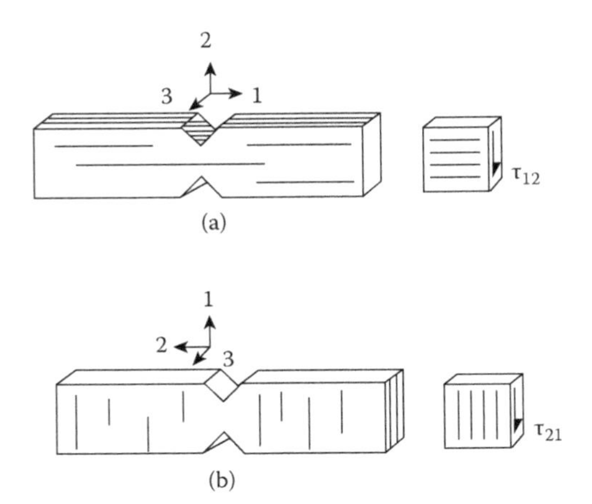

FIGURE 10.3
Iosipescu shear test specimens for in-plane shear testing: (a) 0° specimen and (b) 90° specimen.

practice, it is likely to produce shear strengths closer to the true shear strength of the composite material because premature failures are less likely to occur. That is, the cross-ply laminate is likely to produce more accurate (and in this case higher) shear strengths. However, note that presently the 0° orientation unidirectional specimen is still much more commonly used, in part because a unidirectional laminate is more likely to be available for testing. This may change if the use of cross-ply laminates and back-out factors to determine unidirectional lamina compressive strength, as discussed in Section 9.8 of Chapter 9, increases in popularity.

When a strain gage is attached to one (or both) faces of the specimen in the central region between the notches, a complete shear stress–shear strain curve can be obtained. These attractive features, along with the relatively small specimen size and the general ease of performing the test, have made the Iosipescu shear test method popular.

The standard Iosipescu specimen is shown in Figure 10.2b. The top and bottom edges must be carefully machined to be flat, parallel to each other, and perpendicular to the faces of the specimen to avoid out-of-plane bending and twisting when the load is applied (see Figure 10.2a). The geometry of the notches is less critical (Walrath and Adams 1984). The standard fixture, shown in Figure 10.4, can accommodate a specimen thickness up to 12.7 mm. ASTM D5379 suggests a thickness of 3 to 4 mm. For most composite materials, it is convenient and economical to form the V-shaped notches using a shaped grinding wheel. The notch on one edge of a stack of specimens can be formed, the stack turned over, and the other notch formed. If shear strain is to be measured, a two-element strain gage rosette with the elements oriented ±45° relative to the specimen longitudinal axis can be attached to the central test section region, such as shown in Figure 10.5, and the rosette wired in a

FIGURE 10.4
Photograph of an Iosipescu shear test fixture with specimen installed. (Photograph courtesy of Wyoming Test Fixtures, Inc.)

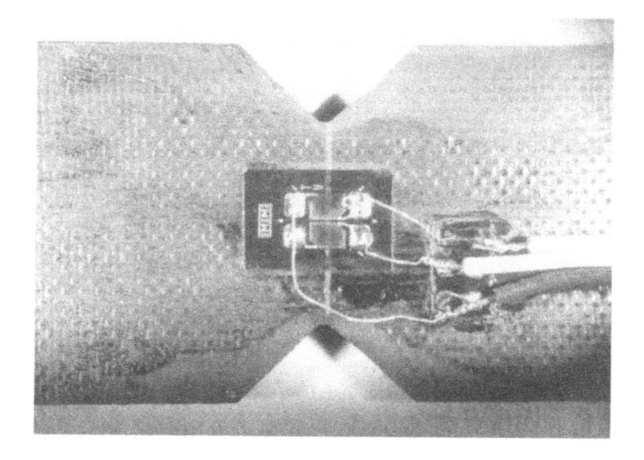

FIGURE 10.5
A ±45° biaxial strain gage rosette bonded to an Iosipescu shear test specimen.

half-bridge circuit. A single-element gage oriented at either plus or minus 45° can be used and wired in a quarter-bridge circuit. If out-of-plane bending and twisting of the specimen are a concern, back-to-back strain gages can be used to monitor these undesired effects (Morton et al. 1992). However, this is normally not necessary.

The specimen should be centered horizontally in the test fixture using the specimen-centering pin (Figures 10.2a and 10.4). Vertical alignment is achieved by keeping the back face of the specimen in contact with the fixture while the wedge-adjusting screws are finger tightened to close any gap between the specimen and the fixture. Note that these wedges are not clamps and need not be tightened. They are provided to accommodate any tolerance in the width dimension of the specimen.

The upper half of the test fixture is loaded in compression through a suitable adapter, attaching it to the crosshead of the testing machine. A crosshead displacement rate of 2 mm/min is typical. The applied load and strain signals are monitored until the specimen fails.

The (average) shear stress across the notched section of the specimen is calculated using the simple formula

$$\tau = P/A \tag{10.1}$$

where P is the applied force, and A is the cross-sectional area of the specimen between the notches. For a unidirectional composite specimen tested in the 0° orientation, detailed stress analyses (Bergner et al. 1977, Walrath and Adams 1984, Morton et al. 1992, Ho et al. 1994), indicate that an initially nonuniform elastic stress state exists. However, if any inelastic material response occurs, particularly if there is initiation and arrested propagation of a crack parallel to the reinforcing fibers at each notch tip (which may occur well before the ultimate loading is attained), the local stress concentrations are significantly relieved. The stress distribution then becomes more uniform across the cross section of the specimen and increases further the accuracy of Equation (10.1).

Shear strain γ is simply calculated as the sum of the absolute values of the ±45° strain gage readings:

$$\gamma = |\varepsilon(45^\circ)| + |\varepsilon(-45^\circ)| \tag{10.2}$$

or, if only a single-element gage mounted at plus or minus 45° is used,

$$\gamma = 2\,|\varepsilon(45^\circ)| \tag{10.3}$$

The shear modulus G is obtained as the initial slope of the shear stress–shear strain curve.

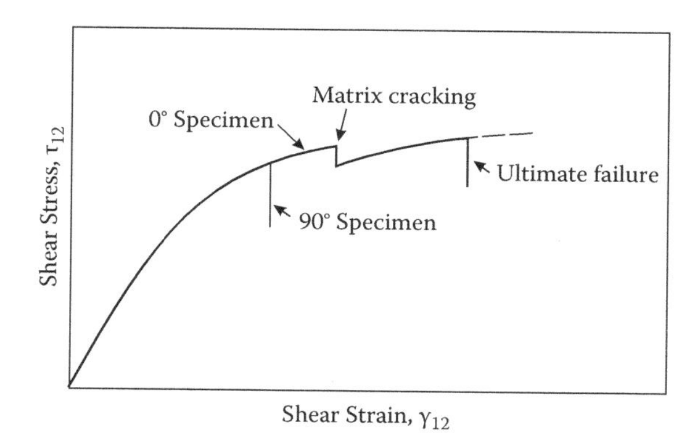

FIGURE 10.6
Schematic stress–strain curves for 0° and 90° Iosipescu shear test specimens.

Premature damage in the form of longitudinal matrix cracks initiating from the notch roots is a common occurrence in 0° unidirectional specimens. Load decreases are observed at about two-thirds of the eventual ultimate load when these cracks initiate and propagate, but they quickly arrest, and the specimen then carries additional load until the true shear failure occurs, which involves multiple matrix cracks parallel to the fibers concentrated in the region of the specimen between the two notches. Because 90° specimens often fail prematurely, particularly for brittle-matrix composites as a result of stress concentrations and induced bending, they may produce a low failure stress, see Figure 10.6. Figure 10.7 shows common failure modes of unidirectional specimens. The 90° specimen usually fails suddenly, parallel to the fibers (Figure 10.7a). The 0° specimen fails in a more gradual manner. As noted, a small load decrease is often observed at approximately two-thirds of the ultimate shear strength (Figure 10.6) because of cracking at the notch root, as indicated in Figure 10.7b. Load drops, relatively close to each other, will occur if the notch root cracks do not happen to occur simultaneously. These are local stress-relieving mechanisms, as discussed previously, and do not represent the shear strength. The stress that results in total failure across the test section, as shown in Figure 10.7c, is the failure stress S_6.

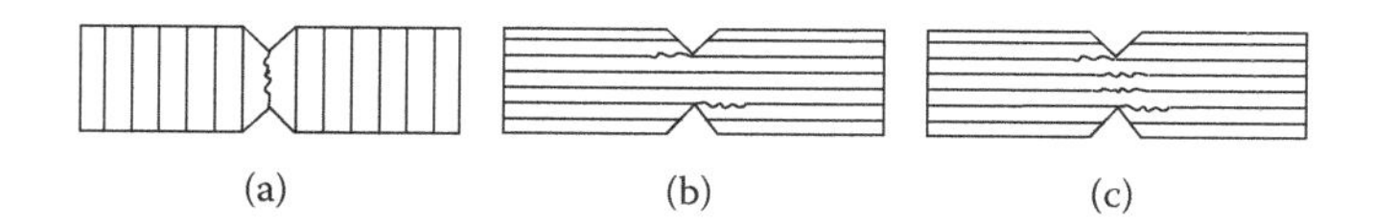

FIGURE 10.7
Failure modes of Iosipescu shear test specimens: (a) matrix cracking in a 90° specimen; (b) and (c) matrix cracking in a 0° specimen.

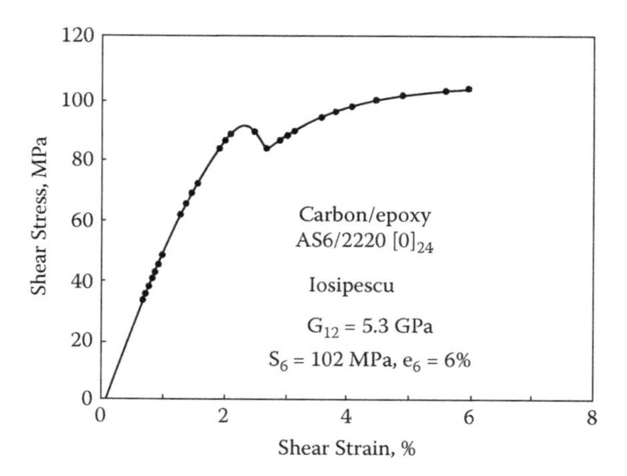

FIGURE 10.8
Shear stress–strain curve for a $[0]_{24}$ carbon/epoxy Iosipescu shear specimen.

Figures 10.8 and 10.9 show typical stress–strain curves for 0° specimens of two different types of composite materials. Note that no load drop is evident in the shear stress–strain curve of Figure 10.9 because the polyphenylene sulfide (PPS) thermoplastic matrix is relatively ductile, relieving the local shear stress concentrations at the notch roots sufficiently to avert local failures. Additional examples of acceptable and unacceptable failure modes are presented in ASTM D5379.

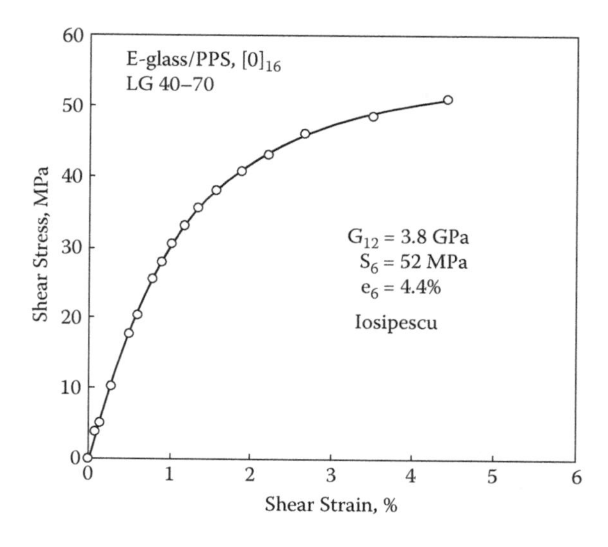

FIGURE 10.9
Shear stress–strain curve for a $[0]_{16}$ glass/PPS Iosipescu shear specimen.

10.3 Two-Rail Shear Test Method

The two-rail shear test is an in-plane shear test method. In this discussion, we restrict the specimen geometry to unidirectional-ply laminates of 0°, 90°, or $[0/90]_{ns}$ fiber orientation. Both the two- and three-rail shear tests are included in the same ASTM Standard D4255 (2007). They are discussed in separate sections here because they utilize different test fixtures and offer different advantages and disadvantages. Presently, these two test methods are used somewhat less frequently than the other three test methods. This is particularly true for the three-rail shear test method, for reasons to be discussed further. However, the two-rail shear test method is given a more prominent position in the present discussion because it has some favorable technical attributes that the two test methods to be discussed next (the $[\pm 45]_{ns}$ tension shear and the short-beam shear test methods) do not exhibit. That is, although presently it is not used as extensively as are the others, it has significant potential for future improvements and hence increased use, as discussed in the following material.

The commonly used tensile-loading version of the two-rail shear test fixture is shown schematically in Figure 10.10a. A compression-loading fixture also exists, and compressive loading can be advantageous in some circumstances. The tensile-loading fixture has had a long history and presumably is based on fixture designs originally developed even earlier (circa 1960) for the shear testing of plywood panels (Boller 1969). As a consequence, this fixture contains some features that are not fully logical for use with composite materials. For example, note that in Figure 7.10a the specimen is loaded at a

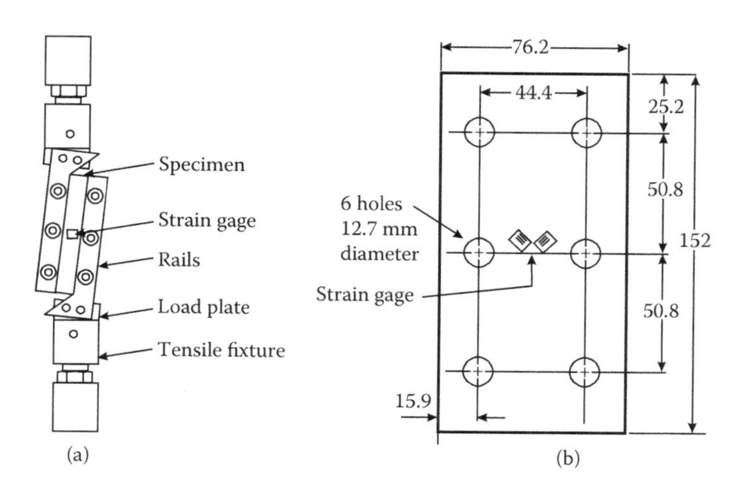

FIGURE 10.10
Two-rail shear test method: (a) fixture configuration and (b) specimen geometry (all in millimeters).

slight angle relative to its axis (indicated as 7° in ASTM D4255). There does not appear to be a technical reason for this; rather, it is probably an artifact of a test fixture for plywood (ASTM 2719 2007) developed in the early 1960s. In that case, because of the type of loading apparatus used, it was convenient to apply the load to the large (610 × 430 mm) plywood test panel slightly off axis.

The two-rail shear test specimen is shown in Figure 10.10b. As indicated, the specimen is 76.2 × 152.4 mm, thus consuming eight times more test material than the Iosipescu shear specimen. Note also that there are six holes in the otherwise-simple rectangular specimen. These are clearance holes for the six bolts that clamp the specimen to the rails. Not only do these holes potentially introduce stress concentrations in the test specimen, but also there is always some inherent concern about making holes in a composite material without introducing auxiliary damage. In addition, for very high-shear-strength composites, the clamping forces have to be very high to avoid slipping of the rails during the test. A bolt torque of 100 N·m, which is a very high torque for the 9.5-mm-diameter bolts, is specified.

Despite these current deficiencies of the two-rail shear test method, there are distinct positive attributes of the method as well. That is, an essentially pure shear loading is applied to the gage section of the specimen (the 12.7-mm-wide portion of the specimen exposed between the rails). The shear stress along the length of the specimen (parallel to the rails) is relatively uniform, except near the ends (which must be at a zero shear stress because they are free surfaces). Some extraneous normal (tensile and compressive) stresses are introduced by the presence of the rails, particularly near the boundaries of the gage section. Finite element analyses have been conducted to characterize these stresses (Bergner et al. 1977, Hussain and Adams 1998), and undoubtedly specimen and fixture modifications can be made to significantly reduce, if not eliminate, them.

When a two-rail shear test is conducted, the specimen must first be properly prepared, heeding in particular the cautions already noted regarding formation of the required six clearance holes in the specimen without introduction of auxiliary damage. When a unidirectional lamina is tested, the fibers can be oriented either parallel (90° orientation) or perpendicular (0° orientation) to the rails. However, a fiber orientation perpendicular to the rails is much preferred because extraneous bending and edge effects have much less influence on the measured properties, and the specimen is much more robust (Hussain and Adams 1998). In fact, testing a $[0/90]_{ns}$ (cross-ply) laminate may be even better; the specimen is even more robust than a unidirectional lamina with fibers oriented perpendicular to the rails. Because there is no shear coupling in the plies of a $[0/90]_{ns}$ laminate (see Chapter 2), this orientation will theoretically produce the same shear properties as a unidirectional composite. In practice, it is likely to produce shear strengths closer to the true shear strength of the composite material because premature failures are less likely to occur. That is, the cross-ply laminate is likely to produce more accurate (and in this case higher) shear strengths. Note that this same general logic was stated in Section 10.2 with reference to the Iosipescu shear test specimen.

The specimen to be tested is clamped between the pairs of rails, as indicated in Figure 10.10a. It is important that the rails do not slip during the test. If they do slip, the clamping bolts can bear against the clearance holes in the specimen, inducing local stress concentrations and leading to premature failure. This, of course, results in an unacceptable test. Most currently commercially available fixtures have thermal-sprayed tungsten carbide particle gripping surfaces (Wyoming Test Fixtures), although ASTM D4255 does not specifically require them. The thermal-sprayed surfaces generally have much better holding power than the other types of grip surfaces listed in the standard.

If shear strain is to be measured, a single-element or a dual-element strain gage is bonded to the specimen at the center of the gage section. If bending or buckling is suspected, back-to-back strain gages can be used, just as discussed for the Iosipescu shear test method in the previous section. Likewise, the calculations of shear stress, shear strain, and shear modulus are also the same; the cross-sectional area in the present case is the length of the specimen parallel to the rails times the specimen thickness.

10.4 Three-Rail Shear Test Method

Although the in-plane shear stress state induced in the three-rail shear test, Figure 10.11, is generally similar to that induced in the two-rail shear test, there also are significant differences between the two test methods. In particular, the test fixtures are quite different. The standard fixture shown is designed to be loaded in compression between the flat platens of a testing machine. Tensile loading is also permissible if the fixture is modified to permit attachment to the base and crosshead of the testing machine. However, in practice this is not commonly done. In fact, the fixture drawings available from ASTM only include the compression-loaded configuration.

The three-rail shear fixture does shear load the specimen along its geometric axis. Nine clearance holes must be cut into the specimen, and the size of the standard specimen is 136 × 152 mm. The (two) gage sections are each 25.4 mm wide. The potential for shear buckling in the gage section is a concern. The methods of specimen gripping, specimen strain gaging, and reducing the test data are essentially the same as for the two-rail shear test method, as detailed in Section 10.3, although the shear force is P/2 for this test.

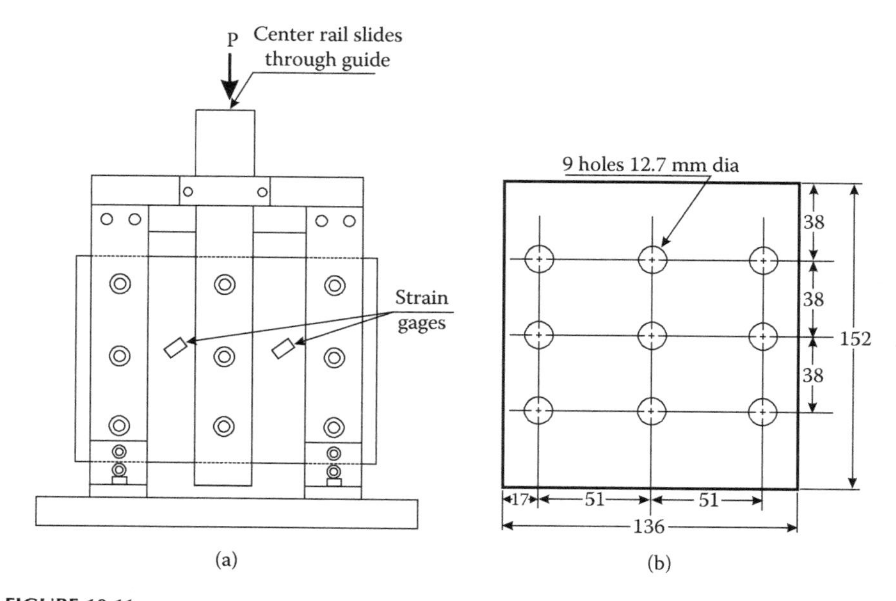

FIGURE 10.11
Three-rail shear test method: (a) fixture configuration and (b) specimen geometry (all in millimeters).

10.5 V-Notched Rail Shear Test Method

The Iosipescu and rail shear test methods discussed in the previous sections were originally developed to characterize unidirectional composites for the purpose of obtaining basic ply shear properties. However, now there is increasing interest in testing laminates of various ply orientations, which typically exhibit significantly higher shear strengths. This includes $[\pm 45]_{ns}$ angle-ply laminates, which are often used as shear panels in structures because of their extremely high shear stiffness and strength.

As discussed in Section 10.2, the Iosipescu specimen is loaded on its edges. If this loading becomes high enough, the edges may crush. This redistributes the local stresses such that the crushing may be arrested, allowing the test to proceed to a valid shear failure in the test section. However, if higher loading to shear failure is required, the crush zone may propagate into the test section between the specimen notches and invalidate the test results.

Likewise, there are difficulties in using the rail shear test methods to test materials with high shear strength, as discussed in Sections 10.3 and 10.4. In particular, slipping of the rails can cause local crushing of the specimen due to bearing of the clamping bolts on the edges of the holes in the specimen.

Thus, the V-notched rail shear test specimen, ASTM D7078 (2005), shown in Figure 10.12 was developed to combine the favorable stress state created

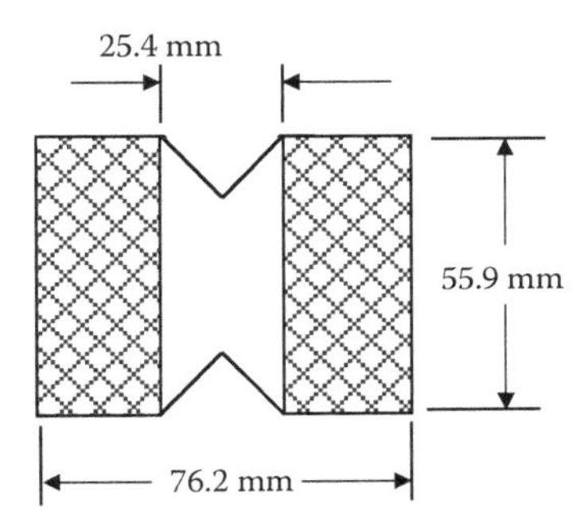

FIGURE 10.12
V-notched rail shear test specimen.

by the V notches with face loading of the specimen (Hussain and Adams 1999, Adams et al. 2003).

The specimen length, 76 mm, is the same as that of the standard Iosipescu specimen. However, the width, 56 mm, is almost three times greater, resulting in a proportionally larger gage section. As for the Isopescu specimen, the V notch may conveniently be machined using a shaped grinding wheel. While less important when testing unidirectional and cross-ply laminates, a larger gage section is very beneficial when general laminates are to be tested (since the failure zone tends to be larger) or when testing coarse-weave fabrics (since it is desired to have several unit-cells within the gage section). Of course, both the Iosipescu and the V-notched rail specimens (and corresponding fixtures) can be scaled up in size if desired.

The resulting test fixture is shown in Figure 10.13. Rather than the clamping bolts passing through holes drilled in the specimens, these bolts push on pairs of clamping plates that sandwich the specimen between them. The standard V-notched rail shear fixture will accommodate specimens up to 12 mm in thickness. Using bolts on both sides permits using one set to center the particular specimen thickness in the fixture while the other set is torqued as required to provide the desired clamping force. The surfaces of the clamping plates in contact with the specimen are coated with tungsten carbide particles to increase friction.

Using this test fixture, shear strengths as high as 500 MPa have been successfully measured for $[\pm 45]_{ns}$ angle-ply laminates (Adams et al. 2002). For comparison, the common limit for successfully testing a unidirectional composite using the Iosipescu fixture is only on the order of 135 MPa.

Unlike the Iosipescu shear fixture, which has built-in features for centering and aligning the specimen, the V-notched rail shear specimen must be manually centered and aligned, both in the length and width directions. This can be done visually with a reasonable degree of accuracy. However, an alignment jig has been developed that makes the process much easier. The use of this jig is illustrated and discussed in ASTM D7078. It is shown installed on a specimen in Figure 10.14.

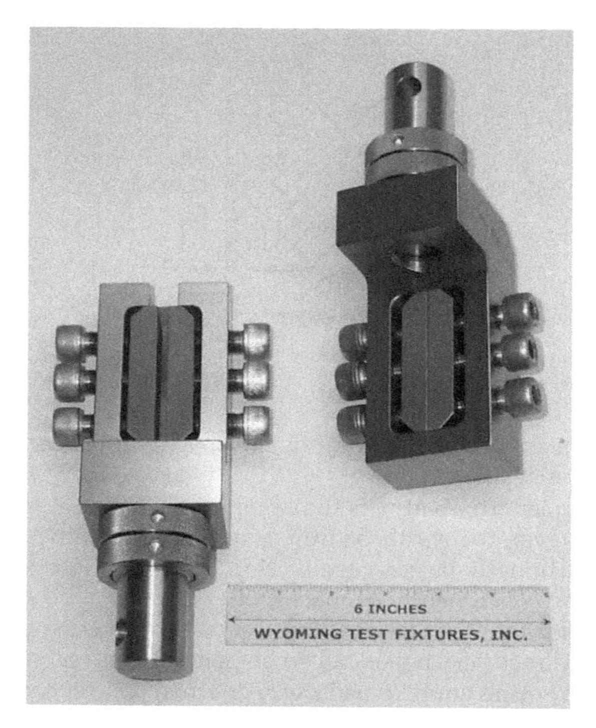

FIGURE 10.13
V-notched rail shear fixture (ASTM D7078).

FIGURE 10.14
V-notched rail shear specimen installation jig.

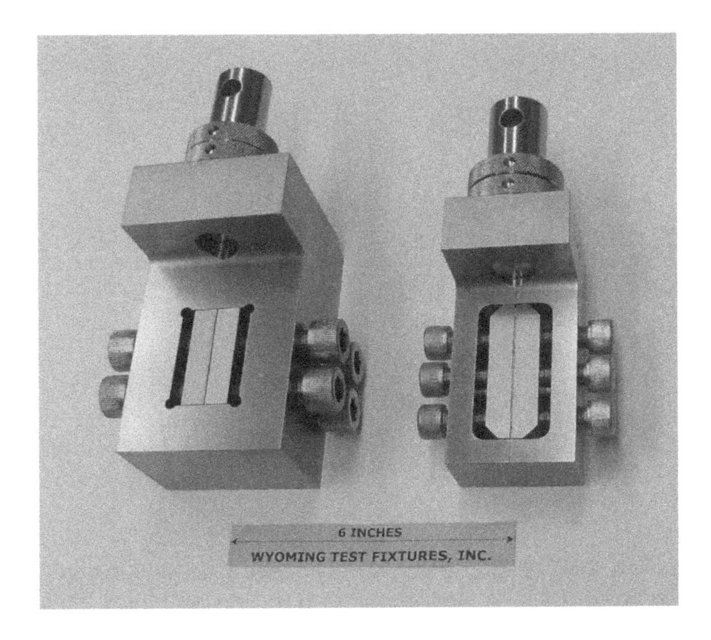

FIGURE 10.15
Combined loading shear (CLS) test fixture.

The strain gage instrumentation to be used and the data reduction are the same as previously discussed for the Iosipescu shear test method. Even though face loading is much more effective than edge loading of the V-notched specimen, there are applications where it is still not adequate. This is leading to the current study of a combined loading shear (CLS) fixture design (Adams 2010). One example is shown in Figure 10.15. In this case, the width of the fixture opening is the same as that of the specimen (56 mm), so that as the face-loaded specimen tends to rotate in the clamped grips due to the applied bending moment, some of the applied load will be transferred to the specimen edges, thus increasing the total loading capability. A modification is to have a width-adjustable fixture so that positive edge-clamping forces can be applied. The CLS could become yet another shear test method, and ASTM standard, in coming years.

10.6 $[\pm 45]_{ns}$ Tensile Shear Test Method

The $[\pm 45]_{ns}$ laminate tensile test is a simple method to determine the shear properties of the unidirectional lamina (see Figure 10.16).

The test specimen is relatively simple to prepare and requires no special test fixture other than standard tensile grips. The test method has been

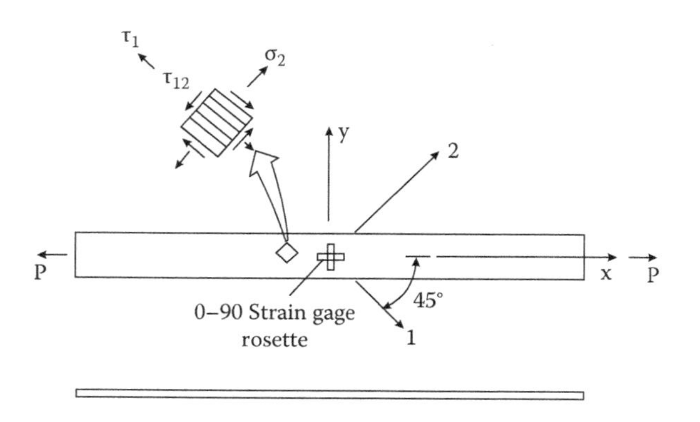

FIGURE 10.16
The $[\pm 45]_{ns}$ tensile specimen for evaluation of the shear stress–strain response of unidirectional composites and the stress state in a ply.

standardized as ASTM D3518 (2007). The primary advantage of this test method is its simplicity of specimen preparation and test. This shear test, however, suffers from the fact that a biaxial tensile stress state coexists with the shear stress. The state of stress in each lamina of the $[\pm 45]_{ns}$ laminate is not pure shear. Each lamina contains tensile normal stresses σ_1 and σ_2 in addition to the desired shear stress τ_{12}. In addition, an interlaminar shear stress τ_{xz} is present near the laminate free edge as discussed in Chapter 2. Biaxial failure theories all indicate interaction between shear and in-plane stress components in determining failure.

Normally, these considerations would lead to the rejection of this test geometry. However, there are mitigating circumstances that reduce these concerns. First, the shear stress–strain responses of many types of composite laminae are nonlinear and may exhibit strain-softening characteristics. Thus, although the biaxial state of stress present in the specimen likely causes the measured value of shear strength to be lower than the true value, the reduction may be small because of the nonlinear softening response. Second, the magnitudes of the interlaminar stresses for laminates containing highly orthotropic plies are a maximum at ply angles of 15° to 25°, and the interlaminar stresses are considerably smaller for 45° ply angles. Therefore, the $[\pm 45]_{ns}$ tensile shear test method may often be reliable in determining lamina shear modulus and strength.

The width of the specimen typically is on the order of 25 mm. End tabs may not be required but are often used. The $[\pm 45]_{ns}$ laminate tension test provides an indirect measure of the in-plane shear stress-strain response of the plies in the material coordinate system. The tensile specimen is instrumented with a 0°/90° biaxial strain gage rosette as shown in Figure 10.16. The specimen is prepared and tested in tension following the procedures outlined for the tension test in Chapter 8.

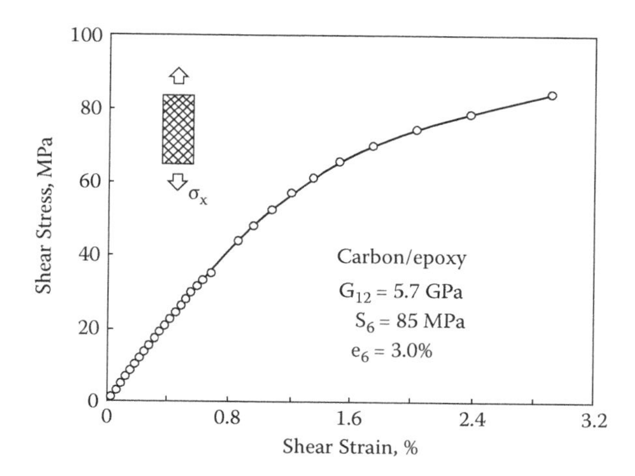

FIGURE 10.17
Shear stress–strain curve obtained from a tensile test of a $[\pm 45]_{2s}$ carbon–epoxy test specimen.

Determination of the intrinsic (lamina) shear properties from the tension test results uses a stress analysis of the $[\pm 45]_{ns}$ specimen (Rosen 1972). The shear stress τ_{12} (Figure 10.12) is simply

$$\tau_{12} = \sigma_x/2 \tag{10.4}$$

where σ_x is the axial stress (P/A). The shear strain is

$$\gamma_{12} = \varepsilon_x - \varepsilon_y \tag{10.5}$$

where ε_x and ε_y are the axial and transverse strains, respectively ($\varepsilon_y < 0$). Hence, the in-plane shear modulus G_{12} is readily determined by plotting $\sigma_x/2$ versus $(\varepsilon_x - \varepsilon_y)$ and establishing the slope of the initial portion of the curve. The ultimate shear stress S_6 is defined as the maximum value of $\sigma_x/2$.

Figures 10.17 and 10.18 show typical shear stress–shear strain curves for the lamina as determined from the laminate tensile test.

10.7 Short-Beam Shear Test Method

The short-beam shear test, ASTM D2344 (2006), is an interlaminar shear test method. A typical short-beam shear test fixture is shown in Figure 10.19. For a unidirectional composite, the τ_{13} component is the shear stress applied, assuming the fibers to be oriented parallel to the beam axis. However, if the

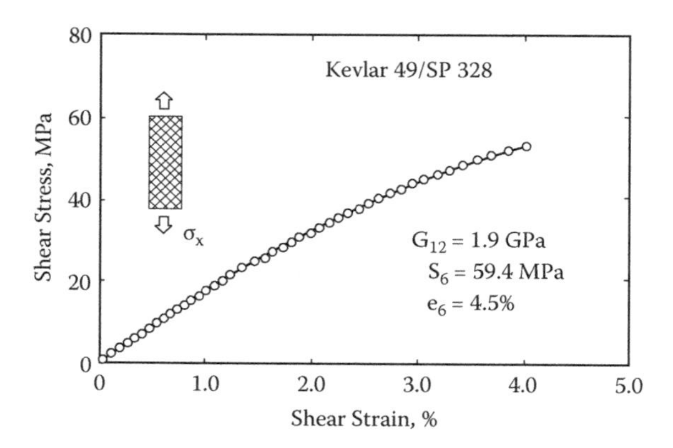

FIGURE 10.18
Shear stress–strain curve obtained from a tensile test of a $[\pm 45]_{2s}$ Kevlar–epoxy test specimen.

shear strengths in the 13 and 12 planes are equal, as is often assumed for a unidirectional composite, this test method can be, and often is, used to determine the in-plane shear strength of a unidirectional lamina. As for the $[\pm 45]_{ns}$ tension shear test method, it is in common use despite some serious limitations. This ASTM standard was titled "Apparent Interlaminar Shear Strength … " for many years, the word *apparent* acknowledging these limitations.

A specimen with a short support span length is loaded in three-point bending. Both bending (flexural) and interlaminar shear stresses are induced in any beam loaded in this manner, as discussed in more detail in Chapter 11. The axial bending stresses are compressive on the surface of the beam where the load is applied and tensile on the opposite surface, varying

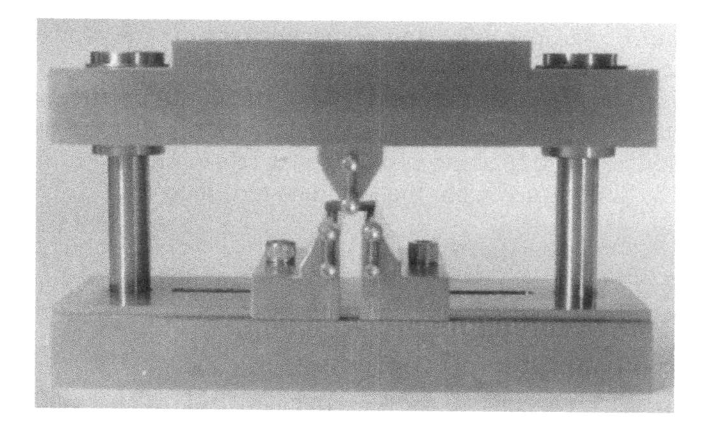

FIGURE 10.19
Short-beam shear test fixture with adjustable support span. (Photograph courtesy of Wyoming Test Fixtures, Inc.)

linearly through the beam thickness if the material response is elastic. By definition, the neutral axis (neutral plane) is where the bending stress passes through zero. It is on this neutral plane that the interlaminar shear stress is theoretically at maximum, varying parabolically from zero on each surface of the beam. Thus, although a combined stress state exists in general, the stress state should be pure shear on the neutral plane. By keeping the span length-to-specimen thickness ratio small (typically 4 or less), the bending stresses can be kept low, promoting shear failures on the neutral plane. Unfortunately, the concentrated loadings on the beam at the loading and support points create stress concentrations throughout much of the short beam (Berg et al. 1972, Whitney and Browning 1985, Kotha and Adams 1998), complicating the stress state. Nevertheless, the ASTM standard assumes a parabolic stress distribution, which for a beam of rectangular cross section results in a maximum shear stress of

$$\tau_{13} = 0.75P/A \tag{10.6}$$

where P is the applied load on the beam, and A is the cross-sectional area of the beam. This assumption is the reason for the use of "apparent" in the previous title of ASTM D2344. Despite these limitations, the short-beam shear test method usually produces reasonable values of shear strength (Adams and Lewis 1997).

The specimen can be very small, consuming a minimal amount of material. For example, when a span length-to-specimen thickness ratio of 4 and a 2.5-mm-thick composite are used, the span length is only 10 mm. If one specimen thickness of overhang is allowed on each end, the total specimen length is still only 15 mm. In addition, minimum specimen preparation time is required because the length and width dimensions, and the quality of the cut edges of the specimen, are not critical. The test fixturing can be relatively simple, and a test can be performed quickly. Another reason the test is both quick and economical is that no strain or displacement measurements are made. For the described reasons, the short-beam shear test is used extensively as a materials screening and quality control test. It has definite advantages for these purposes.

Testing is typically conducted at a crosshead displacement rate of 1–2 mm/min.

11

Lamina Flexural Response

11.1 Introduction

As defined in Chapters 8 through 10, pure, uniform tension, compression, and shear loadings must be individually applied to establish the fundamental strength and stiffness properties of a composite material. A flexure test (i.e., the bending of a beam) typically induces tensile, compressive, and shear stresses simultaneously. Thus, this test is not usually a practical means of determining the fundamental properties of a composite material (Whitney et al. 1984, Adams 2000a).

Nevertheless, flexure tests are popular because of the simplicity of both specimen preparation and testing, as discussed subsequently. Gripping of the specimen, having the need for end tabs, obtaining a pure stress state, avoiding buckling, and most of the other concerns discussed in the previous three chapters are usually nonissues when conducting a flexure test.

Flexural testing can, for example, be a simple method of monitoring quality during a structural fabrication process. The usual objective of a flexure test is to determine the flexural strength and flexural modulus of the beam material. This might be particularly relevant if the component being fabricated is to be subjected to flexural loading in service. However, because of the complex stress state present in the beam, it is typically not possible to directly relate the flexural properties obtained to the fundamental tensile, compressive, and shear properties of the material.

11.2 Testing Configurations

Figure 11.1 indicates the configuration of the ASTM D790 (2010) three-point flexure test. This standard was created in 1970 by the plastics committee within the American Society for Testing and Materials (ASTM) for use with unreinforced and reinforced plastics and electrical insulating materials, as its title suggests. For more than 25 years, until 1996 when it was removed,

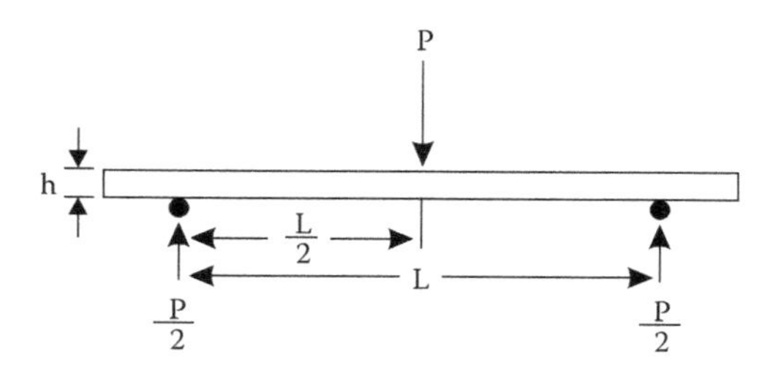

FIGURE 11.1
Three-point flexure loading configuration.

this standard also included four-point flexure. In response to demands by the composite materials community, a new standard, ASTM D6272 (2010), was introduced by the plastics committee specifying four-point flexure. That is, two standards now exist. The composites committee of ASTM has now written its own flexural test standard specifically for composite materials, ASTM D7264 (2007), which includes both three-point and four-point flexure. The three standards differ sufficiently in detail that it is advisable to refer to all three for guidance. It is unfortunate, but understandable because of its long existence, that many experimentalists still (incorrectly) only quote ASTM D790 as the governing standard for all flexural testing.

Analysis of a macroscopically homogeneous beam of linearly elastic material (Timoshenko 1984) shows that an applied bending moment is balanced by a linear distribution of normal stress σ_x, as shown in Figure 11.2. For the three-point flexural loading shown, the top surface of the beam is in compression, while the bottom surface is in tension. Assuming a beam of rectangular cross section, the midplane contains the neutral axis and is under zero bending stress. The shear force will be balanced by a distribution of interlaminar shear stress τ_{xz}, varying parabolically from zero on the free surfaces to a maximum at the center as shown (Timoshenko 1984). For three-point flexure, the shear stress is constant along the length of the beam and directly proportional to the applied force *P*. However, the flexural stress, in addition to being directly proportional to *P*, is zero at each

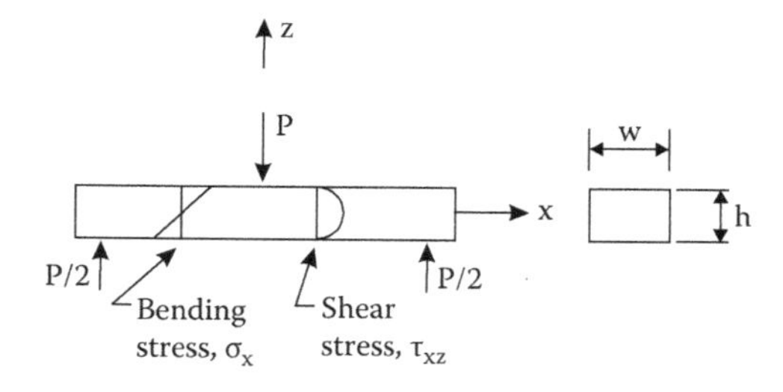

FIGURE 11.2
Stresses in a beam subjected to three-point flexure.

end support and maximum at the center. The stress state is highly dependent on the support span length-to-specimen thickness ratio (L/h). Beams with small L/h ratios are dominated by shear. As discussed in Chapter 10, a short-span ($L/h = 4$) three-point flexure test is used for interlaminar shear strength determination. Beams with long spans deflect more and usually fail in tension or compression. Typically, composite materials are stronger in tension than compression. Also, a concentrated load is applied at the point of maximum compressive stress in the beam, which tends to introduce local stress concentrations. Thus, the composite beam usually fails in compression at the midspan loading point.

Although testing of a unidirectional composite is the primary subject of the present chapter, laminates of various other orientations can also be tested in flexure, as discussed in more detail in Chapter 15.

Flexural testing of a unidirectional composite is generally limited to beams with the fibers aligned parallel to the beam axis. That is, 0° flexural properties are determined. Beams with fibers oriented perpendicular to the beam axis almost always fail in transverse tension on the lower surface because the transverse tensile strength of most composite materials is less than the transverse compressive strength, usually by a factor of three or more. In fact, the transverse flexure test has been suggested as a simple means of obtaining the transverse tensile strength of a unidirectional composite (Adams et al. 1990).

As do the other two flexural test standards, ASTM D7264 requires that a sufficiently large support span-to-specimen thickness ratio be chosen, "such that failures occur in the outer fibers of the specimens, due only to the bending moment." The standard recommends support span-to-thickness ratios of 16:1, 20:1, 32:1, 40:1, and 60:1, indicating that, "as a general rule support span-to-thickness ratios of 16:1 are satisfactory when the ratio of the tensile strength to shear strength is less than 8 to 1." High-strength unidirectional composites can have much higher strength ratios, requiring correspondingly higher support span-to-thickness ratios. For example, ASTM D7264 suggests a ratio of 32:1 for three-point and four-point flexure. Although longer spans promote flexural failure, they also result in larger deflections. Excessive deflections lead to geometric nonlinear effects, which should be avoided.

Per ASTM D7264, the diameter of the loading noses and supports should be 6 mm. The other two ASTM standards have a slightly different requirement, (viz., 10-mm diameters). This perhaps confirms experimental observations that the test results obtained are not strongly influenced by the specific diameters used as long as the diameters are not so small that local bearing damage of the composite material occurs (Adams and Lewis 1995).

Four-point flexural loading is typically conducted in either "third-point" or "quarter-point" loading, as shown in Figure 11.3. For quarter-point loading, the loading points are each positioned one-quarter of the support span length from the respective support and hence are one-half of the support span length from each other. For third-point loading, the loading points are each positioned one-third of the support span length from the respective

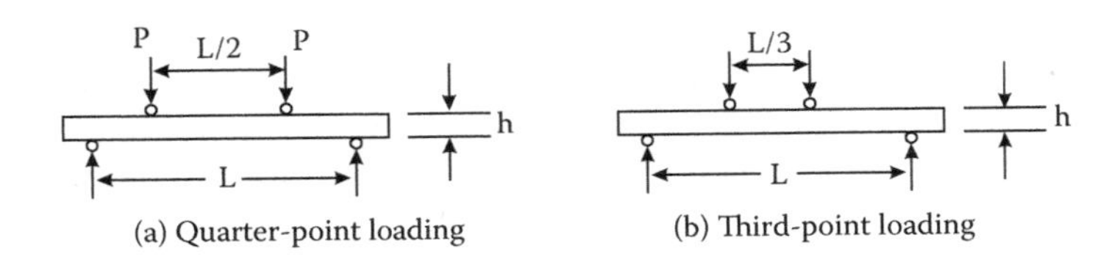

FIGURE 11.3
Four-point flexure test configurations: (a) quarter-point loading and (b) third-point loading.

support and hence are also one-third of the support span length from each other. ASTM D7264 includes only quarter-point loading.

Although the specimen thickness can be arbitrary (as long as the recommended span-to-thickness ratio is maintained), ASTM D7264 suggests a specimen that is 100 mm long, 2.4 mm thick, and 13 mm wide and a 76.8-mm span for materials of high-strength. This results in a specimen overhang of 11.6 mm at each end. Suggestions are given for other types of composites as well.

11.3 Three- versus Four-Point Flexure

As noted in the previous section, either three- or four-point flexure can be, and is, used. The various ASTM standards make no specific recommendations concerning when to use each test. In fact, there is no clear advantage of one test over the other, although there are significant differences. Figure 11.4 indicates the required loadings and the corresponding bending moment M and transverse shear force V distributions in the beam for each of the loadings.

For three-point flexure, the maximum bending moment in the beam, and hence the location of the maximum tensile and compressive flexural stresses, is at midspan and is equal to $M_{max} = PL/4$. For four-point flexure with loading at the quarter points, attaining the same maximum bending moment, that is, $M_{max} = PL/4$, requires twice the testing machine force (i.e., $2P$); this maximum bending moment is constant over the entire span $L/2$ between the applied loads (Figure 11.4b). For four-point flexure with loading at the one-third points, attaining the same maximum bending moment, that is, $M_{max} = \frac{3}{4}P(L/3) = PL/4$, requires 50% more testing machine force (i.e., $1.5P$) than for three-point flexure. This maximum bending moment is constant over the entire span $L/3$ between the applied loads.

Thus, for quarter-point loading the force exerted by the testing machine must be twice as high, and for third-point loading the force must be one and one-half times as high, as for three-point flexure. Normally, this is not in itself a significant factor because the loads required to fail a beam in flexure are not extremely high. More significant is the magnitude of the required force at the loading point(s). It is realized that the highest flexural stresses

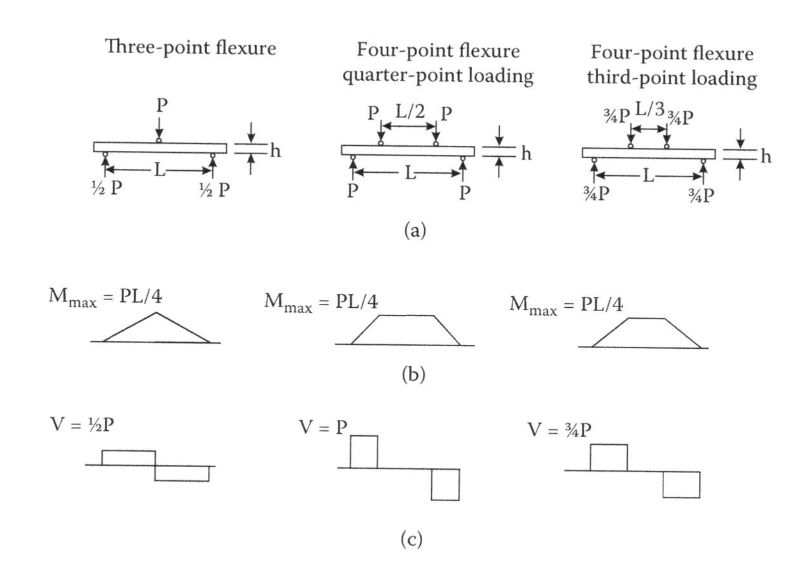

FIGURE 11.4
Required loadings for equal maximum bending moment in various beam configurations, with corresponding vertical shear force distributions: (a) loading diagrams, (b) moment diagrams, and (c) shear diagrams.

occur at these locations, and these stresses will be magnified by the stress concentration factors at the load introduction points. As noted in Figure 11.4, for both three-point flexure and four-point flexure with quarter-point loading, the maximum concentrated force on the beam is P. However, for four-point flexure with third-point loading, the load applied is only $\frac{3}{4}P$. This indicates an advantage for third-point loading.

There are other considerations, however. The transverse shear force V, and hence the interlaminar shear stress in the beam, also varies with the type of loading, as indicated in Figure 11.4. For three-point loading, the shear force is equal to $\frac{1}{2}P$ and is constant over the entire support span. For four-point flexure, quarter-point loading, the maximum shear force is equal to P and exists only over the end quarters of the beam. For four-point flexure, third-point loading, the maximum shear force is equal to $\frac{3}{4}P$ and exists only over the end thirds of the beam. To avoid shear failure, it is desirable to keep the ratio of interlaminar shear stress to flexural stress sufficiently low. As explained, this is normally achieved by increasing the span length-to-specimen thickness ratio. For a given specimen thickness, three-point flexure ($V = \frac{1}{2}P$) would be preferred because it minimizes the required support span length, followed by four-point flexure, third-point loading ($V = \frac{3}{4}P$).

One additional consideration should be noted, although it is usually of lesser importance. As discussed in Section 11.5, it may also be desired to determine the flexural modulus from the measured deflection of the beam.

The shear stresses in the beam contribute to the total deformation in proportion to the product of the shear force and the length over which it acts. The entire length of the beam in three-point flexure is subjected to the shear force $\frac{1}{2}P$, one-half of the four-point flexure; the quarter-point loading beam is subjected to the shear force P and two-thirds of the four-point flexure; and the third-point loading beam is subjected to the shear force $\frac{3}{4}P$. Hence, the net shear deformation is the same for all three cases. This consideration is important when beam deflection is used to determine modulus.

In summary, four-point flexure with third-point loading appears to be a good overall choice. However, each of the three loading modes (Figure 11.4) has some individual advantages. Thus, all three are used, with three-point flexure the most common, perhaps only because it requires the simplest test fixture. A typical test fixture, with adjustable loading and support spans such that it can be used for both three- and four-point flexure, is shown in Figure 11.5.

As an aside, note that three-point flexure is commonly (although perhaps erroneously) referred to as three-point loading, and likewise four-point

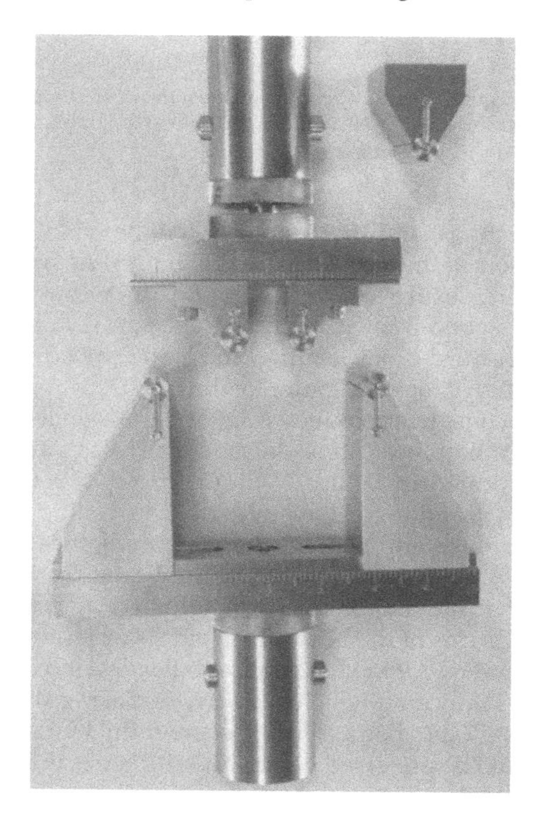

FIGURE 11.5
Photograph of a flexure test fixture with interchangeable three- and four-point loading heads and adjustable spans. (Photograph courtesy of Wyoming Test Fixtures, Inc.)

flexure is referred to as four-point loading. As a result, this has become accepted terminology, included in all three ASTM standards.

11.4 Specimen Preparation and Flexure Test Procedure

The flexure specimen is simply a strip of test material of constant width and thickness. As noted in Section 11.2, for a unidirectional $[0]_n$ composite under flexure, the suggested dimensions in ASTM D 7264 are support 76.8 mm span length, 100 mm specimen total length, 13 mm specimen width, and 2.4 mm specimen thickness. Suggested tolerances on these dimensions are also given in the standard.

Although all three ASTM standards specify the use of a deflection-measuring device mounted under the midspan of the specimen, occasionally a strain gage is used instead. One longitudinal strain gage can be mounted at the midspan on the tension side (bottom surface) of the specimen. The test fixture support span is to be set according to the beam thickness, specimen material properties, and fiber orientation, as discussed previously. ASTM D7264 specifies that the specimen is to be loaded at a crosshead rate of 1 mm/min, although a crosshead rate between 1 and 5 mm/min is common.

The beam deflection δ is measured using a calibrated linear variable differential transformer (LVDT) at the beam midspan. Alternatively, the beam displacement may be approximated as the travel of the testing machine crosshead if the components of this travel that are due to the machine compliance and to the indentations of the specimen at the loading and support points are subtracted. If a strain gage is used, the specimen is to be placed in the fixture with the strain gage on the tension side of the beam and centered at midspan. The strain or displacement readings may be recorded continuously or at discrete load intervals. If discrete data are recorded, the load and strain-displacement readings should be taken at small load intervals, with at least 25 points in the linear response region (so that an accurate flexural modulus can be determined). The total number of data points should be enough to accurately describe the complete beam response to failure.

11.5 Data Reduction

From classical beam theory (Timoshenko 1984), the tensile and compressive stresses at the surfaces of the beam at the location where the bending moment is a maximum is

$$\sigma_{max} = \frac{M_{max}(h/2)}{I} \qquad (11.1)$$

where M_{max} is the maximum bending moment, h is the thickness of the beam, and $I = wh^3/12$ is the moment of inertia of a beam of rectangular cross section, with w being the beam width.

For three-point flexure, for example, the maximum bending moment occurs at the midlength of the beam. The maximum value is given by

$$M_{max} = PL/4 \tag{11.2}$$

Substitution of M_{max} and I into Equation (11.1) gives

$$\sigma_{max} = \frac{3PL}{2wh^2} \tag{11.3}$$

This equation enables construction of a stress–strain plot if load versus strain has been recorded.

A reasonable approximation for most materials is that the modulus in tension is the same as in compression. If a strain gage is used, the initial slope of the flexural stress–strain plot can be obtained using a linear least-square fit. Figures 11.6 and 11.7 show typical stress–strain curves obtained from three-point flexure tests.

When the three-point flexure specimen is not strain gaged, the flexural modulus may be determined from a plot of load P versus center deflection δ as

$$E_f = \frac{L^3}{4wh^3} \frac{\Delta P}{\Delta \delta} \tag{11.4}$$

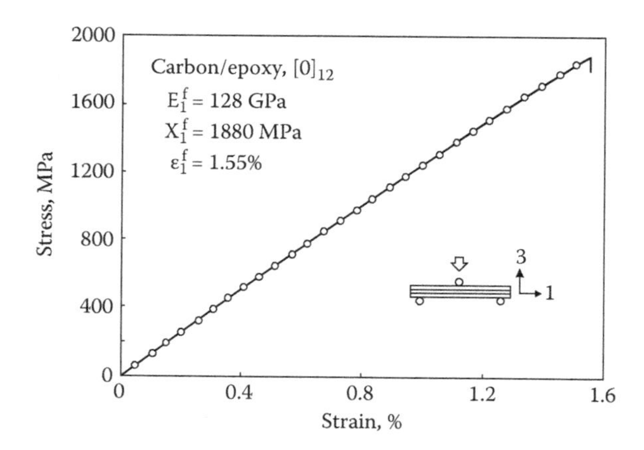

FIGURE 11.6
Flexural stress–strain response of a $[0]_{12}$ carbon–epoxy test specimen.

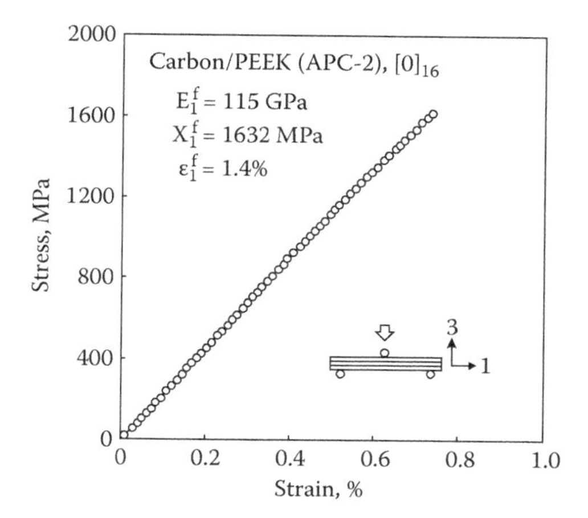

FIGURE 11.7
Flexural stress–strain response of a $[0]_{16}$ carbon–PEEK (polyetheretherketone) test specimen.

This relation, however, assumes that shear deformation is negligible. For $[90]_n$ beams (fibers perpendicular to the beam axis), shear deformation is generally insignificant, and Equation (11.4) should be accurate. For $[0]_n$ beams, however, the results of Zweben (1990) shown in Figure 11.8 illustrate that certain unidirectional composites require relatively long support spans to

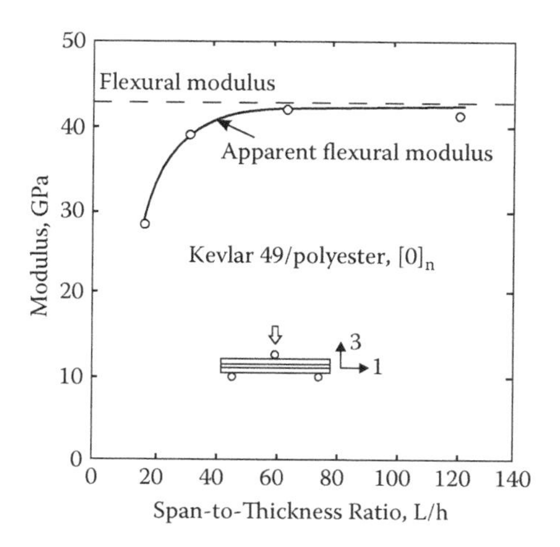

FIGURE 11.8
Apparent flexural modulus of a $[0]_n$ Kevlar 49–polyester specimen as a function of span-to-thickness ratio. (From Zweben, C., 1990, in *Delaware Composites Design Encyclopedia*, Vol. 1, Technomic, Lancaster, PA, pp. 49–70).

minimize the influence of shear deformation and thus produce an accurate flexural modulus.

If a flexural test is conducted for which the shear deformation component is significant, the following equation may be used to evaluate the flexural modulus:

$$E_f = \frac{L^3}{4wh^3}\left(1 + \frac{6h^2E_f}{5L^2G_{13}}\right)\frac{\Delta P}{\Delta\delta} \quad (11.5)$$

where G_{13} is the interlaminar shear modulus. The second term in the parentheses of Equation (11.5) is a shear correction factor that may be significant for composites with a high axial modulus and a low interlaminar shear modulus. The flexural and shear moduli (E_f and G_{13}, respectively), however, may not be known prior to the test. For use of Equation (11.5), the flexural modulus E_f may be replaced by the tensile modulus E_1, and the out-of-plane shear modulus can be approximated as the in-plane shear modulus G_{12} if the fibers are oriented along the beam axis. When E_1 and G_{12} are not known, the flexural modulus can be evaluated using Equation (11.4) for tests conducted on a slightly loaded beam over a range of span lengths until a constant value is achieved at long span lengths, as suggested in Figure 11.8.

Additional details of data reduction, including for four-point flexure, can be found in the three ASTM flexural testing standards cited in this chapter.

12

Lamina Off-Axis Tensile Response

12.1 Introduction

The off-axis tension test of unidirectional composites has received considerable attention by the composites community. "Off-axis" here refers to the material axes (1–2) being rotated through an angle θ with respect to the specimen axis and direction of loading (Figure 12.1). The off-axis specimen is typically 230 mm long and between 12.7 and 25.4 mm wide. A thickness of eight plies is common (0.127-mm ply thickness).

The off-axis tension test is rarely used to determine basic ply properties. Most commonly, the purpose of this test is to verify material properties determined in tension, compression, and shear, as discussed in Chapters 8–10, using the transformed constitutive relations discussed in Chapter 2. Testing of specimens at off-axis angles between 10° and 20° produces significant shear in the principal material system. Consequently, the 10° off-axis test has been proposed as a simple way to conduct a shear test (Chamis and Sinclair 1977). The test has been used also to verify biaxial strength criteria because, as will be discussed, uniaxial loading will lead to a combined state of stress in the principal material system.

12.2 Deformation and Stress in an Unconstrained Off-Axis Specimen

Because of the off-axis configuration of the specimen, the in-plane response is characterized by a fully populated compliance matrix, as shown in Equation (2.16):

$$\begin{bmatrix} \varepsilon_x \\ \varepsilon_y \\ \gamma_{xy} \end{bmatrix} = \begin{bmatrix} \bar{S}_{11} & \bar{S}_{12} & \bar{S}_{16} \\ \bar{S}_{12} & \bar{S}_{22} & \bar{S}_{26} \\ \bar{S}_{16} & \bar{S}_{26} & \bar{S}_{66} \end{bmatrix} \begin{bmatrix} \sigma_x \\ \sigma_y \\ \tau_{xy} \end{bmatrix} \tag{12.1}$$

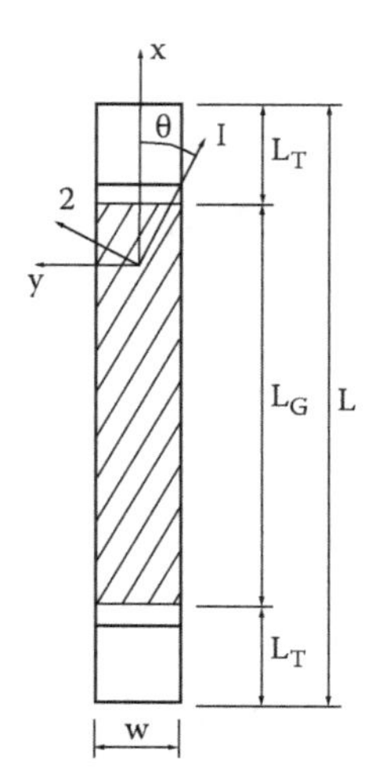

FIGURE 12.1
Geometry of the off-axis tensile coupon.

where the x-y system is defined in Figure 12.1, and expressions for the transformed compliance elements S_{ij} are given in Appendix A.

For an ideal, uniformly stressed off-axis tensile coupon, the only stress acting is σ_x, ($\sigma_y = \tau_{xy} = 0$), and Equations (12.1) give the following state of strain in the specimen:

$$\begin{bmatrix} \varepsilon_x \\ \varepsilon_y \\ \gamma_{xy} \end{bmatrix} = \sigma_x \begin{bmatrix} \bar{S}_{11} \\ \bar{S}_{12} \\ \bar{S}_{16} \end{bmatrix} \tag{12.2}$$

Consequently, the off-axis coupon subjected to a uniform uniaxial state of stress thus exhibits shear strain (γ_{xy}) in addition to the axial and transverse strains, ε_x and ε_y, (Figure 12.2).

A set of material properties may be evaluated based on measurement of axial stress (σ_x) and axial, transverse, and shear strains ε_x, ε_y, γ_{xy}, respectively. It is customary to determine the axial stiffness E_x and Poisson's ratio ν_{xy} of the off-axis specimen:

$$E_x = \frac{\sigma_x}{\varepsilon_x} \tag{12.3}$$

$$\nu_{xy} = -\frac{\varepsilon_y}{\varepsilon_x} \tag{12.4}$$

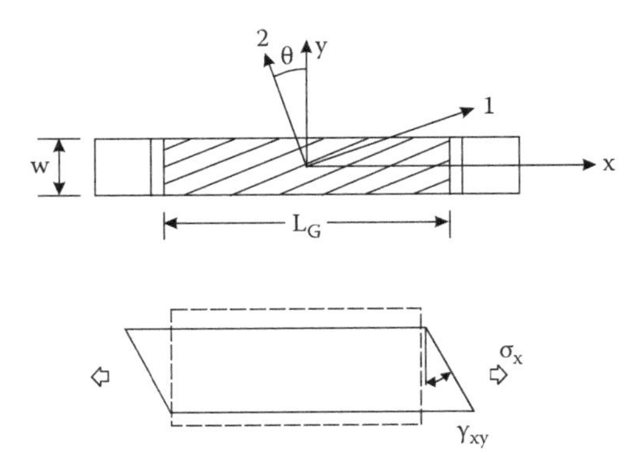

FIGURE 12.2
Off-axis coupon under uniform axial stress.

In addition, a shear coupling ratio η_{xy} that quantifies coupling between shear and axial strains is defined according to

$$\eta_{xy} = \frac{\gamma_{xy}}{\varepsilon_x} \tag{12.5}$$

The off-axis tension test may also be used to determine the in-plane shear stiffness G_{12} in the principal material coordinate system. This property, according to Equation (2.9), is defined by

$$G_{12} = \frac{\tau_{12}}{\gamma_{12}} \tag{12.6}$$

Consequently, determination of G_{12} requires determination of shear stress and strain in the principal material coordinate system. Equations (2.12) and (2.14) yield

$$\tau_{12} = -mn\sigma_x \tag{12.7}$$

where $m = \cos\theta$ and $n = \sin\theta$. The shear strain is obtained from Equations (2.14)

$$\gamma_{12} = 2mn(\varepsilon_y - \varepsilon_x) + (m^2 - n^2)\gamma_{xy} \tag{12.8}$$

where the strain ε_x is directly measured, and the transverse strain ε_y and shear strain γ_{xy} are determined as subsequently explained.

The properties E_x, ν_{xy}, η_{xy}, and G_{12} can be evaluated from test data using procedures detailed further in this chapter. The mechanical properties so determined can be compared to theoretical values calculated from the compliance relations, Equations (12.2), and the definitions in Equations (12.3–12.5):

$$E_x = \frac{1}{\bar{S}_{11}} \tag{12.9a}$$

$$\nu_{xy} = \frac{-\bar{S}_{12}}{\bar{S}_{11}} \tag{12.9b}$$

$$\eta_{xy} = \frac{\bar{S}_{16}}{\bar{S}_{11}} \tag{12.9c}$$

If the principal (basic) material properties (E_1, E_2, ν_{12}, and G_{12}) are known from previous tests (Chapters 8–10), it is possible to calculate the off-axis properties E_x, ν_{xy}, and η_{xy} using Equations (A.1) (Appendix A) and compare those to the experimentally determined values. In addition, the shear modulus G_{12} measured according to Chapter 10 may be compared to G_{12} measured in the off-axis test [Equation (12.6)].

12.3 Influence of End Constraint

As first pointed out by Halpin and Pagano (1968), most testing machines used in testing laboratories employ rigid grips that constrain the shear deformation illustrated in Figure 12.2. As a result, the specimen assumes a shape schematically illustrated in Figure 12.3. To quantify the influence of gripping on the response of an off-axis tension specimen, Halpin and Pagano performed a stress analysis of a constrained coupon and obtained the following expressions for the shear strain and longitudinal strain at the specimen centerline:

$$\gamma_{xy} = \bar{S}_{16}C_2 - \bar{S}_{66}C_0 w^2/4 \tag{12.10a}$$

$$\varepsilon_x = \bar{S}_{11}C_2 - \bar{S}_{16}C_0 w^2/4 \tag{12.10b}$$

with

$$C_0 = \frac{12\bar{S}_{16}\varepsilon_0}{3w^2(\bar{S}_{11}\bar{S}_{66} - \bar{S}_{16}^2) + 2\bar{S}_{11}L_G^2} \tag{12.11a}$$

$$C_2 = \frac{C_0}{12\bar{S}_{16}}\left(3\bar{S}_{66}w^2 + \bar{S}_{11}L_G^2\right) \tag{12.11b}$$

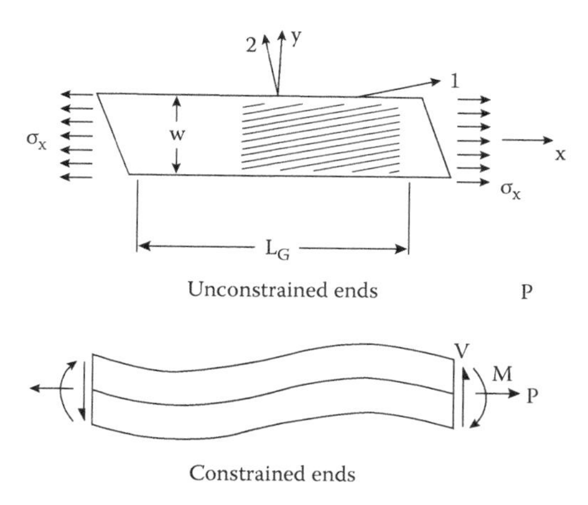

FIGURE 12.3
Influence of gripped end regions on deformation of off-axis specimen. (From Halpin, J.C., and Pagano, N.J., 1968, *J. Compos. Mater.*, 2, 18–31.)

where $\varepsilon_0 = \Delta L/L_G$ (elongation/gage length), and w and L_G are specimen width and gage length, respectively (Figure 12.1).

On the basis of this analysis, it is possible to derive an expression for the apparent axial modulus, including end constraint:

$$(E_x)_a = \frac{E_x}{1-\xi} \tag{12.12}$$

in which E_x is the modulus for an unconstrained off-axis specimen. The parameter ξ is given by

$$\xi = \frac{1}{\bar{S}_{11}}\left[\frac{3\bar{S}_{16}^2}{3\bar{S}_{66} + 2\bar{S}_{11}\left(L_G/w\right)^2}\right] \tag{12.13}$$

Examination of these equations reveals that $\xi \to 0$ and $(E_x)_a \to E_x = 1/\bar{S}_{11}$ when $L_G/w \to \infty$.

Similarly, Pindera and Herakovich (1986) derived an expression for the apparent Poisson's ratio $(\nu_{xy})_a$:

$$(\nu_{xy})_a = \nu_{xy}\frac{1-\dfrac{3}{2}\left(\dfrac{\bar{S}_{26}}{\bar{S}_{11}}\right)\beta}{1-\dfrac{3}{2}\left(\dfrac{\bar{S}_{16}}{\bar{S}_{12}}\right)\beta} \tag{12.14}$$

where β is given by

$$\beta = \frac{\left(\frac{w}{L_G}\right)^2 \left(\frac{\bar{S}_{16}}{\bar{S}_{11}}\right)}{1 + \frac{3}{2}\left(\frac{w}{L_G}\right)^2 \left(\frac{\bar{S}_{66}}{\bar{S}_{11}}\right)} \tag{12.15}$$

The apparent shear coupling ratio of the specimen subjected to end constraint is

$$(\eta_{xy})_a = \frac{\bar{S}_{16}}{\bar{S}_{11}}\left[1 + \frac{3}{2}\left(\frac{w}{L_G}\right)^2 \left(\frac{\bar{S}_{66}}{\bar{S}_{11}} - \left(\frac{\bar{S}_{16}}{\bar{S}_{11}}\right)^2\right)\right]^{-1} \tag{12.16}$$

Note that when the length-to-width ratio $L_G/w \rightarrow \infty$, the apparent shear coupling ratio approaches $\bar{S}_{16}/\bar{S}_{11}$, as given by Equation (12.9c). Hence, long and narrow specimens should produce more accurate test results.

Pindera and Herakovich (1968) examined the influence of end constraint on the evaluation of G_{12} from the off-axis tension specimen using the elasticity solution of Halpin and Pagano (1986) and found that the procedure outlined in Section 12.2 leads to error in G_{12}. The main source of error is the neglect of the shear stress τ_{xy} in Equation (12.7). The proper transformation is (Pindera and Herakovich 1986)

$$\tau_{12} = -mn\sigma_x + (m^2 - n^2)\tau_{xy} \tag{12.17}$$

This equation, combined with the definition of G_{12} [Equation (12.6)], yields an expression for the correct value of the shear modulus:

$$G_{12} = (G_{12})_a \frac{1 + \frac{3(m^2 - n^2)}{2mn}\beta}{1 - \beta(\bar{S}_{16}/\bar{S}_{11})} \tag{12.18}$$

where $(G_{12})_a$ is the apparent shear modulus determined from Equation (12.6), and β is defined in Equation (12.15). As $(w/L_G) \rightarrow \infty$, $\beta \rightarrow 0$, $(G_{12})_a$ approaches G_{12}.

The expressions for apparent off-axis properties may be used to correct measured values or to design the off-axis specimen for minimum error resulting from end constraint. An obvious way to reduce the error is to use specimens with large aspect ratios, L_G/w. As mentioned in the beginning of this chapter, specimens are typically 230 mm long and between 12.7 and 25.4 mm wide. For 38-mm-long tabs at the ends, this corresponds to aspect ratios between 6 and 12.

For a carbon–polyimide composite specimen with an aspect ratio of 10 and an off-axis angle of 10°, Pindera and Herakovich (1986) found an error in E_x of about 2–4%. The error in shear modulus G_{12} is larger, approximately 12–15%. For more accurate determination of G_{12}, Pindera and Herakovich (1986) recommend use of 45° coupons with an aspect ratio of at least 10.

12.4 Off-Axis Tensile Strength

The off-axis tension test has been used to examine theories proposed for the prediction of failure of composites under combined stress. In such studies, slender specimens are used for strength measurements to avoid the complications of end constraint effects discussed. This leads to a biaxial stress state in the on-axis system (see Figure 12.4) as given by Equations (2.12) and (2.14):

$$\begin{bmatrix} \sigma_1 \\ \sigma_2 \\ \tau_{12} \end{bmatrix} = \sigma_x \begin{bmatrix} m^2 \\ n^2 \\ -mn \end{bmatrix} \tag{12.19}$$

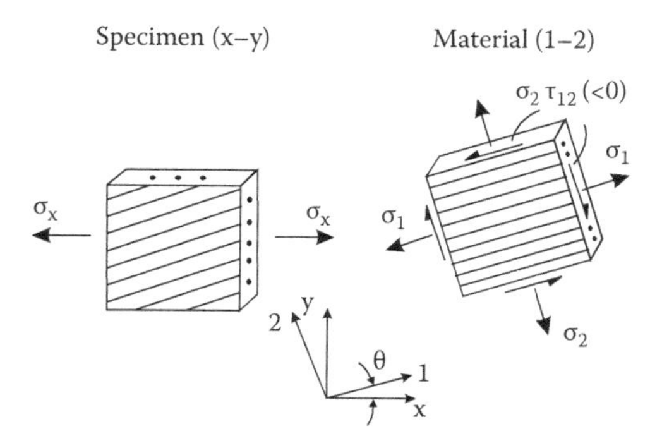

FIGURE 12.4
State of stress in the specimen for an off-axis tension test.

Experimental studies conducted on on-axis and off-axis specimens (Pipes and Cole 1973, Pindera and Herakovich 1981, Chatterjee et al. 1993c) showed that the off-axis specimen under tension fails along planes parallel to the fibers except for zero and very small angles, where failure involves fiber fractures. To predict the failure stress of the off-axis tension specimen, the on-axis stresses given by Equations (12.19) are substituted into the failure criterion of choice (see Section 2.6). The maximum stress and strain criteria (see Section 2.6) yield three equations for the ultimate stress σ_x^{ult}, and the appropriate strength is identified by the least of the three values. Substitution of the stresses given by Equations (12.19) into the Tsai-Wu failure criterion, Equation (2.44), yields a quadratic equation in σ_x^{ult}

$$A\left(\sigma_x^{ult}\right)^2 + B\sigma_x^{ult} - 1 = 0 \tag{12.20}$$

The solution of Equation (12.20) yields two roots: the positive associated with the tensile strength and the negative associated with the compressive strength of the off-axis specimen.

As mentioned in the beginning of this chapter, the 10° off-axis tension test has been proposed for measuring the in-plane shear strength (S_6) of unidirectional composites (Chamis and Sinclair 1977). However, because failure occurs under the influence of normal stresses σ_1 and σ_2 (Figure 12.4), where σ_2 and τ_{12} combined generate failure of planes parallel to the fibers, this test is not recommended for measuring the shear strength (Chatterjee et al. 1993c).

12.5 Test Procedure

1. Prepare off-axis tension coupons from a unidirectional, six- to eight-ply-thick panel. The specimens should be about 230 mm long and between 12.5 and 25 mm wide. Select at least three different off-axis angles (e.g., 15°, 30°, and 60°). Use the same tolerances as for the tension specimen discussed in Chapter 8 and bond end tabs as described in Chapter 4.
2. The off-axis test specimen is instrumented with a three-element strain gage rosette with one of the elements aligned with the coupon axis (x-direction in Figure 12.1), one element at 45°, and one element at –45°.

3. Measure the specimen cross-sectional dimensions (average six measurements).
4. Mount the specimen in a properly aligned and calibrated test frame. Set the crosshead rate at about 0.5 to 1 mm/min.
5. Monitor the load-strain response of the specimen (all three elements). Take strain readings at small load intervals to collect at least 25 data points in the linear region. Load the specimen to failure.

12.6 Data Reduction

12.6.1 Elastic Properties

Modulus, Poisson's ratio, and shear coupling ratio may be determined from measured stress–strain data according to Equations (12.3)–(12.5). The axial strain ε_x is obtained directly from the axially oriented strain gage. Transverse strain and shear strain ε_y and γ_{xy}, respectively, are obtained from the ±45° gages using Equations (2.13).

$$\varepsilon(45^\circ) = (\varepsilon_x + \varepsilon_y + \gamma_{xy})/2 \tag{12.21a}$$

$$\varepsilon(-45^\circ) = (\varepsilon_x + \varepsilon_y - \gamma_{xy})/2 \tag{12.21b}$$

Combining Equations (12.21) yields

$$\varepsilon_y = \varepsilon(45^\circ) + \varepsilon(-45^\circ) - \varepsilon_x \tag{12.22}$$

$$\gamma_{xy} = \varepsilon(45^\circ) - \varepsilon(-45^\circ) \tag{12.23}$$

The modulus E_x is determined from the initial slope of the curve of σ_x versus ε_x. Poisson's ratio ν_{xy} is obtained by plotting the negative of the strain ε_y versus ε_x and determining the slope of the line. The shear coupling ratio η_{xy} is determined from the slope of a plot of shear strain versus axial strain, as shown in Figure 12.5 for a 30° off-axis carbon–epoxy specimen. Note that the experimentally determined properties are apparent because they are influenced by the constraints imposed by the grips.

Figures 12.6–12.8 show experimentally obtained off-axis modulus, Poisson's ratio, and shear coupling ratio versus off-axis angle for carbon–fiber

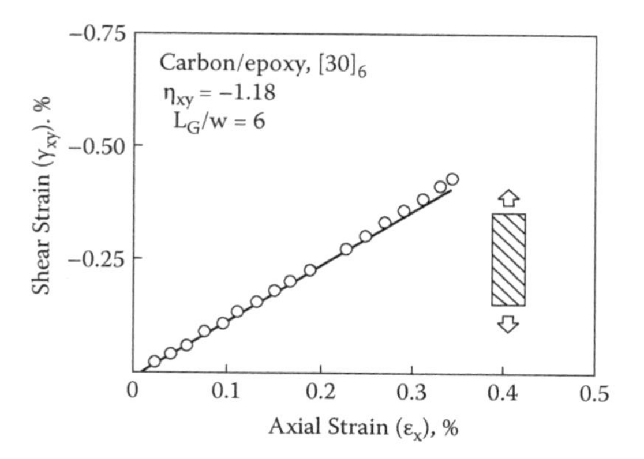

FIGURE 12.5
Shear strain versus axial strain for a $[30]_6$ carbon–epoxy composite.

composites. Shown in these graphs are reduced data (apparent), theoretical curves calculated neglecting the end constraints [Equations (12.9)], and curves corrected for end constraint according to Section 12.3. It is observed that the apparent modulus and Poisson's ratio are larger than the unconstrained values, whereas the magnitude of the apparent shear coupling ratio is reduced because of end constraints. The expressions in Section 12.3 incorporating end constraints due to finite aspect ratio bring the results in close agreement with measured data.

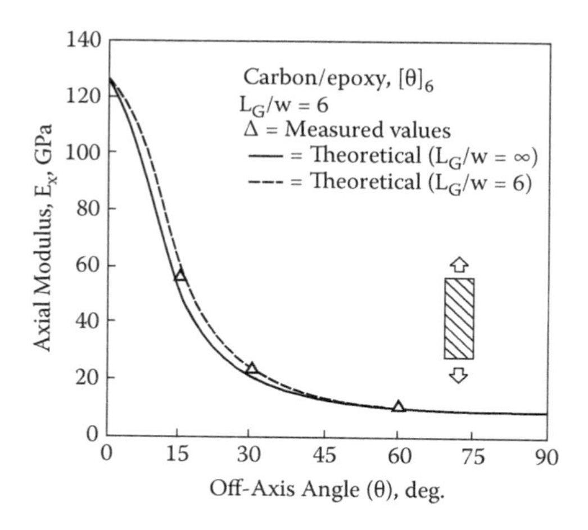

FIGURE 12.6
Modulus versus off-axis angle for a carbon–epoxy composite.

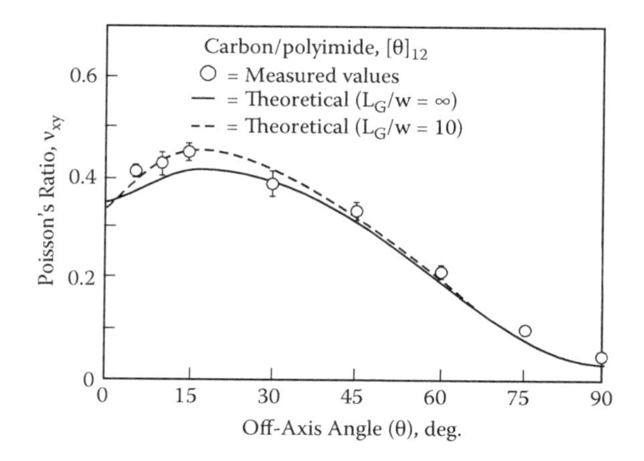

FIGURE 12.7
Poisson's ratio versus off-axis angle for a carbon–polyimide composite.

The shear modulus G_{12}, determined for a carbon–polyimide composite using Equation (12.6) and corrected using Equation (12.18), is shown versus off-axis angle in Figure 12.9. For angles up to 30°, the end constraint will increase the apparent shear modulus. The correction brings the value closer to measured data. This graph emphasizes the need to use off-axis specimens of angles 45° or greater when determining the shear modulus.

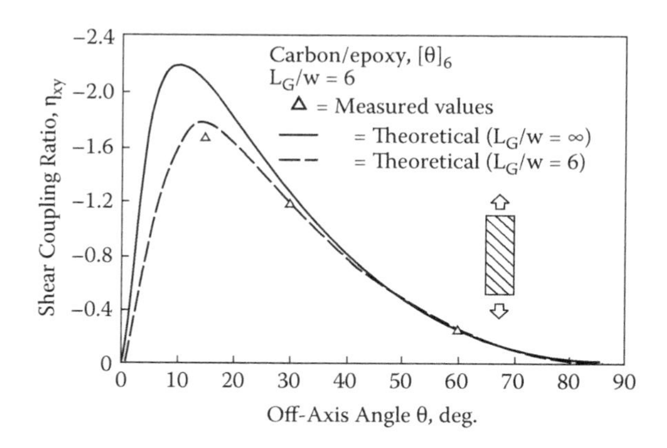

FIGURE 12.8
Shear coupling ratio versus off-axis angle for a carbon–epoxy composite.

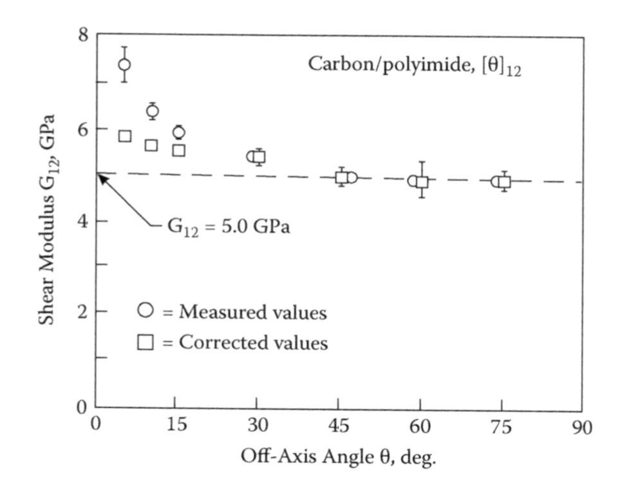

FIGURE 12.9
Shear modulus versus off-axis angle for a carbon–polyimide composite. (From Pindera, M.J., and Herakovich, C.T., 1986, *Exp. Mech.*, 26, 103–112.)

TABLE 12.1

Basic Ply Mechanical Properties for Carbon–Epoxy and Carbon–Polyimide Composites Considered in Figures 12.5–12.10

Material	E_1 (GPa)	E_2 (GPa)	ν_{12}	G_{12} (GPa)
Carbon/epoxy	126	10.0	0.30	5.2
Carbon/polyimide	137	9.79	0.35	5.0

TABLE 12.2

Experimental Off-Axis Tensile Strength Data for a Carbon–Epoxy Composite

Angle, θ (deg.)	Tensile Strength (MPa)
5	780
15	305
30	112
60	65

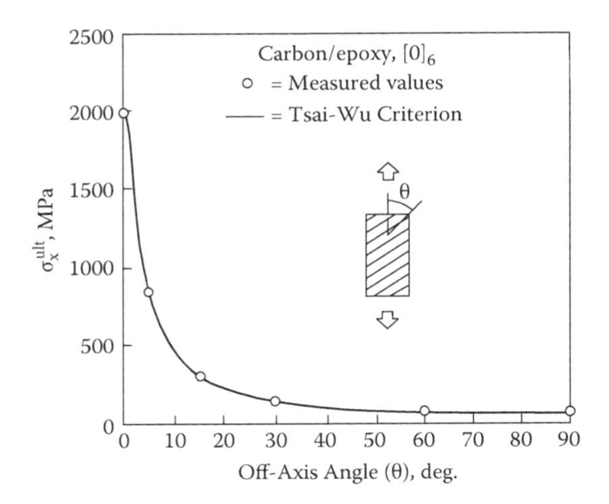

FIGURE 12.10
Tensile failure stress versus off-axis angle for a carbon–epoxy composite.

As a reference, basic ply properties for the carbon–epoxy and carbon–polyimide composites considered here are listed in Table 12.1.

12.6.2 Tensile Strength of Off-Axis Specimen

Table 12.2 presents experimental off-axis tensile strength data for a carbon–epoxy composite. Figure 12.10 shows experimentally determined and predicted [Equation (2.44)] failure stress versus off-axis angle for this composite. Excellent agreement is noted.

13

Lamina and Laminate Thermoelastic Response

13.1 Introduction

The dimensions of most materials change as the temperature is changed. Thermal expansion is defined as the change of dimensions of a body or material as a result of a temperature change. Such a property is very important in the application of composite materials in structures that undergo temperature changes, such as engine parts and space structures. The material property constant describing this phenomenon is the coefficient of thermal expansion (CTE), indicated by the symbol α and defined as

$$\alpha = \frac{\Delta\varepsilon}{\Delta T} \tag{13.1}$$

where $\Delta\varepsilon$ is the increment of strain measured for an unconstrained material subject to a temperature change ΔT. Thus, the in-plane thermal deformations of an orthotropic composite lamina are, from Equation (13.1),

$$\alpha_1 = \frac{\Delta\varepsilon_1}{\Delta T} \tag{13.2a}$$

$$\alpha_2 = \frac{\Delta\varepsilon_2}{\Delta T} \tag{13.2b}$$

where $\Delta\varepsilon_1$ and $\Delta\varepsilon_2$ are the thermally induced strains in the principal material directions over the temperature interval ΔT.

Several methods have been devised for measuring thermal expansion coefficients of composite materials (Adams 2000b). Basically, this determination requires a controlled-temperature chamber and a deformation measuring device.

American Society for Testing and Materials (ASTM) Standards E228 (2011), D696 (2008), and E831 (2012) propose using a dilatometer or thermomechanical analysis (TMA) apparatus for materials with CTE values greater than $5 \times 10^{-6}/°C$. ASTM Standard E289 (2010) describes a method based on interferometry, which permits determination of the CTE of materials with extremely small thermal expansion coefficients, as low as $10^{-8}/°C$. For composites, strain gages have been successfully used for measuring thermal expansion coefficients as low as about $10^{-5}/°C$ (Freeman and Campbell 1972, Whitney et al. 1984, Yaniv et al. 1987). Strain gages are readily available and require no special apparatus except for the strain-reading instrument. Further, orthotropy in thermal expansion requires measurement of more than one strain component. Multiple strain gages may be bonded to the surface of a specimen or embedded between the plies in a laminate. Here, we restrict attention to surface-bonded strain gages.

The test specimen used for determining the CTEs of a unidirectional lamina or woven fabric ply or laminate using strain gages should be a flat panel. Although the in-plane dimensions are not critical, a commonly used specimen size is 50 × 50 (mm) (Figure 13.1). The thickness of the panel is

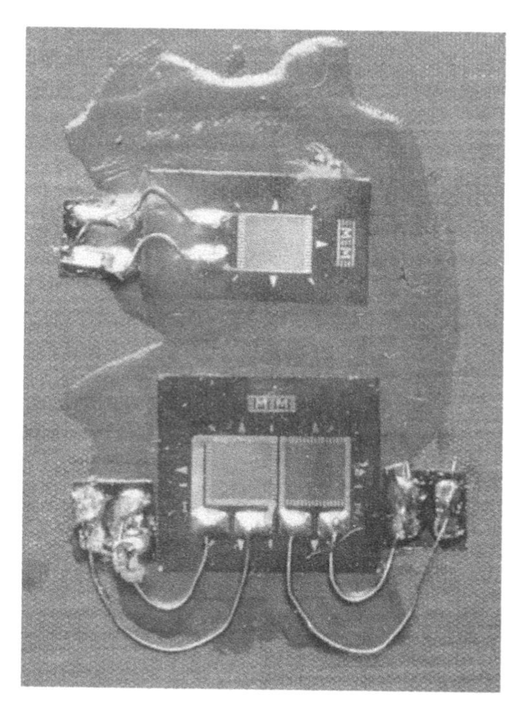

FIGURE 13.1
Typical carbon–epoxy specimen fitted with strain gages (bottom) and temperature sensor (top).

commonly about 1 mm. Panels that are too thin are easily cracked if they are unidirectional, and panels that are too thick require a long soaking time to achieve thermal equilibrium (uniform temperature). The temperature can be monitored with a temperature sensor or a thermocouple. The temperature range should be selected with regard to the type of strain gages and sensors used and the temperature capability of the resin in the composite. For a typical 175°C-cure epoxy–matrix composite, a suitable temperature range is 20–150°C. Because moisture induces dimensional changes in many resin systems, it is important to dry the specimens in an oven at 70°C until the weight stabilizes before measuring the thermal strains.

As discussed by Adams (2000b), an important limitation of strain gages is their upper use temperature. Maximum-accuracy gages can only be used up to 65°C. Other types of strain gages may be used to extend this range. Another important limitation of the use of strain gages is undesirable local reinforcement of the test material by the metal foil strain gage. At elevated temperatures, the transverse stiffness of unidirectional polymer–matrix composites will decrease as a result of softening of the matrix resin. Local reinforcement of the tested material by the gage would lead to apparent CTE values that are too small.

The adhesive bond between the strain gage and specimen is also an important factor. Because the matrix is typically a polymer resin, the adhesive will soften at elevated temperatures and exhibit viscoelastic creep or stress relaxation effects that would influence the strain readings. Selection of strain gage adhesive to match the anticipated temperature range is thus important.

13.2 Temperature Gage Sensing System

To monitor temperature, it is convenient to use a resistance gage circuit, which allows the experimenter to monitor the temperature of a test specimen while taking strain readings. The temperature gage (Micro Measurements Type ETG-50B or equivalent) shown in Figure 13.1 consists of a sensing grid of high-purity nickel foil that is bonded to the specimen by standard strain gage techniques using a high-temperature adhesive (M-Bond 600 or equivalent). This temperature sensor exhibits a linear change in resistance with temperature.

After the gage has been properly mounted and wired, it is connected to the gage scanning equipment along with the other strain gages being used. A special resistance network (Micro Measurements Type LST-10F-350D or equivalent) may be incorporated in the circuit, which modifies the gage signal, producing a direct readout of the temperature in °C or °F. The temperature readout is set to room temperature before testing, and then the proper gage factor is set. Lead wires exposed to temperatures greater than about 75°C should be protected with a high-temperature plastic wrap. Many laboratories use thermocouples as an alternative to temperature gages. One or several thermocouples may be attached to the specimen.

13.3 Temperature Compensation and CTE Measurements

In nonisothermal applications of strain gages, techniques must be employed to compensate for changes in the performance characteristics of the gages resulting from a change in temperature (Dally and Riley 1991). The change in performance of the gage relates to the following:

- The gage dimensions change with temperature.
- The resistance of the gage changes with temperature.
- Transverse strain sensitivity of the gage (Chapter 4) will induce an error in the measurements.
- The gage factor may change with temperature.

Consider a strain gage bonded to a composite specimen. For a given temperature change $\Delta T = T - T_0$, where T_0 is the initial temperature, the relative change in gage resistance $\Delta R/R$ may be expressed as

$$\Delta R/R = (\alpha_c - \alpha_g)S_g\Delta T + \gamma\Delta T \tag{13.3}$$

where α_c and α_g are the CTEs of the composite and the gage, respectively; γ is the temperature coefficient of resistivity of the gage material; and S_g is the gage factor, here assumed to be constant. For a large temperature change, it is also necessary to take into account the temperature dependence of the gage factor. Thus, the gage will be subjected to a strain mismatch of $(\alpha_c - \alpha_g)$ ΔT. If the coefficient γ is not zero, the strain measuring system will record an apparent strain that physically does not exist. To correct the apparent strain reading, a common temperature compensation method includes a reference gage, identical to the one bonded to the composite, mounted on a substrate with known CTE. For the gage bonded to the composite, Equation (13.3) gives

$$\Delta R_1/R = (\alpha_c - \alpha_g)S_g\Delta T + \gamma\Delta T \tag{13.4a}$$

and for the gage bonded to the reference substrate,

$$\Delta R_2/R = (\alpha_r - \alpha_g)S_g\Delta T + \gamma\Delta T \tag{13.4b}$$

where R is the gage resistance, α_r is the CTE of the reference substrate. Combining Equations (13.4) yields

$$\alpha_c = \alpha_r + \frac{\Delta R_1 - \Delta R_2}{RS_g\Delta T} \tag{13.5}$$

Hence, the CTE for the composite is

$$\alpha_c = \alpha_r + (\varepsilon_c - \varepsilon_r)/\Delta T \tag{13.6}$$

where ε_c and ε_r are the strain readings for the composite and the reference substrate.

If the gages are connected in a Wheatstone half bridge, the changes in resistance of the two gages ΔR_1 and ΔR_2 are subtracted (Figure 13.2). Thus, the output voltage of the bridge is directly proportional to the difference $(\varepsilon_c - \varepsilon_r)$.

The choice of reference material should be determined by the anticipated magnitude of α_c. If the values of α_c and α_r are very close, the apparent strain measured $\varepsilon_c - \varepsilon_r$ will be very small, and the sensitivity of the measurement will suffer. A common reference material is quartz (with $\alpha_r \approx 0.56 \times 10^{-6}/°C$). Measurements of CTEs less than $0.56 \times 10^{-6}/°C$ using strain gages may thus suffer from poor accuracy. In addition, the accuracy of the gage and strain measuring system, typically $\pm 2 \times 10^{-6}$, may further limit the accuracy of the measurement of small CTEs.

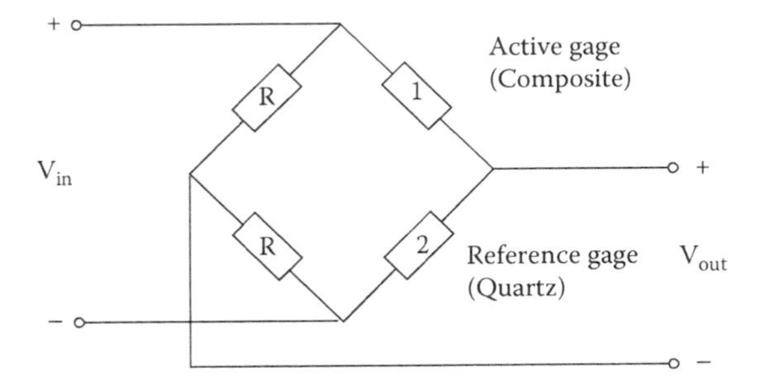

FIGURE 13.2
Wheatstone half-bridge circuit.

To measure the generally very small CTEs in the fiber direction of unidirectional composites (see Table 1.2) with sufficient accuracy, it may be necessary to use the interferometry techniques mentioned in the beginning of this chapter.

13.4 Measurement of Thermal Expansion

1. Bond two strain gages (Micro Measurements Type WK-06-125AC or equivalent) and one temperature sensor (or thermocouple) to the composite specimen. Align the strain gages (Figure 13.1) parallel to the principal material directions. Use a high-temperature strain gage adhesive cured according to adhesive specifications. Locate the strain gages and temperature sensor near the specimen center and on the same side to minimize the possible influence of thermal gradients.
2. Place the composite specimen and the reference material inside a laboratory oven (near the center). The gage lead wires inside the oven should be protected by a temperature-resistant coating such as Teflon.
3. Connect the strain gages and the temperature sensor (or thermocouple or both) to the recording system.
4. Raise the oven temperature slowly to 150°C. Monitor the oven thermometer, making strain and temperature measurements at regular temperature intervals. After a temperature of 150°C is reached, reduce the oven temperature slowly to room temperature. Take strain and temperature measurements also during this cooldown period. Occasionally, some composites display hysteresis during the first cycle but stabilize during subsequent cycles. Multiple temperature cycles are therefore desirable.

13.5 Data Reduction for CTEs

From the apparent strain measured via the half bridge $\varepsilon_A = \varepsilon_c - \varepsilon_r$, the actual strain ε_c is determined from Equation (13.6):

$$\varepsilon_c = \alpha_r \Delta T + \varepsilon_A \tag{13.7}$$

Plot ε_c versus the temperature T or the change in temperature $\Delta T = T - T_0$, where T_0 is the initial temperature of the specimen. Figure 13.3 shows typical thermal strain data for a carbon–epoxy specimen on heating (Whitney et al. 1984). To determine the CTE in the actual temperature range, evaluate the slope of the strain-versus-temperature plot. Figure 13.4 shows examples of plots of thermal strains versus temperature for Kevlar (E.I. du Pont de

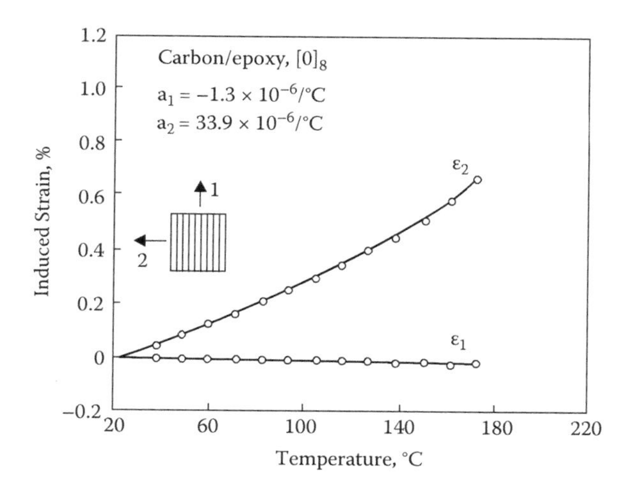

FIGURE 13.3
Thermal expansion strains for a carbon–epoxy composite. (From Whitney, J.M., Daniel, I.M., and Pipes, R.B., 1984, *Experimental Mechanics of Fiber Reinforced Composite Materials,* rev. ed., Society for Experimental Mechanics, Prentice-Hall, Englewood Cliffs, NJ.)

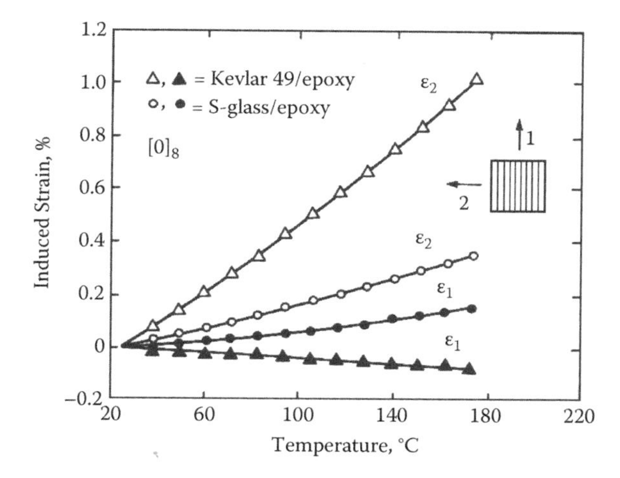

FIGURE 13.4
Thermal expansion strains for Kevlar–epoxy and S-glass–epoxy composites. Kevlar–epoxy: $\alpha_1 = -4.0 \times 10^{-6}/°C$, $\alpha_2 = 57.6 \times 10^{-6}/°C$. S-glass–epoxy: $\alpha_1 = 6.6 \times 10^{-6}/°C$, $\alpha_2 = 19.7 \times 10^{-6}/°C$. (From Whitney, J.M., Daniel, I.M., and Pipes, R.B., 1984. *Experimental Mechanics of Fiber Reinforced Composite Materials,* rev. ed., Society for Experimental Mechanics, Prentice-Hall, Englewood Cliffs, NJ.)

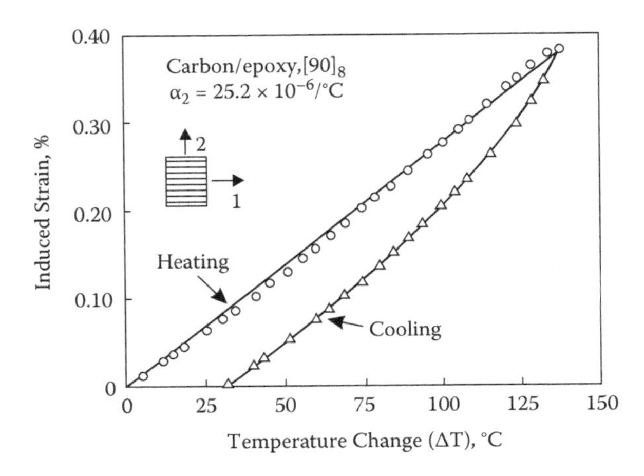

FIGURE 13.5
Thermal expansion response in the transverse direction for a carbon–epoxy composite showing hysteresis on cooling.

Nemours and Company)–epoxy and S-glass–epoxy composites. The determination of CTE discussed here inherently assumes linear expansion over the temperature range considered. For polymers at temperatures above their glass transition temperature T_g, the CTE is larger than at temperatures below T_g. Such phenomena and other factors make the expansion-versus-temperature curve more complex and sometimes nonlinear. For most cases, it is common to specify a temperature range of interest and calculate CTE for this range using a linear least-squares fit.

Hysteresis is often observed in heating and cooling cycles (Figure 13.5). Hysteresis is generally thought to be a result of viscoelastic creep and stress relaxation effects in the adhesive that bonds the gage to the specimen that are magnified by the increased temperature. It is also possible that residual thermal stresses in the composite will relax at elevated temperatures, which may change the dimensions. Higher rates of temperature change appear to produce more hysteresis, indicating that the material is not in thermal equilibrium. However, at lower temperatures, the slopes of the heating and the cooling curves are consistent. It is recommended to measure the coefficients of thermal expansion below the glass transition temperature of the polymer to minimize hysteresis.

13.6 Analysis of Thermal Expansion of Symmetric and Balanced Laminates

The thermoelastic response of unsymmetric laminates may be very complex (Hyer 1981). For the particular case of unsymmetric laminates, bending–extension coupling, Equations (2.44) and (2.45) indicate the existence of out-of-plane deflections for a laminate subject to a temperature change. Hyer (1981) and Dang and Hyer (1998) have performed very detailed experiments and analysis of warping deformations of unsymmetric composite laminates during cooling from elevated (cure) temperatures. For symmetric laminates, however, it can be shown that the bending–extension coupling disappears, [B] = [0]. For a balanced laminate, $A_{16} = A_{26} = 0$. Hence, a symmetric and balanced laminate behaves as a homogeneous orthotropic material in a macroscopic sense. Typical balanced symmetric laminates are $[0/\pm45/90]_s$, $[0/\pm45]_s$, and $[0/90]_s$.

For a symmetric and balanced laminate, $[b] = [c] = [0]$, and Equation (2.51) yields

$$\begin{bmatrix} \varepsilon^\circ \\ \kappa \end{bmatrix} = \begin{bmatrix} a & 0 \\ 0 & d \end{bmatrix} = \begin{bmatrix} N^T \\ M^T \end{bmatrix} \tag{13.8}$$

where $[N^T]$ and $[M^T]$ are the thermally induced forces and moments given by Equations (2.37) and (2.38). For such laminates, it may also be shown that the thermal moment resultants vanish, $[M^T] = [0]$, and the thermal in-plane shear force resultant $N_{xy}^T = 0$. Equations (13.8) then yield

$$[\varepsilon^0] = [a][N^T] \tag{13.9a}$$

$$[\kappa] = [0] \tag{13.9b}$$

Hence, a symmetric laminate does not warp due to a temperature change [Equation (13.9b)]. The expanded form of Equation (13.9a) is

$$\begin{bmatrix} \varepsilon_x \\ \varepsilon_y \\ \gamma_{xy} \end{bmatrix} = \begin{bmatrix} a_{11} & a_{12} & 0 \\ a_{12} & a_{22} & 0 \\ 0 & 0 & a_{66} \end{bmatrix} \begin{bmatrix} N_x^T \\ N_y^T \\ 0 \end{bmatrix} \tag{13.10}$$

Consequently, $\gamma_{xy} = 0$, which shows that a balanced laminate will not deform in shear due to the temperature change.

The thermal expansion coefficients of the laminate α_x and α_y are defined by

$$\alpha_x = \frac{\varepsilon_x}{\Delta T} \tag{13.11a}$$

$$\alpha_y = \frac{\varepsilon_y}{\Delta T} \tag{13.11b}$$

where ΔT is the temperature change from the reference state. Combining Equations (13.10) and (13.11) yields

$$\alpha_x = \left(a_{11}N_x^T + a_{12}N_y^T\right)/\Delta T \tag{13.12a}$$

$$\alpha_y = \left(a_{12}N_x^T + a_{22}N_y^T\right)/\Delta T \tag{13.12b}$$

where the compliance elements a_{ij} are obtained by inversion of the stiffness matrix [A].

To determine the laminate thermal expansion coefficients, the effective thermal forces $[N_x^T, N_y^T]$ and the stiffnesses A_{11}, A_{12}, and A_{22} are calculated using Equations (2.37) and (2.35a), respectively. Such calculation requires knowledge of the basic ply (lamina) mechanical properties and thermal expansion coefficients. The calculation of the laminate thermal expansions is quite involved. It is recommended that a computer code be used.

13.6.1 Example of Analysis of Laminate Thermal Expansion

Thermal expansion strains ε_x and ε_y for a quasi-isotropic $[0/\pm45/90]_s$ carbon–epoxy laminate are shown in Figure 13.6. The experimental values of the CTEs α_x and α_y obtained from the initial slopes of the curves shown in Figure 13.6 are as follows: $\alpha_x = 3.54\times10^{-6}/°C$, and $\alpha_y = 3.50\times10^{-6}/°C$.

Calculation of the CTEs (α_x and α_y) for the quasi-isotropic laminate was done using Equations (13.12) and the following ply data:

$$E_1 = 140\ GPa \quad \alpha_1 = -0.7\times10^{-6}/°C$$

$$E_2 = 10.3\ GPa \quad \alpha_2 = 31.2\times10^{-6}/°C$$

$$\nu_{12} = 0.29 \quad G_{12} = 15.5\ GPa$$

The results are

$$\alpha_x = \alpha_y = 3.30\times10^{-6}/°C$$

Good agreement between experiments and the laminate analysis is observed.

Further experimental CTE data for an IM6 carbon–3501 epoxy laminate with a layup $[0/\pm60/0]_s$ are listed in Table 13.1. It is observed that the data are reasonably consistent on heating and cooling. Note also the significant

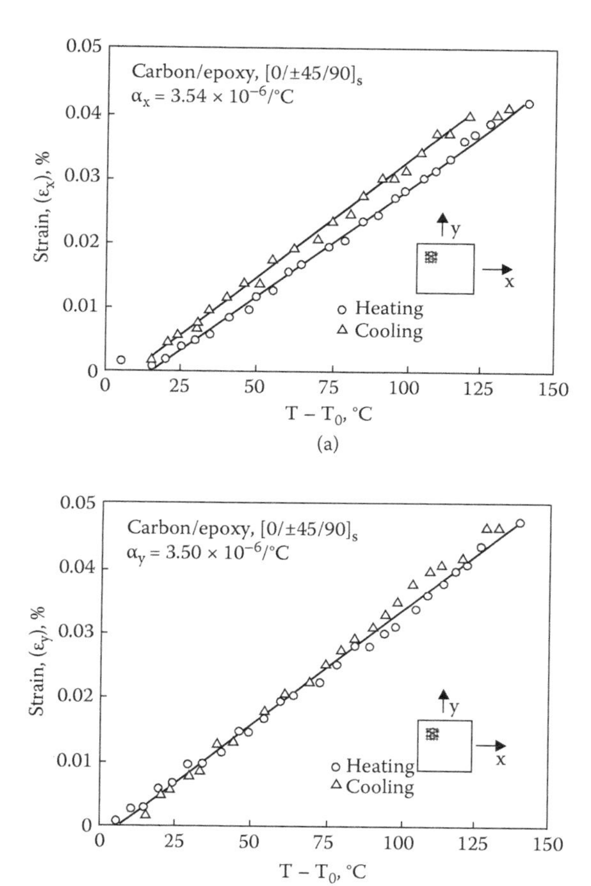

FIGURE 13.6
Thermal expansion strains for a quasi-isotropic $[0/\pm45/90]_s$ carbon–epoxy laminate: (a) ε_x and (b) ε_y.

TABLE 13.1

Measured and Predicted Thermal Expansion Coefficients (in units of 10^{-6}/°C for a $[0/\pm60/0]_s$ carbon–epoxy laminate)

CTE	Measured	Predicted
α_x	1.97 (H)[a] 2.04 (C)[a]	1.68
α_y	3.15 (H) 3.41 (C)	3.94

[a] H and C represent heating and cooling, respectively.

difference between α_x and α_y for this layup. The unidirectional ply properties are

$$E_1 = 160\ GPa \quad \alpha_1 = 0.64 \times 10^{-6}/°C$$

$$E_2 = 9.2\ GPa \quad \alpha_2 = 28.1 \times 10^{-6}/°C$$

$$\nu_{12} = 0.33 \quad G_{12} = 5.24\ GPa$$

The predicted and experimentally measured CTEs listed in Table 13.1 can be compared. The analysis somewhat underpredicts α_x and overpredicts α_y.

14

Through-Thickness Tension and Compression Testing

14.1 Introduction

Composite materials have largely been used in relatively thin laminate forms (i.e., as two-dimensional structural elements). Even so, delaminations and free-edge effects have been of concern. Now, as applications of composites continue to expand, there is increasing interest in the use of increased-thickness laminates as well and determination of how they respond to through-the-thickness stresses. This includes sandwich structures as well as solid laminates. Thus, it has become necessary to develop and standardize corresponding through-thickness material property characterization tests. Numerous instances of premature failure of laminated composites have been reported due to the low through-the-thickness tensile strength. Kedward et al. (1989) report values as low as 3% of the in-plane tensile strength of laminates.

For solid laminates, interlaminar tensile testing offers particular challenges. If a thin, flat laminate is to be tested, one major difficulty is properly gripping the thin specimen. If the specimen is bonded between blocks, the adhesive must be stronger than the laminate itself to obtain a valid test. And, the applied normal stress must be induced uniformly over the surface area of the specimen. Two specimen geometries currently being used (flatwise and curved beam) to measure the interlaminar tensile strength are discussed in the following sections. In addition, a test method for determining the through-thickness compressive strength is discussed.

14.2 Flatwise Through-Thickness Tensile Test

Until relatively recently, there has been no standardized test method for through-the-thickness tensile testing of solid laminates, and relatively few test data are available. An obvious approach is to adhesively bond the solid

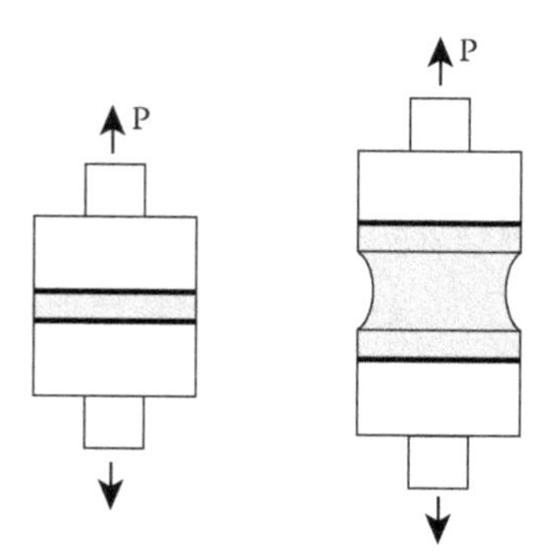

FIGURE 14.1
Typical flatwise through-thickness tensile specimens.

laminate between two metal blocks, usually steel, which are then loaded in tension normal to the plane of lamination. Guidance has been taken from American Society for Testing and Materials (ASTM) C297 (2010), a standard specifically developed for testing sandwich panels discussed in Chapter 19. While the sandwich specimen is square in plan-form, the flatwise composite specimen has a circular cross section. Figure 14.1 illustrates the ASTM D7291 (2007) standard flatwise test specimens and end tab assembly.

Test specimens with a symmetric and balanced, quasi-isotropic layup are recommended. The specimen should be at least 2.5 mm thick. The specimen should be bonded to cylindrical end tabs made from titanium or steel. Aluminum is not recommended because its relatively low modulus leads to excessive deformation of the tabs during testing. Figure 14.2 shows a drawing of the end tab.

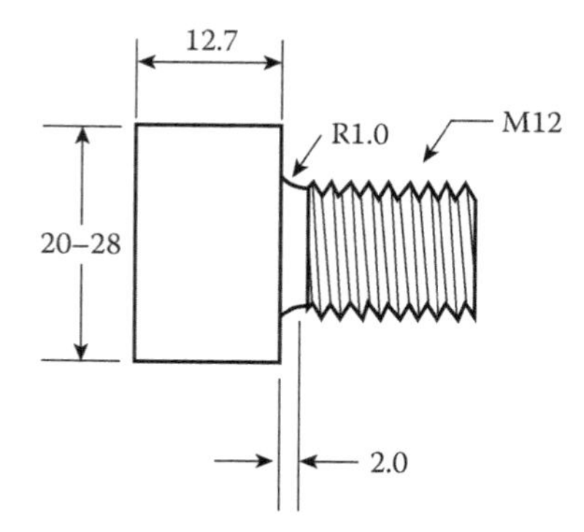

FIGURE 14.2
Drawing of cylindrical end tab. All dimensions are given in millimeters.

The diameter of the end tab ranges from 20 to 28 mm. The nominal diameter of the tab where the specimen is bonded to the tab is 25 mm. But, the end tabs can be reused after machining until they reach a diameter of 20 mm. The end tab thickness should be at least 12.7 mm as indicated. The threads on the end tabs allow attachment to the loading arrangement in the test machine. The threads also provide a method to attach constant-diameter bushings for the alignment of the specimen and a pair of end tabs in a special alignment and bonding fixture explained in Chapter 4, Section 4.10. The key factor in the short gage test is alignment. It is extremely important that the specimen thickness is uniform and that the bonding surfaces are flat and parallel and perpendicular to the direction of load application. After bonding of the end tabs (Chapter 4, Section 4.10), the specimen–tab assembly must be machined to obtain the required concentricity. Low-stress grinding or turning techniques combined with water coolant are appropriate.

For determination of the through-thickness modulus E_3, ASTM D7291 specifies a minimum specimen thickness of 6 mm and bonding of strain gages with an active gage length of 1.5 mm (although larger gages may be more suitable provided the specimen has long enough gage length). For composite laminates, the strain gage should cover a minimum of three laminate plies. The standard suggests using either two strain gages at locations that are 180° apart or three strain gages 120° apart around the cylindrical surface of the specimen, at the center of the gage section. Note that using only two gages may not allow detection of bending effects in the test geometry.

The specimen–tab assembly should be attached to the loading fixture in the test machine. The main factor is that the load is applied in the center of the assembly. The testing should be performed at a slow crosshead speed of 0.1 mm/min. Record load and displacement to collect at least 500 data points prior to rupture of the specimen. Determine the maximum load at failure.

Inspect the failure surfaces. Acceptable modes of failure are within the test specimen at least one ply thickness from the bonding surfaces. ASTM D7291 indicates four types of failure: (a) single plane within the specimen, (b) multiple planes within the specimen, (c) partly through the specimen and partly through the bonding surface, and (d) bondline failure. The only acceptable failure modes are (a) and (b). For the spool specimen (Figure 14.1), failure is most likely in the gage region, although failure sometimes occurs close to the tabs. Again, such failures are acceptable as long as they do not involve adhesive bondline failure.

The through-the-thickness tensile strength X_3^T is defined as the maximum applied load divided by the minimum cross-sectional area of the specimen. It is noteworthy to mention that since the Young's moduli and Poisson's ratios of the end block material and the laminate often differ significantly, a complex state of stress can be induced into the laminate or sandwich through the adhesive, and the average stress calculation (force/area) may be inaccurate. Further, interlaminar stress concentrations are also induced at the edges of the laminate and possibly singularities due to the bimaterial composite adhesive–end block interface. These issues are important considerations in interpreting strength and stiffness data developed in this test method.

14.3 Curved-Beam Tests for Interlaminar Tensile Strength

An alternative method, albeit indirect, for determining the through-thickness (interlaminar) tensile strength of a solid laminate is to subject a curved beam to an opening bending moment. Stress analysis of a curved beam under opening moment loading (discussed further in this chapter) reveals that, in addition to bending stresses, there will be an interlaminar tensile stress present that may cause the desired delamination mode of failure. To induce an opening bending moment in the curved section, the beam specimens may be loaded in tension as shown in Figure 14.3. The beam may be of uniform curvature along its length (Figure 14.3a) or have straight segments joined to a relatively sharp central curvature (Figure 14.3b). The 90° angle specimen shown in Figure 14.3b is most popular since the location of failure is well defined.

For the tensile-loaded curved-beam specimens (Figure 14.3), it is realized that the curved section is not under pure moment loading. It is clear that the region is under tension due to the applied load P, but most sections of

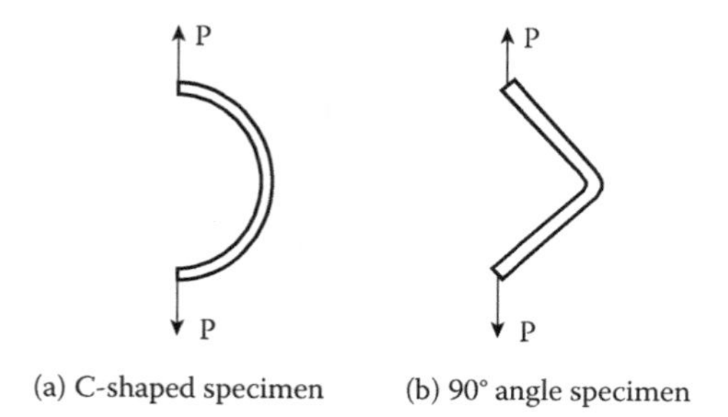

FIGURE 14.3
Curved-beam tensile specimens.

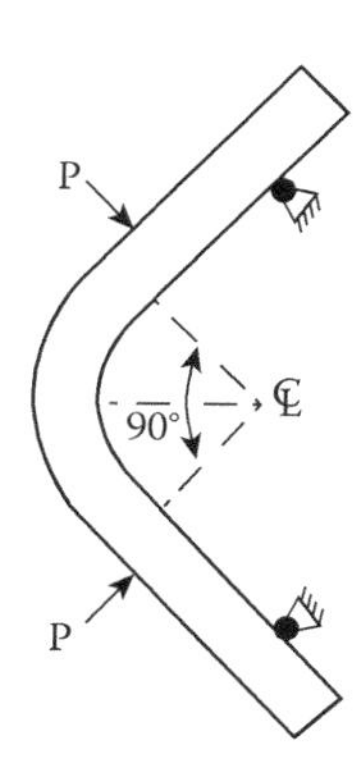

FIGURE 14.4
Curved-beam specimen loaded in four-point flexure.

the specimen are also subject to transverse shear forces, thereby making the failure analysis less clear cut.

Curved-beam specimens may also be loaded in flexure as shown schematically in Figure 14.4. Pure bending produces a favorable state of stress in the curved section of the specimen, free from shear stresses. Furthermore, the radial tensile stress does not vary along the curved region.

Figure 14.5 shows a typical failure of a curved-beam specimen. The delaminations shown were induced by the through-the-thickness tensile stresses developed in the curved region of the laminate.

Both tensile and flexure loading of curved beams may be appropriate. Due to the reasons discussed, however, only the flexure test has become standardized, as ASTM D6415 (2006).

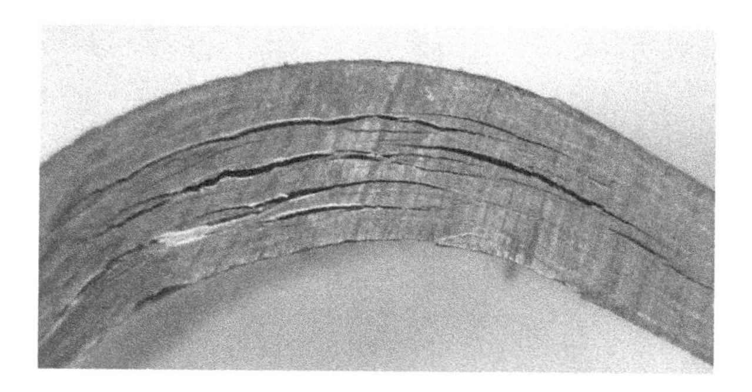

FIGURE 14.5
Delamination failure of a curved-beam specimen.

14.4 Flexure Test of Curved Specimens

The four-point test specimen shown schematically in Figure 14.4 is more clearly depicted in Figure 14.6. The curved beam is a unidirectional $[0]_n$ laminate with fibers directed along the straight sections and around the curved section. The beam is of uniform thickness h and width b. The straight sections, of length L, connect the curved section of inner radius r_i to form a smooth 90° bend. When the test is utilized to test multiaxial laminates, failures may result from free-edge stresses and matrix cracking. Composites with three-dimensional (3D) reinforcement fiber architecture may fail by in-plane failure. Consequently, for all types of composites, the failure process should be carefully examined to ensure that the specimens fail in the desired mode shown in Figure 14.5. The actual dimensions of the test specimens, according to ASTM D6415, are a width b of 25 mm, a thickness h of 6.4 ± 0.2 mm, and a length L of the straight sections of at least 50 mm. The inner radius r_i is specified as 6.4 mm.

A procedure for manufacturing the curved specimens is outlined in Chapter 4. The curved beam interlaminar strength is extremely sensitive to fiber volume fraction and void content. Consequently, the test results may reflect manufacturing quality as much as material properties. The machined edges of the specimens may be polished to aid in visually detecting damage initiation and growth during the test. Alternatively, the edges in the curved section may be painted white to aid visual inspection.

14.4.1 Four-Point Flexure Test Procedure

Figure 14.7 shows the ASTM D6415 four-point flexure test fixture. Machine drawings of the fixture are provided in the ASTM D6415 standard. The cylindrical loading and support rollers should be in the range from 6 to 10 mm in diameter (D) mounted on roller bearings. The horizontal distance between the centers of the top loading rollers ℓ_t should be 75 ± 2 mm, and

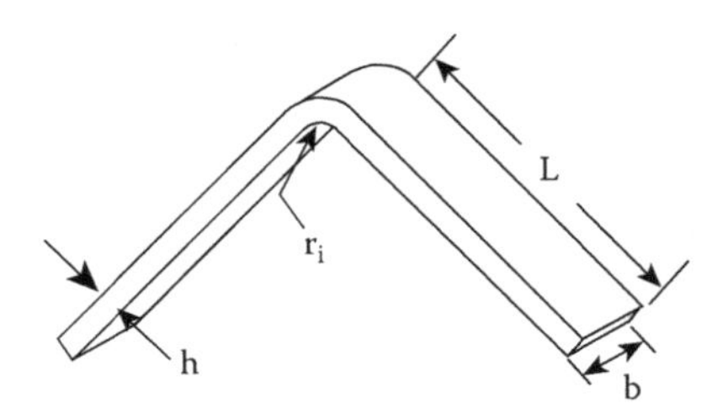

FIGURE 14.6
Four-point flexure curved-beam test specimen.

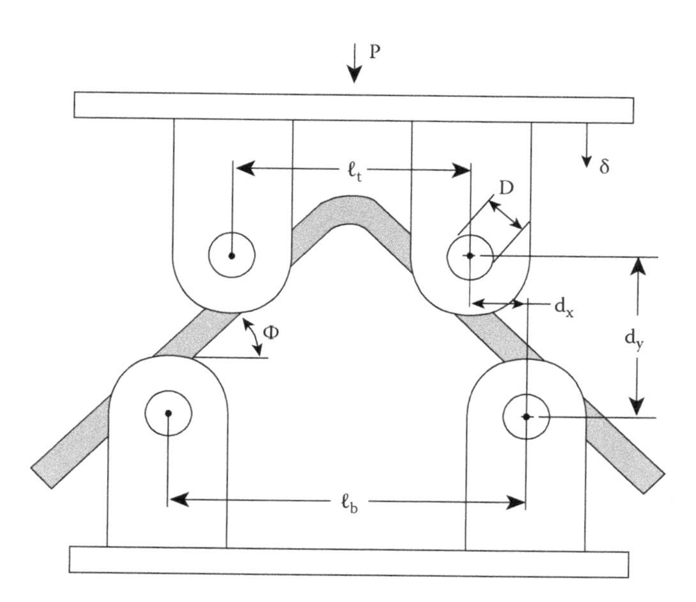

FIGURE 14.7
Four-point flexure fixture for curved-beam testing.

the horizontal distance between the bottom support rollers ℓ_b should be 100 ± 2 mm. Figure 14.8 shows a photo of an actual test rig.

Measure the thickness of the specimen at several locations in the curved regions. Also determine the variation in thickness, which should not exceed 5% of the nominal thickness. Record the average thickness for each leg and for the curved section. Also, measure the angle between the two legs (2φ in Figure 14.7). The angle should be 90° ± 3° but may deviate from the targeted 90° angle because of the "spring-in effect" discussed in Section 4.10 of Chapter 4.

The four-point-bending fixture is mounted in the testing machine. Align the top and bottom parts of the fixture such that all loading bars are parallel, and that the two parts are centered. Place the specimen on the bottom part of the fixture so that the crest of the curved region becomes centered between the support rollers. Lower the upper part of the fixture until the upper loading rollers make contact with the upper surface of the specimen.

Load the specimen at a constant rate of 0.5 mm/min and record the load P and crosshead displacement (δ). As the specimen bends, observe the curved region at the top of the specimen for damage. Delamination initiation is typically followed by rapid propagation and growth, as well as formation of additional delaminations as shown in Figure 14.5. A schematic load-displacement curve for this test is shown in Figure 14.9.

The various peaks indicate onset of delamination, each at a different interface. The vertical drops of the load indicate rapid growth of the delamination

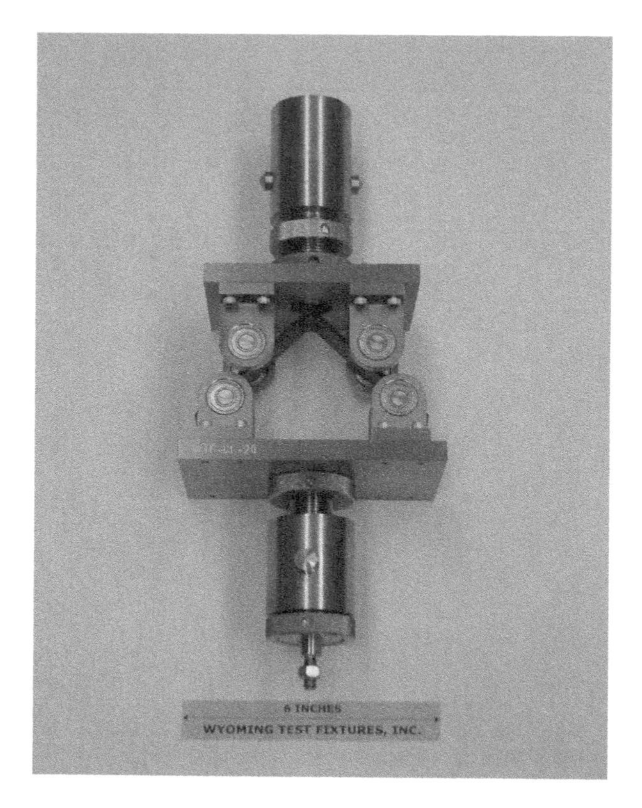

FIGURE 14.8
Photograph of four-point flexure fixture for curved-beam testing. (Courtesy of Wyoming Test Fixtures.)

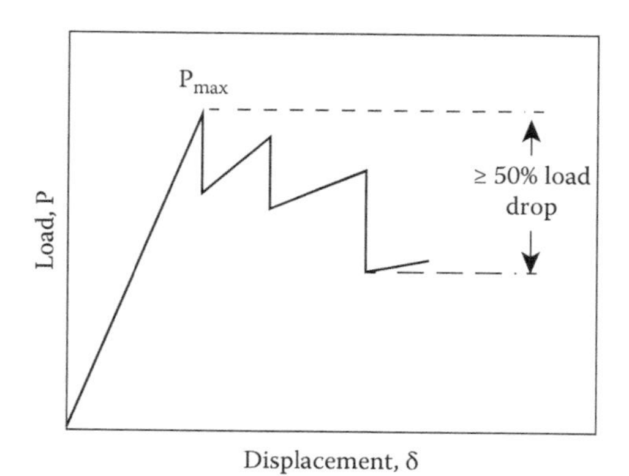

FIGURE 14.9
Schematic load-displacement curve for a curved-beam specimen loaded in four-point flexure.

around the circumference of the curved region (see Figure 14.5). The test is to be stopped after the load drops about 50% of the maximum load P_{max}. ASTM D6415 recommends testing of at least five replicate specimens unless valid results can be achieved with fewer specimens.

14.4.2 Determination of Through-Thickness Tensile Strength

The stress state in a curved-beam specimen under pure bending consists of circumferential (hoop) stress (σ_θ) and through-thickness radial stress σ_r. For opening moments, σ_θ is compressive on the outer radius r_0 and tensile at the inner radius r_i. σ_r is zero at both these two positions with a maximum at a radius r_m, which points to a position below the center of the cross section.

The stresses are independent of angular position because the section is under pure bending. Hence, the only loading entering into the stress calculations is the bending moment M, given by ASTM D6415 as

$$M = \frac{P}{2\cos\varphi}\left(\frac{d_x}{\cos\varphi} + (D+h)\tan\varphi\right) \tag{14.1}$$

where P is the load applied; d_x is the horizontal distance between the centers of adjacent loading and support rollers (see Figure 14.6), that is, $d_x = (\ell_b - \ell_t)/2$, D is the diameter of the loading and support rollers; and h is the specimen thickness. The angle φ (defined in Figure 14.7) will change from its initial value ($\varphi = 45°$) during the test due to bending deformation of the specimen. Hence, to obtain a more accurate value of the failure moment from Equation (14.1), the value of φ at failure should be used. ASTM 6415 suggests a kinematic analysis based on the vertical distance d_y between the loading and support rollers (Figure 14.7). The initial value of this distance is given by

$$d_i = d_x \tan\varphi_i + \frac{D+h}{\cos\varphi_i} \tag{14.2}$$

where φ_i is the initial angle of the unloaded specimen. The distance d_y after load application becomes

$$d_y = d_i - \delta \tag{14.3}$$

where δ is the crosshead displacement. Using trigonometric functions, the angle is given by

$$\varphi = \sin^{-1}\left(\frac{d_y\sqrt{d_x^2 + d_y^2 - D^2 - 2Dh} - d_x(D+h)}{d_x^2 + d_y^2}\right) \tag{14.4}$$

The only parameter in this equation changing during loading is d_y as given in Equation (14.3). The interlaminar tensile strength X_3^T is obtained using Lekhnitskii's (1968) analysis:

$$X_3^T = \frac{-M_{\max}}{br_0^2 g}\left[1-\left(\frac{1-\rho^{\kappa+1}}{1-\rho^{2\kappa}}\right)\left(\frac{r_m}{r_0}\right)^{\kappa-1}-\left(\frac{1-\rho^{\kappa-1}}{1-\rho^{2\kappa}}\right)\rho^{\kappa+1}\left(\frac{r_0}{r_m}\right)^{\kappa+1}\right] \quad (14.5)$$

where

$$g = \frac{1-\rho^2}{2}-\left(\frac{\kappa}{\kappa+1}\right)\frac{(1-\rho^{\kappa+1})^2}{\left(1-\rho^{2\kappa}\right)}+\left(\frac{\kappa\rho^2}{\kappa-1}\right)\frac{(1-\rho^{\kappa-1})^2}{\left(1-\rho^{2\kappa}\right)} \quad (14.6)$$

and with

$$\kappa = \sqrt{\frac{E_\theta}{E_r}} \qquad \rho = \frac{r_i}{r_0} \qquad r_m = \left[\frac{(1-\rho^{\kappa-1})(\kappa+1)(\rho r_0)^{\kappa+1}}{(1-\rho^{\kappa+1})(\kappa-1)r_0^{-(\kappa-1)}}\right]^{\frac{1}{2\kappa}} \quad (14.7a)\text{–}(14.7c)$$

E_θ and E_r in Equation (14.7a) are the moduli of the composite specimen in the hoop and radial directions, respectively. They may be approximated by E_1 and E_2 for the unidirectional flat composite laminate.

14.5 Tensile Test of Curved Specimens

The tensile loaded 90° angle specimen is shown in Figure 14.10. As discussed previously, this test specimen suffers from a more complex stress state involving shear and superimposed tensile stresses, and the element where the radial stress σ_r is maximum is under biaxial loading. For this and several other reasons, this specimen has not been accepted as an ASTM standard. This test, however, has been demonstrated to be useful for quality assurance of laminated composites long before the four-point configuration appeared in 2006, as reported by Jackson and Martin (1993). We therefore include a brief review of their work in this chapter.

Tensile load of magnitude P is applied via steel end tabs bolted to the specimen ends. This configuration allowed the specimen to be tested in a standard tensile test machine. Carbon–epoxy unidirectional composite specimens with 16, 24, and 48 plies were laid up over a rounded corner of an aluminum block. The inner radius r_i ranged from 3.2 to 8.5 mm for the specimens. The length of the straight portions of the specimen was 8 cm, and the specimen width

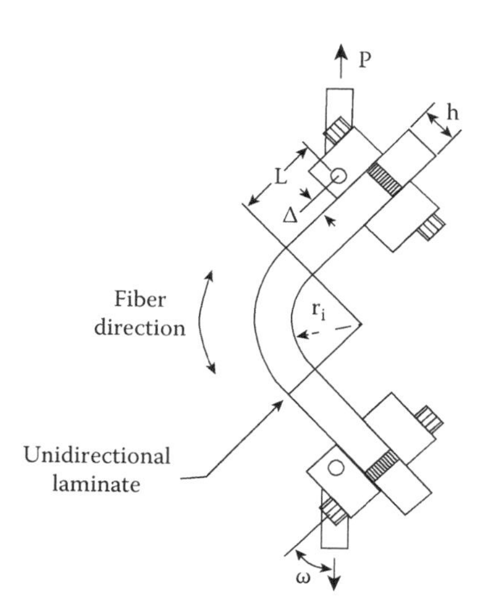

FIGURE 14.10
Tensile test of the 90° angle specimen. (Jackson, W.C., and Martin, R.H., 1993, *An Interlaminar Tensile Strength Test*, ASTM Spec. Tech. Pub. 1206, American Society for Testing and Materials, West Conshohocken, PA, pp. 333–354.)

b was 12.7 or 25.4 mm, nominally. The length of the loading arm L, defined in Figure 14.10, was 25.4 mm in most cases. The specimens were loaded at a crosshead rate of 0.5 mm/min until failure. At failure, the load dropped about 40–50% from its peak. After failure, circumferential delamination cracks were observed in the curved segment, indicating tensile failure.

The 16-ply specimens tended to have the most cracks, up to seven on each side. Most specimens had one to three cracks, and at different locations on opposite sides of the same specimen. Stress analysis was conducted using finite-element analysis and a modified Lekhnitskii (1968) solution to accommodate the tensile load applied to the beam specimen (Figure 14.10). This procedure led to an expression for the radial stress that is too extensive to reproduce here; furthermore, their analysis does not provide the radius where σ_r is maximum. An approximate solution for isotropic curved beams was provided:

$$(\sigma_r)_{\max} = \frac{3P\left(L\cos\omega - \left(\Delta + \frac{h}{2}\right)\sin\omega\right)}{2bh\sqrt{r_i r_0}} \tag{14.8}$$

where ω is the angle between the line of load application and the normal to the straight sections of the specimen (see Figure 14.10). Note that the $\Delta + h/2$ term in Equation (14.8) represents the offset of the loading center from the neutral axis of the straight sections. At the beginning of the test, the nominal value of ω is 45°. Similar to the four-point flexure specimen discussed in Section 14.4, the angle w will change from its initial value of 45° during the test as the specimen deforms in bending, but there is no simple kinematics analysis to update this angle. For the carbon–epoxy specimens examined by Jackson and Martin (1993), the maximum stress obtained from Equation (14.7) is about 10% off the finite-element result.

Test results from a relatively large range of specimens revealed that the through-thickness tensile strength X_3^T fell into two groups: one "high," with an average of about 80 MPa, and one "low," about 30 MPa on average. The results, however, displayed much scatter. Part of the large scatter is attributed to the highly localized region of high stress in the tensile test curved beams. The low strength category was explained by presence of a macroscopic flaw in the specimen. Such flaws were more common in thicker (48=ply) laminates. The thinner laminates failed overall at higher stress due to smaller flaws.

15.6 Through-Thickness Compressive Test

The through-thickness compressive strength X_C^T is defined as maximum applied load divided by the cross-sectional area. There is no standardized test for through-thickness compression of a solid laminate, although there is ASTM Standard C365 (2009) for compression testing sandwich core materials. The basic approach is to load the specimen between two compression platens. The ASTM standard specifies that one platen be fixed and the other mounted on a spherical seat. However, two fixed platens can be used if they are properly aligned to have faces parallel to each other and perpendicular to the loading axis. Since solid laminates are typically much stiffer than core materials, there is a greater chance of nonuniform loading; thus, incorporating a spherical seat is advisable.

Since the through-thickness compressive strength X_3^C of a solid laminate can be much greater than the corresponding tensile strength X_3^T, specimens of small cross section are typically used. Specimens, either round or square in cross section, with cross-sectional dimensions of 12 to 25 mm are common. For example, for a material with a through-thickness compressive strength of 200 MPa, an applied force of over 130 kN would be required to fail a 25-mm square specimen. Specimen buckling is a consideration when applying a compressive loading on long unsupported specimens. Thus, the maximum specimen length is typically kept below two to three times the cross-sectional dimensions of the specimen.

Just as for through-thickness tension testing discussed in Section 14.2, the available test material may be thin. For compressive loading, this is much less of a problem since individual pieces of the test panel can be stacked up to provide whatever total thickness is desired. These pieces need not be adhesively bonded together other than for ease of handling. A small amount of a soft adhesive on the edges is adequate. As an alternative, the stack can be wrapped with adhesive tape. If, for some reason, it is desirable, or necessary, to adhesively bond the faces together, a soft adhesive (e.g., rubber cement or similar adhesive that will not chemically react with the composite), used in very thin bond lines, should be used.

During the compression testing, the layers adjacent to the specimen ends will be constrained by the platen surfaces due to differential transverse strains (the Poisson's ratio effect discussed in Section 14.2), which strengthens the material. Thus, end constraint can be beneficial in the sense that it forces the failure away from the specimen ends where the load is applied. Failure typically occurs at a position two or three layers away from the loading platens where a more uniform state of axial compressive stress exists. Thus, reduction in cross section of the compression specimen is not required.

As for through-the-thickness tensile testing discussed in Section 14.2, because the solid laminate being compressed is typically relatively thin, the crosshead rate should be kept low. A crosshead displacement rate of 0.1 mm/min is recommended. The compression strength X_3^C is defined as the maximum applied load divided by the cross-sectional area.

15

Laminate Mechanical Response

15.1 Introduction

The basic mechanical tests used to characterize a lamina (i.e., tension, compression, and shear) must be performed to experimentally characterize a laminate as well. Sometimes, a flexural test is also performed on a laminate beam specimen. However, the complexity of the mechanical behavior of a laminated plate (Figure 15.1) is considerably greater than that of the lamina discussed in previous chapters. Because the laminate generally includes off-axis plies (a unidirectional composite consisting of multiple unidirectional laminae is sometimes also referred to as a laminate), the stress state in a given ply is biaxial. Moreover, at free edges, a fully three-dimensional stress state develops (Pagano and Pipes 1971, 1973, Pipes et al. 1973), and edge delamination may occur. As shown in the works cited previously, however, the free-edge interlaminar stresses decay to small magnitudes within a boundary-layer region, which extends only about one laminate thickness in from the free edge. Classical laminated plate theory, reviewed in Chapter 2, is an accurate predictor of the stress state in the remainder of the laminate and of the overall structural behavior of the laminate. In this chapter, all analysis is therefore based on classical laminated plate theory.

The mechanical response of the composite laminate is typically monitored using the same strain instrumentation as previously discussed for lamina response to establish similar stiffness, strength, and strain-to-failure properties. The specimen configurations and sizes are usually similar, if not identical. Correspondingly, when appropriate, the specimens must be end tabbed, strain gaged, and then loaded to ultimate failure.

Specimens investigated in this chapter are limited to symmetric and balanced laminates. Symmetric means that for each ply above the midsurface of the laminate, there is an identical ply of the same orientation at the same distance below the midsurface. Balanced means that for each angle ply with an orientation angle θ with respect to the x-axis (Figure 15.1), there is an identical ply oriented at $-\theta$ with respect to the x-axis. Note that for balanced laminates this identical ply can be anywhere within the laminate. Typical examples of symmetric and balanced laminates are $[0/\pm45/90]_s$, $[0_2/\pm45]_s$, and $[0/90]_s$.

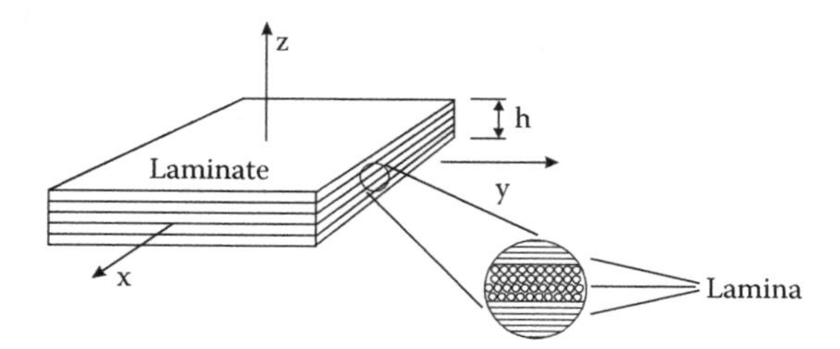

FIGURE 15.1
Element of a laminate.

15.2 Analysis of Extensional and Shear Stiffness Properties of Laminates

Consider the element of a symmetric and balanced laminate subjected to in-plane extensional and shear forces shown in Figure 15.2. The stiffness relation for an element of the symmetric and balanced laminate is given by Equation (2.33):

$$\begin{bmatrix} N_x \\ N_y \\ N_{xy} \end{bmatrix} = \begin{bmatrix} A_{11} & A_{12} & 0 \\ A_{12} & A_{22} & 0 \\ 0 & 0 & A_{66} \end{bmatrix} \begin{bmatrix} \varepsilon_x \\ \varepsilon_y \\ \gamma_{xy} \end{bmatrix} \tag{15.1}$$

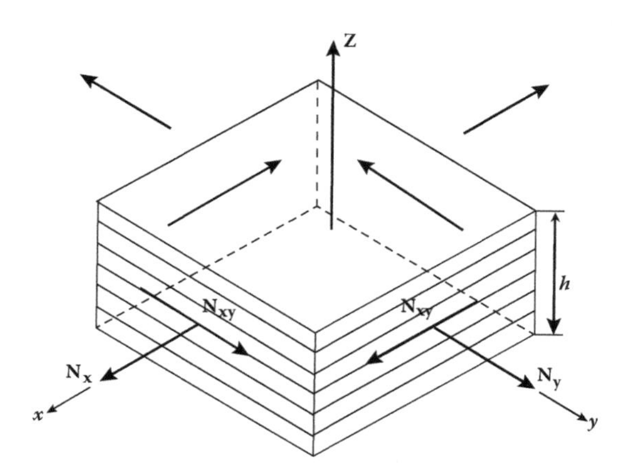

FIGURE 15.2
Symmetric and balanced laminate element under in-plane forces.

where the laminate stiffnesses A_{ij} are defined in terms of the ply stiffnesses $\bar{Q}_{ij}$ and ply thickness h_{ply} in Equation (2.35a):

$$A_{ij} = \sum_{k=1}^{N} (\bar{Q}_{ij})_k (z_k - z_{k-1}) \tag{15.2}$$

where N is the number of plies in the laminate.

For the purpose of establishing the effective laminate stiffness properties E_x, E_y, ν_{xy}, ν_{yx}, and G_{xy}, where symbols E, ν, and G represent extensional modulus, Poisson's ratio, and shear modulus, respectively, it is most convenient to use the inverted (compliance) form of Equation (15.1):

$$\begin{bmatrix} \varepsilon_x \\ \varepsilon_y \\ \gamma_{xy} \end{bmatrix} = \begin{bmatrix} a_{11} & a_{12} & 0 \\ a_{12} & a_{22} & 0 \\ 0 & 0 & a_{66} \end{bmatrix} \begin{bmatrix} N_x \\ N_y \\ N_{xy} \end{bmatrix} \tag{15.3}$$

The compliance elements a_{ij} are obtained from the stiffness elements A_{ij} defined in Equation (15.1) as follows:

$$a_{11} = \frac{A_{22}}{A_{11}A_{22} - A_{12}^2}, \quad a_{12} = \frac{-A_{12}}{A_{11}} \quad \text{(15.4a), (15.4b)}$$

$$a_{22} = \frac{A_{11}}{A_{11}A_{22} - A_{12}^2}, \quad a_{66} = \frac{1}{A_{66}} \quad \text{(15.4c), (15.4d)}$$

To proceed with the derivation of effective mechanical properties of the laminate element, it is useful to define the following average stresses and strains for uniaxial and pure shear loadings of the element shown in Figure 15.2:

$$\sigma_x = N_x/h \tag{15.5a}$$

$$\sigma_y = N_y/h \tag{15.5b}$$

$$\tau_{xy} = N_{xy}/h \tag{15.5c}$$

where h is the total laminate thickness (Figure 15.2) $(h = Nh_{ply})$.

Here, each load N_x, N_Y, and N_{xy} is supposed to be applied one at a time to be consistent with the mechanical testing described further in this chapter. The effective stiffnesses are then given by the following:

For $N_x \neq 0$ and $N_y = N_{xy} = 0$ (uniaxial tension or compression),

$$E_x = \frac{\sigma_x}{\varepsilon_x} \tag{15.6a}$$

$$\nu_{xy} = \frac{-\varepsilon_y}{\varepsilon_x} \tag{15.6b}$$

For $N_y \neq 0$ and $N_x = N_{xy} = 0$ (uniaxial tension or compression),

$$E_y = \frac{\sigma_y}{\varepsilon_y} \tag{15.7a}$$

$$\nu_{yx} = \frac{-\varepsilon_x}{\varepsilon_y} \tag{15.7b}$$

For $N_{xy} \neq 0$ and $N_x = N_y = 0$ (pure shear),

$$G_{xy} = \frac{\tau_{xy}}{\gamma_{xy}} \tag{15.8}$$

Combination of Equations (15.3) and (15.5)–(15.8) yields the following set of effective engineering properties:

$$E_x = \frac{1}{ha_{11}}, \qquad \nu_{xy} = \frac{-a_{12}}{a_{11}} \qquad \text{(15.9a), (15.9b)}$$

$$E_y = \frac{1}{ha_{22}}, \qquad \nu_{yx} = \frac{-a_{12}}{a_{22}} \qquad \text{(15.9c), (15.9d)}$$

$$G_{xy} = \frac{1}{ha_{66}} \tag{15.9e}$$

Note that only four of the five stiffnesses are independent ($E_x \nu_{yx} = E_y \nu_{xy}$), consistent with Equation (2.8) for a homogeneous orthotropic material.

In terms of the stiffness elements A_{ij}, the stiffness properties are given by

$$E_x = \frac{A_{11}A_{22} - A_{12}^2}{hA_{22}}, \qquad \nu_{xy} = \frac{A_{12}}{A_{22}} \qquad \text{(15.10a), (15.10b)}$$

$$E_y = \frac{A_{11}A_{22} - A_{12}^2}{hA_{11}}, \qquad \nu_{yx} = \frac{A_{12}}{A_{11}} \qquad \text{(15.10c), (15.10d)}$$

$$G_{xy} = \frac{A_{66}}{h} \qquad \text{(15.10e)}$$

15.2.1 Laminate Strength Analysis

Strength analysis of the laminate under tension, compression, and shear loading is performed as outlined in Section 2.5. The ply stresses are calculated for a unit load, for example, $N_x = 1$, and for the actual temperature and moisture conditions experienced and substituted into the selected failure criterion. N_x is then increased until that failure criterion is satisfied and first-ply failure occurs. Many current designs do not allow first-ply failure.

Failure of a ply may or may not lead to laminate failure. If the remaining plies of the laminate are able to carry the load redistributed at first-ply failure, further loading may be applied to the laminate. A conservative estimate of the laminate behavior after first-ply failure is based on the assumption that certain of the elastic stiffness properties are removed from the laminate constitutive relation. For example, matrix cracking in a unidirectional composite ply, illustrated schematically in Figure 15.3, leaves the fibers intact, and the ply still can carry load in the fiber direction. Jones (1999) presents a laminate failure analysis in which the modulus E_1 remains intact while the following transverse properties of the cracked ply are assumed to be close to zero:

$$E_2 = \nu_{12} = G_{12} \approx 0 \qquad (15.11)$$

From this assumption, a new stiffness matrix, Equation (15.1), for the laminate containing ply cracks is determined, and further load is applied until the next ply failure occurs. This procedure may be repeated until last-ply failure occurs, indicating total failure. More elaborate analyses of damaged composites have been presented, for example, by Highsmith and Reifsnider (1982) and Hashin (1985).

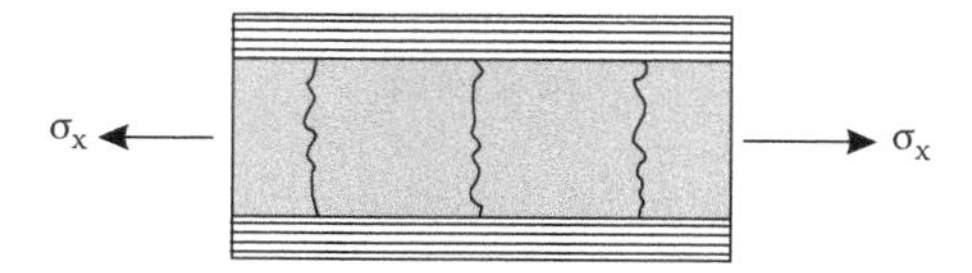

FIGURE 15.3
Matrix cracking of a ply in a laminate.

15.3 Analysis of Laminate Beam Flexure

Unlike for the in-plane load cases discussed in Section 15.2, the flexural response of a laminate, including a balanced, symmetric laminate, is a function of the stacking sequence of the plies.

Different from the in-plane extensional stiffness matrix [Equation (15.1)], the bending stiffness matrix of a general symmetric and balanced laminate may contain the bending/twisting coupling stiffnesses D_{16} and D_{26}. We consider a cross section of a laminate beam subject to bending moment M_x per unit width of the beam, that is,

$$M_x = \frac{M}{b} \tag{15.12}$$

where M is the moment at this section, and b is the width of the beam. The B matrix for such a laminate is zero. For such a loading, Equation (2.34) becomes

$$\begin{bmatrix} M_x \\ 0 \\ 0 \end{bmatrix} = \begin{bmatrix} D_{11} & D_{12} & D_{16} \\ D_{12} & D_{22} & D_{26} \\ D_{16} & D_{26} & D_{66} \end{bmatrix} \begin{bmatrix} \kappa_x \\ \kappa_y \\ \kappa_{xy} \end{bmatrix} \tag{15.13}$$

For analysis of bending of a laminated beam, it is useful to consider Equation (15.13) in the inverted (compliance) form:

$$\begin{bmatrix} \kappa_x \\ \kappa_y \\ \kappa_{xy} \end{bmatrix} = \begin{bmatrix} d_{11} & d_{12} & d_{16} \\ d_{12} & d_{22} & d_{26} \\ d_{16} & d_{26} & d_{66} \end{bmatrix} \begin{bmatrix} M_x \\ 0 \\ 0 \end{bmatrix} \tag{15.14}$$

To develop a beam formulation, it is customary to assume that the deflection w is a function of x only, which is reasonable for long and narrow beams:

$$w = w(x) \tag{15.15}$$

Combining the definition of κ_x ($\kappa_x = -\partial^2 w/\partial x^2$) with Equations (15.14) and (15.15) yields

$$\frac{d^2 w}{dx^2} = -d_{11} M_x \tag{15.16}$$

This equation is identical to the classical differential equation governing bending of isotropic and homogeneous beams (Gere and Timoshenko 1997):

$$\frac{d^2w}{dx^2} = \frac{-M}{EI} \tag{15.17}$$

where the moment of inertia for the rectangular cross section ($b \times h$) is given by

$$I = \frac{bh^3}{12} \tag{15.18}$$

Equations (15.16)–(15.18) yield an expression for the effective flexural modulus of the laminated beam:

$$E_x^f = \frac{12}{d_{11}h^3} \tag{15.19}$$

Similarly, the flexural modulus for a beam cut perpendicular to the x-axis is

$$E_y^f = \frac{12}{d_{22}h^3} \tag{15.20}$$

15.4 Test Specimen Preparation

Symmetric and balanced laminate panels should be prepared. Examples of such laminates are $[0/\pm 45/90]_s$, $[0_2/\pm 45]_s$, and $[0_2/90_2]_s$. A typical specimen preparation procedure is to bond end tabs if required (Chapter 4) and machine the specimens to the required dimensions. It is critical that the specimens be taken from the fabricated panel in the orientations desired. For example, the $[0_2/\pm 45]_s$ laminate becomes a $[90_2/\mp 45]_s$ laminate if cut from the panel perpendicular to the direction intended.

Measure the cross-sectional dimensions (average six measurements) and check for parallelism of the edges and of the end tab surfaces. If required, apply strain gages as appropriate for the specific test to be performed.

15.5 Test Procedures

Many test procedure details are common to all types of testing to be performed. These details include, for example, correctly installing the specimen in the grips or test fixture and ensuring that proper alignment is attained and proper gripping–clamping forces are used when appropriate. Testing machine settings, particularly load range, strain instrumentation calibrations, and data acquisition systems, should be checked.

15.5.1 Tension Test Procedures

A laminate tensile specimen is typically about 225 mm long and 25 mm wide. For an axial tensile test, two strain gages are mounted at the center of each specimen, one in the longitudinal direction and one in the transverse direction, so that both the axial stiffness E_x and major Poisson's ratio ν_{xy} can be determined.

Mount the specimen in mechanical wedge grips or hydraulic grips and follow the test procedure outlined in Chapter 8. Set the crosshead rate at about 0.5 to 1 mm/min. Avoid unprotected eyes in the test area. The strain readings may be recorded continuously or at discrete load intervals. If discrete data are recorded, a sufficient number of data points are required to adequately reproduce the stress–strain behavior. A total of 40 to 50 points, with at least 25 data points in the linear response region, are desirable to establish the total stress–strain response. Monitor all specimens to failure. Plot the data for subsequent reduction.

15.5.2 Compression Test Procedures

Because only relatively thin laminates are usually available for testing, gross buckling of the specimen must be prevented. Thus, some type of lateral restraint fixturing is required. It is important to ensure that this fixturing does not create a redundant load path. As discussed in Chapter 9, each compression test method has its own unique fixture. In all cases proper specimen installation in the fixture is critical. Specimen dimensions are typically the same as for lamina testing.

Whether the specimen is end loaded or shear loaded, specimen tabs may be required to prevent either end crushing or gripping surface damage. However, because laminates typically have lower compressive strengths than unidirectional composites, a particular test may not require tabs. It is important to recognize when tabs are not needed because their use increases specimen preparation cost and potentially introduces additional stress concentrations and loading abnormalities, as discussed in Chapter 9.

15.5.3 Shear Test Procedures

The Iosipescu and rail shear test methods discussed in Chapter 10 are applicable for laminates. However, the in-plane shear failure modes for laminates are quite different from those for unidirectional composites. Unidirectional composites subjected to shear loadings fail in a clean manner parallel to the fibers. Laminates do not have these weak through-the-specimen-thickness planes of failure because the individual plies are at different orientations. The resulting failure path can be tortuous and the damage zone extensive. It is important that the test fixture geometry not inhibit the failure mode. The dimensions of the Iosipescu and short-beam shear specimens are the same as for lamina testing.

15.5.4 Flexural Test Procedures

As previously noted in Section 15.3, the ply stacking sequence of the laminate influences the composite stiffness properties measured in a flexural test. It also influences the flexural strength. Thus, it is important to verify that the orientation of the laminate being tested is that intended.

Because both the flexural and shear stiffnesses depend on the laminate orientation being tested, a different span length-to-specimen thickness ratio may be required than would be used to test a unidirectional composite of the same material. The same guidelines as indicated in Chapter 11 apply, however.

15.6 Data Reduction

To determine the laminate properties, recall the following definitions:

E_x Initial slope of the stress–strain curve ($\Delta\sigma_x/\Delta\varepsilon_x$)
ν_{xy} Negative ratio of the transverse to longitudinal strains ($-\varepsilon_y/\varepsilon_x$)
σ_x^{ult} Ultimate load/cross-sectional area
G_{xy} Initial slope of the shear stress–strain curve ($\Delta\tau_{xy}/\Delta\gamma_{xy}$)
τ_x^{ult} Ultimate shear force/shear area
E_x^{flex} Initial slope of the flexural stress–strain curve

Determine the modulus and Poisson's ratio by least-squares fits of the initial slopes of σ_x versus ε_x and σ_x versus ε_y. The transverse tensile and compressive properties may be determined accordingly for loadings in the y direction. The shear and flexural moduli are determined by least-squares fits of the initial slopes of the corresponding stress–strain curves.

Figure 15.4 shows typical tensile test results for a $[0/\pm45/90]_s$ carbon–epoxy laminate. Data reduction provides the following mechanical properties of the laminate: $E_x = 56.5\ GPa$, $\nu_{xy} = 0.34$, and $\sigma_x^{ult} = 626\ MPa$.

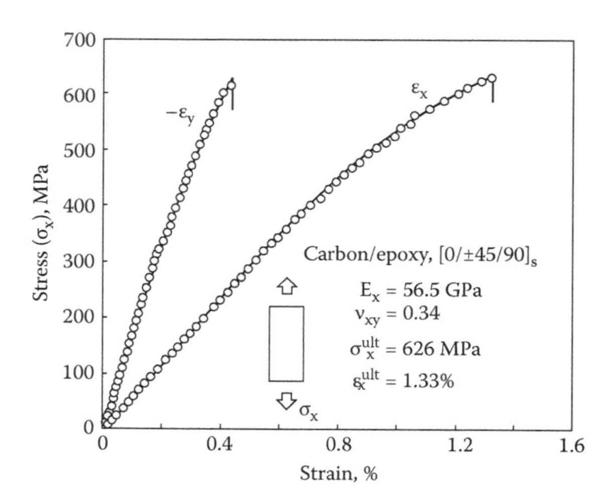

FIGURE 15.4
Tensile stress–strain response for a $[0/\pm45/90]_s$ carbon–epoxy laminate test specimen.

15.7 Example of a Typical Analysis: Axial Tensile Response of a Laminate

To analyze the tensile response of the laminate, the unidirectional ply properties are required. Consider the following set of carbon–epoxy ply properties:

$$E_1 = 140 \text{ GPa} \qquad X_1^T = 1950 \text{ MPa} \qquad S_6 = 85 \text{ MPa}$$

$$E_2 = 10.3 \text{ GPa} \qquad X_1^C = 1500 \text{ MPa} \qquad \alpha_1 = -0.7 \times 10^{-6}/°\text{C}$$

$$\nu_{12} = 0.29 \qquad X_2^T = 48 \text{ MPa} \qquad \alpha_2 = 31.2 \times 10^{-6}/°\text{C}$$

$$G_{12} = 5.15 \text{ GPa} \qquad X_2^C = 130 \text{ MPa} \qquad h_{\text{ply}} = 0.127 \text{ mm}$$

Laminated plate theory gives, with these lamina properties, $E_x = 54.2$ GPa and $\nu_{xy} = 0.31$, in good agreement with measured values listed in Figure 15.4. Nonlinear behavior is noted in the response curve at stress levels beyond approximately 400 MPa. This may be due to ply cracking, which renders the laminate more compliant.

Failure of the laminate was analyzed using both the maximum stress and Tsai-Wu strength criteria (see Section 2.6). Stress–strain levels for ply failures were calculated on the basis of the ply properties listed and a temperature change $\Delta T = -150°\text{C}$. The results are listed in Table 15.1. It is observed that

TABLE 15.1

Calculated Ply Failure Stresses and Strains in the Laminate Coordinate System for a $[0/\pm45/90]_s$ Carbon–Epoxy Laminate ($\Delta T = -150°C$)

	σ_x (MPa)	
Ply Angle (degrees)	**Maximum Stress Criterion**	**Tsai-Wu Criterion**
90	113	102
±45	231	220
0	768	738

the 90° ply is predicted to fail first. The ±45° plies then fail, followed by last-ply failure of the 0° plies (fiber failure). Note that the 90° ply is predicted to fail at stresses significantly lower than the experimentally measured ultimate stress, $\sigma_x^{ult} = 626$ MPa. Last-ply failure, on the other hand, is predicted to occur at stresses somewhat greater than the experimentally determined value. The ply failures preceding ultimate laminate failure were neglected in the calculations. Incorporation of ply failures by degradation of transverse properties should result in more realistic ultimate strength predictions. The differences between the ply failure predictions as a result of the choice of failure criterion will also be noted.

Note that the analyses of laminate ultimate strength reviewed here are highly approximate because the damage introduced after the occurrence of first-ply failure is not considered or is considered in an ad hoc manner. The prediction of first-ply-failure stress is more reliable because the laminate is undamaged to this point. The first-ply-failure criterion is appropriate for many current designs that do not allow ply cracks in any part of a structure.

16

Strength of Laminates with Holes

16.1 Introduction

Experiments have shown that the tensile and compressive strengths of a composite laminate containing a hole or notch depend on hole or notch size. Because of the complexity of the fracture process in notched laminates, most strength models are semiempirical. In this chapter, some of the more commonly accepted and computationally simple strength models (i.e., the point and average stress criteria developed by Whitney and Nuismer in 1974) are discussed. In addition, a modification of the point stress criterion (PSC), proposed by Pipes et al. (1979), is described.

Reasons for the substantial tensile and compressive strength reductions of composites because of holes and notches are the brittleness of the material and the large stress concentration factors brought about by the anisotropy of the material. These strength reductions are not necessarily the same for tensile and compressive loading because the failure modes are different.

As discussed in Chapter 2, the stress concentration factor for a plate containing a circular hole of radius R (Figure 16.1) is

$$K = \frac{\sigma_x(R,0)}{\bar{\sigma}_x} \tag{16.1}$$

where the x-axis is oriented along the loading direction, and the y-axis is oriented transverse to the loading direction. R is the hole radius, and $\bar{\sigma}_x$ is the average normal stress applied at the horizontal boundaries of the plate (Figure 16.1). For an infinite plate (i.e., where $L,w \to \infty$), Lekhnitskii (1968) derived the following expression (see also Karlak 1977):

$$K_\infty = 1 + \sqrt{2\left(\sqrt{E_x/E_y} - v_{xy} + E_x/(2G_{xy})\right)} \tag{16.2}$$

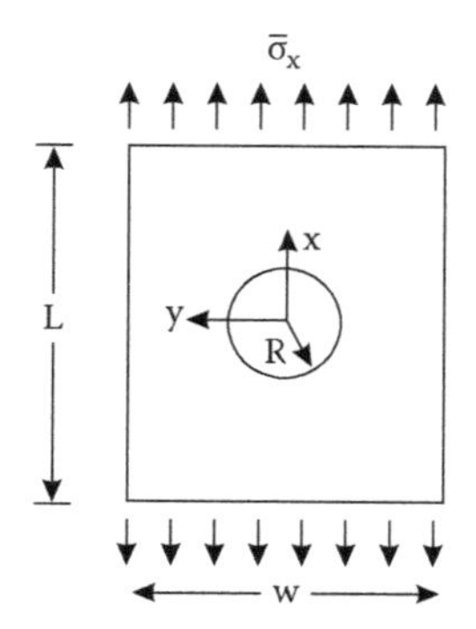

FIGURE 16.1
Finite-size plate containing a hole of diameter $D = 2R$ subject to uniaxial tension.

where E_x, E_y, ν_{xy}, and G_{xy} are the effective engineering elastic constants of the laminated plate. Note that the x-axis is oriented along the loading direction, and the y-axis is oriented transverse to the loading direction.

It is observed from Equation (16.2) that the stress concentration factor for an infinite plate is independent of hole radius R. For an ideally brittle infinite plate, the notched strength (defined as failure load/gross section area) would thus be

$$\sigma_N = \sigma_0/K_\infty \tag{16.3}$$

where σ_0 is the strength of the plate without a hole. Experiments, however, show that the strength reduction of composite plates containing large holes is much greater than that observed for small holes (Whitney and Nuismer 1974, Pipes et al. 1979). Such a difference for large plates cannot be explained by a net area reduction. Consequently, there must be factors other than the stress concentration factor controlling the notched strength. Of course, it is well known that the state of stress at the free edge within a multiaxial laminate is complicated and is not well represented by the effective modulus approach taken in this work. For this and reasons likely due to damage accumulation prior to propagation, the approach discussed next must be considered a first-order approach to understanding these phenomena.

Consideration of the normal stress distribution over the ligament ($x = 0$) for the homogeneous plate adjacent to the hole reveals some interesting features. The approximate stress distribution in an infinite plate containing a circular hole according to Konish and Whitney (1975) is

$$\sigma_x(0,y) = \frac{\sigma_x(\infty)}{2}[2 + \xi^2 + 3\xi^4 - (K_\infty - 3)(5\xi^6 - 7\xi^8)] \tag{16.4}$$

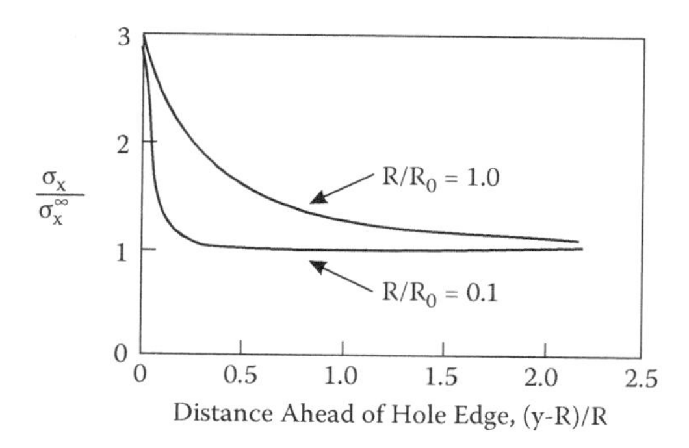

FIGURE 16.2
Normal stress distributions ahead of the hole edge for quasiisotropic plates containing holes of two sizes.

where $\xi = R/y$, and $\sigma_x(\infty)$ is the far-field applied normal stress. Figure 16.2 shows distributions of stress σ_x across a ligament for quasiisotropic plates containing holes of two sizes ($R/R_0 = 0.1$ and 1.0), where R_0 is a reference radius. It is observed that the volume of material subject to a high stress is much larger for the plate with a larger hole, thus increasing the opportunity for damage accumulation and stress redistribution. This may explain the reduced notched strength with increasing hole size.

16.2 Point and Average Stress Criteria

The point and average stress criteria (Whitney and Nuismer 1974) incorporate the hole size effect in two computationally simple fracture criteria.

For the point stress criterion (PSC), the laminate is assumed to fail in tension when the stress, σ_x, at a distance, d_0, ahead of the hole edge ($x = 0$, $y = R + d_0$) reaches the unnotched laminate strength, σ_0.

$$\text{PSC: } \sigma_x(0, R + d_0) = \sigma_0 \tag{16.5a}$$

For the average stress criterion (ASC), the laminate is assumed to fail in tension when the stress, σ_x, averaged over a region spanning from the hole edge ($y = R$) to $y = R + \sigma_0$, reaches the unnotched laminate strength, σ_0.

$$\text{ASC: } \frac{1}{a_0} \int_R^{R+a_0} \sigma_x(0, y)\,dy = \sigma_0 \tag{16.5b}$$

16.2.1 Point Stress Criterion

Combination of the PSC [Equation (16.5a)] and the expression for the stress distribution [Equation (16.4)] yields

$$\frac{\sigma_N}{\sigma_0} = \frac{2}{2+\lambda^2+3\lambda^4-(K_\infty-3)(5\lambda^6-7\lambda^8)} \tag{16.6}$$

where

$$\lambda = \frac{R}{R+d_0} \tag{16.7}$$

Note that for very large holes, d_0 is small compared with R, and Equation (16.6) gives

$$\frac{\sigma_N}{\sigma_0} = 1/K_\infty \tag{16.8}$$

Consequently, the notched strength ratio for a plate with a large hole is given by the inverse of the stress concentration factor. Furthermore, a notch-insensitive laminate is characterized by a large d_0 in comparison to R. For that case, $\lambda \approx 0$ in Equation (16.6), and $\sigma_N/\sigma_0 \approx 1.0$.

The PSC thus contains two parameters (d_0, σ_0) that have to be determined by experiment. Having established d_0 and σ_0, the PSC allows for strength predictions of laminates containing holes of arbitrary size. Figure 16.3 shows σ_N/σ_0 plotted versus hole size for a unidirectional $[0]_n$ boron/aluminum composite (Karlak 1977). Reasonable agreement with experimental data is observed.

16.2.2 Average Stress Criterion

Substitution of the stress distribution [Equation (16.4)] into the ASC [Equation (16.5b)] yields, after integration, the following expression for the notched laminate strength:

$$\frac{\sigma_N}{\sigma_0} = \frac{2}{(1+\delta)(2+\delta^2+(K_\infty-3)\delta^6)} \tag{16.9}$$

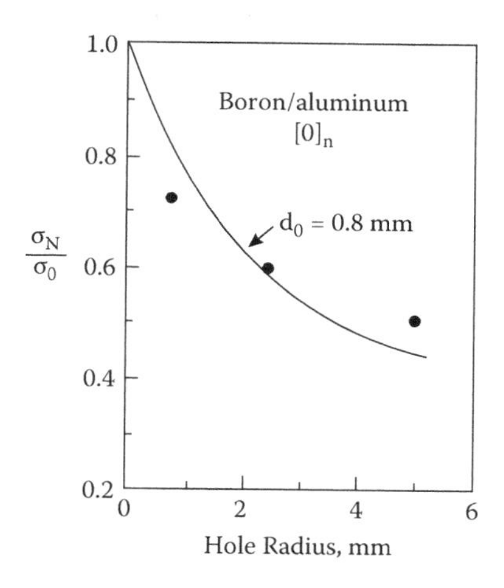

FIGURE 16.3
Experimental data on notched strength of a boron–aluminum composite and predictions based on the point stress criterion. (From R.F. Karlak, *Proceedings of a Conference on Failure Models in Composites (III)*, American Society for Metals, Chicago, 1977. With permission.)

with

$$\delta = \frac{R}{R + a_0} \tag{16.10}$$

Figure 16.4 shows experimental strength data for a $[0_2/\pm 45]_s$ carbon–epoxy laminate (Daniel 1982). Experimental results are in good agreement with the ASC with $a_0 = 5$ mm.

16.2.3 Modification of PSC

To improve the accuracy of notched strength predictions using the PSC, Pipes et al. 1979, following Karlak 1977, let the characteristic distance d_0 [Equation (16.6)] become a power function of hole radius:

$$d_0 = (R/R_0)^m/C \tag{16.11}$$

where m is an exponential parameter, R_0 is a reference radius, and C is the notch sensitivity factor. In essence, this model adds one more parameter (the exponent m) to the PSC. The reference radius may arbitrarily be chosen as $R_0 =$ 1 mm. The parameter λ [Equation (16.7)] then becomes

$$\lambda = 1/(1 + R^{m-1}C^{-1}) \tag{16.12}$$

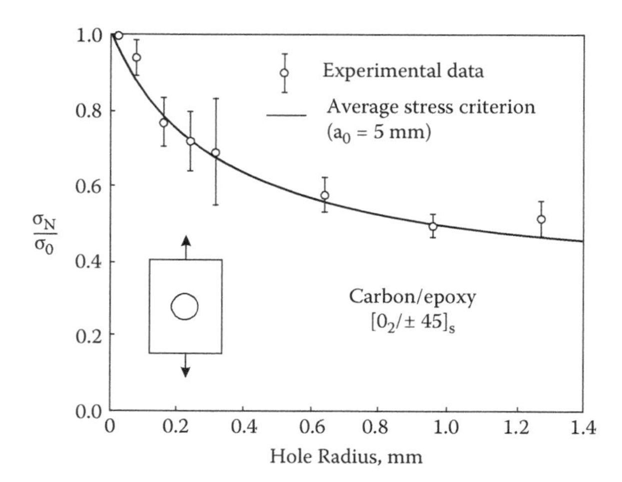

FIGURE 16.4
Notched strength data and predictions based on the average stress criterion for a notched $[0_2/\pm45]_s$ carbon/epoxy laminate. (From Konish, H.J., and Whitney, J.M., 1975. *J. Compos. Mater.*, 9, 157–166.)

Figures 16.5 and 16.6 display the influences on notched strength σ_N/σ_0 of the parameters m and C. Figure 16.5 shows that the exponential parameter affects the slope of the notch sensitivity curve, while Figure 16.6 shows that the parameter C shifts the curves along the log R axis without affecting the shape of the curves. The admissible ranges for the parameters are $0 \leq m < 1$

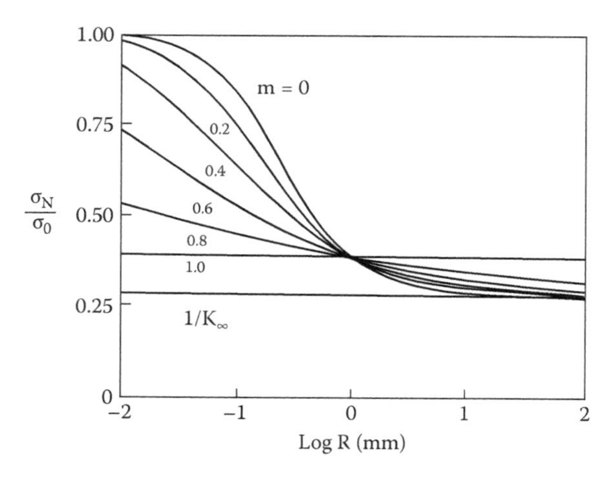

FIGURE 16.5
Influence of exponent m on notched strength, $C = 10.0$ mm^{-1}. (From Pipes, R.B., Wetherhold, R.C., and Gillespie, J.W., Jr., 1979. Notched strength of composite materials, *J. Compos. Mater.*, 13, 148–160.).

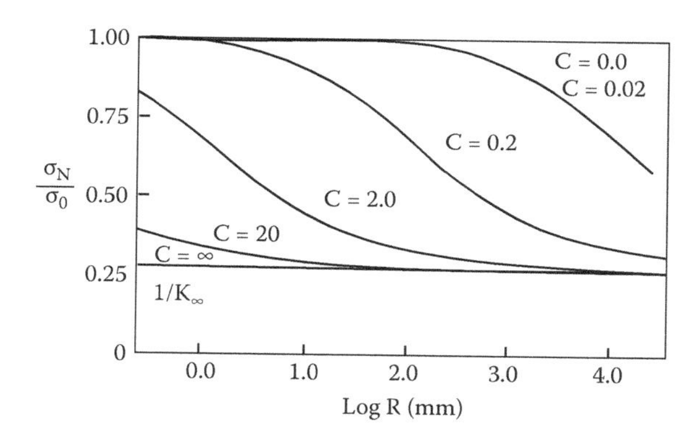

FIGURE 16.6
Influence of notch sensitivity factor C (mm^{-1}) on notched strength, $m = 0.5$. (From Pipes, R.B., Wetherhold, R.C., and Gillespie, J.W., Jr., 1979. Notched strength of composite materials, *J. Compos. Mater.*, 13, 148–160.).

and $C \geq 0$. A notch-insensitive laminate is characterized by a large d_0 in comparison to R. This corresponds to $m \to 1$ and $C \to 0$.

Figure 16.7 shows excellent agreement between predicted and measured notched strength versus hole radius for two quasi-isotropic carbon–epoxy laminates with $[\pm 45/0/90]_s$ and $[90/0/\pm 45]_s$ layups and the magnitudes of the fitting parameters m and C determined as outlined in Section 16.4.

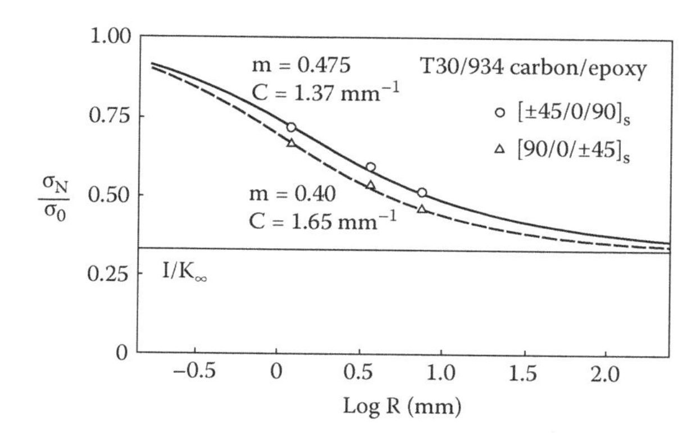

FIGURE 16.7
Notched strength data for $[\pm 45/0/90]_s$ and $[90/0/\pm 45]_s$ carbon–epoxy laminates. (From Pipes, R.B., Wetherhold, R.C., and Gillespie, J.W., Jr., 1979. Notched strength of composite materials, *J. Compos. Mater.*, 13, 148–160.)

16.3 Test Specimen Preparation

Although any laminate configuration can be used, most commonly $[0/\pm45/90]_{ns}$ (quasi-isotropic) laminates have been studied. Laminates with higher or lower percentages of 0° plies are tested when of specific interest to the intended design application (i.e., greater or lesser axial stiffness laminates). However, it should be noted that notch sensitivity is a function of laminate configuration, and these tests are carried out for the purpose of determining relative notch sensitivity of the given laminate of choice.

Often, the same specimen configuration is used for both tensile and compressive open-hole tests. One commonly used specimen size is 305 mm in length and 38 mm in width. An open-hole compression test method utilizes a specimen that is only 75 mm in length and 25 mm in width, to be discussed. If the test facilities permit, it is strongly recommended that wide specimens be used to accommodate a large range of hole sizes and better approximate an infinitely wide specimen. Daniel (1982), for example, used 127-mm-wide laminates and hole diameters ranging from 6.4 to 25.4 mm. Specimen thickness is not critical and is somewhat dependent on the specific laminate configuration to be tested. A specimen thickness on the order of 2.5 to 5 mm is commonly used. The diameter of the hole in the specimen, which is to be centered at the midlength of the specimen, can also be arbitrarily selected. However, to be discussed, the ratio of specimen width to hole diameter influences the magnitude of the stress concentration induced.

Unless a laminate with a high percentage of 0° plies is to be tested, tabs are not usually necessary. If aggressively serrated tensile wedge grips are used, it may be necessary to protect the open-hole tension specimen surfaces with one or more layers of emery cloth, an (unbonded) layer of plastic sheet material (approximately 1 to 2 mm thick), or similar padding material. The open-hole compression test methods typically involve the use of some type of special fixture to prevent specimen buckling, as will be discussed further. These fixtures are usually designed for use with an untabbed specimen.

Measure the cross-sectional dimensions (average six measurements) and check for parallelism of the edges and of the end-tab surfaces if used (see Chapter 8 for tensile specimen tolerances). If a series of tests is to be conducted for various hole sizes, divide the specimens into groups by hole size. Note also that specimens without holes must be tested to determine the unnotched strength σ_0. At least three specimens should be assigned to each group, although five specimens is more common. At least three hole diameters should be investigated, for example, D = 3, 5, and 7 mm. Machine the holes as specified in Section 4.2.

16.4 Tensile Test Procedure and Data Reduction

The specimens should be mounted and tested in a properly aligned and calibrated testing machine with mechanical wedge action or hydraulic grips. Set the crosshead rate at about 0.5 to 1 mm/min. Record the load-versus-crosshead displacement to detect the ultimate load and any anomalous load-displacement behavior. If a strain gage is used, place it midway between the hole and the end tab. Make sure eyes are protected in the test area. Load all specimens to failure. Figure 16.8 shows a carbon–epoxy open-hole tension specimen after testing.

Notched strength σ_N is calculated based on the gross cross-sectional area ($A = wh$). A typical set of unnotched and notched strength data for $[0/45/90]_s$ carbon–epoxy coupons is given in Table 16.1. Because the analysis discussed here is restricted to plates with an infinite width-to-hole diameter ratio, a comparison between experimental data and the notch strength model requires correction for the finite width of the specimen. A common way to correct the data is to multiply the experimental notched strength with a correction factor K/K_∞, where K is the stress concentration factor for an orthotropic plate of finite width; that is,

$$\sigma_N(\infty) = \sigma_N(w)\frac{K}{K_\infty} \tag{16.13}$$

FIGURE 16.8
Carbon–epoxy open-hole tensile specimen tested to failure.

TABLE 16.1

Unnotched and Notched Strength Data for 38 mm wide $[0/\pm45/90]_s$ Carbon–Epoxy Coupons

Notch Radius (mm)	Strength (MPa)
0	607 (σ_0)
1.6	437
2.5	376
3.3	348

where $\sigma_N(w)$ is the experimental strength for a plate of width w, and $\sigma_N(\infty)$ is the notched strength for an infinite plate. A closed-form expression for K, however, does not exist, and K has to be determined numerically (Ogonowski 1980, Bathe 1982). Table 16.2 gives finite width correction factors as a function of width-to-hole diameter ratio (w/D) for various carbon–epoxy layups (Gillespie and Carlsson 1988). Note that $K/K_\infty > 1$, which means that finite-width specimens exhibit larger stress concentration than infinitely wide specimens ($w/D \geq 8$). Consequently, it is expected that a hole in a finite-width specimen will be more detrimental in terms of strength than a hole in an infinitely wide specimen. A common approximation, which is also reasonably accurate for composite laminates with $w/D > 4$ (Gillespie and Carlsson 1988) (see also Table 16.2) is to use an isotropic expression (Peterson 1974, Nuismer and Whitney 1975) for K/K_∞:

$$\frac{K}{K_\infty} = \frac{2+(1-(D/w))^3}{3(1-(D/w))} \tag{16.14}$$

To allow comparison of the data with the PSC [Equation (16.6)], the experimental data (Table 16.1) are corrected for the finite width, using Equation (16.14), i.e., the experimental ratio σ_N/σ_0 is multiplied by K/K_∞. Then, to solve

TABLE 16.2

Finite Width Correction Factor (K/K_∞) for Carbon–Epoxy (AS4/3501-6) Layups $E_1 = 125$ GPa, $E_2 = 9.9$ GPa, $\nu_{12} = 0.28$, and $G_{12} = 5.5$ GPa

Layup	K_∞^a	$w/D = 2$	3	4	6	8	10
$[0/\pm45/90]_s$	3.00	1.4340	1.1495	1.0736	1.0260	1.0107	1.0037
$[0_2/\pm45]_s$	3.48	1.3725	1.1291	1.0632	1.0216	1.0093	1.0031
$[0_4/\pm45]_s$	4.07	1.3226	1.1109	1.0577	1.0172	1.0095	1.0041
$[0_6/\pm45]_s$	4.44	1.2992	1.1006	1.0472	1.0152	1.0102	1.0051
$[\pm45]_s$	2.06	1.6425	1.2379	1.1215	1.0442	1.0180	1.0062
Equation (16.14)		1.417	1.148	1.076	1.031	1.017	1.011

[a] Determined from Equation (16.2).

Source: Gillespie and Carlsson, 1988. *Compos. Sci. Technol.*, 32, 15–30.

TABLE 16.3

Corrected Notched Strength Data (Table 16.1) and Values of λ Determined Using the Newton-Raphson Method

R (mm)	σ_N/σ_0^a	λ
1.6	0.73	0.5998
2.5	0.64	0.6867
3.3	0.62	0.7052

[a] Corrected using Equation (16.14).

for the parameter λ in Equation (16.6) an iterative method must be applied such as Muller's method (Muller 1956) or Newton-Raphson's method (Hornbeek 1975, Conte and de Boor 1965). From the definition of λ [Equation (16.12)], it is observed that only the root, $0 < \lambda < 1$, is required.

For illustrative purposes, the notched strength data listed in Table 16.1 were corrected according to Equation (16.14), and the corresponding λ values were determined with the Newton-Raphson method and are listed in Table 16.3.

To obtain the parameters m and C, Equation (16.12) may be written as

$$-\log(1/\lambda - 1) = \log C + (1 - m)\log R \qquad (16.15)$$

By plotting $-\log(1/\lambda - 1)$ versus $\log R$, the slope and the intercept at $\log R = 0$ can be obtained by the least-squares method. The slope is equal to $1 - m$, and the intercept is equal to $\log C$. Figure 16.9 shows $-\log(1/\lambda - 1)$ plotted versus $\log R$ for the data of Table 16.3.

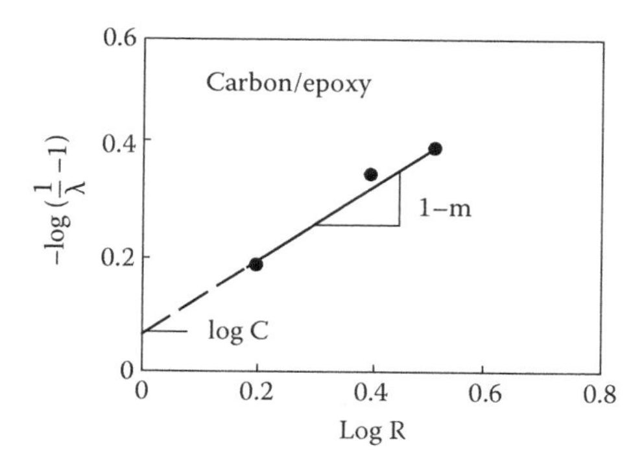

FIGURE 16.9
Determination of the parameters m and C for $[0/\pm45/90]_s$ carbon–epoxy laminate specimens, $m = 0.36$ and $C = 1.16\ \text{mm}^{-1}$.

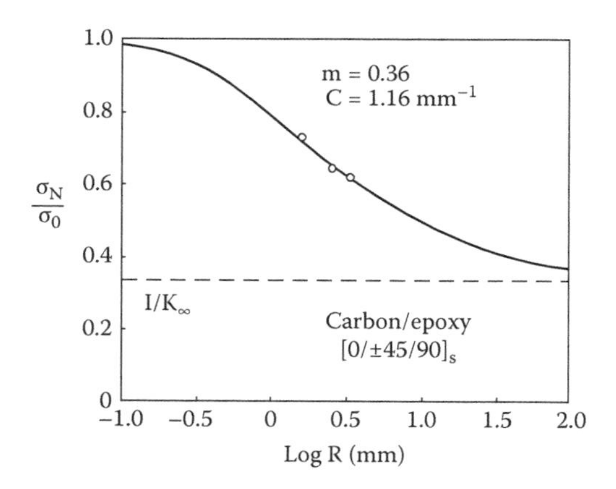

FIGURE 16.10
Notched strength of $[0/\pm45/90]_s$ carbon–epoxy specimens.

As an illustration of the goodness of the fit for the parameters m and C, the predicted curve of σ_N/σ_0 is plotted versus log R in Figure 16.10 along with the (corrected) experimental data. Within the limited range of experimental data, excellent agreement is observed. Once the parameters m and C are established, it is possible to predict the notched strength for any hole size and coupon width (within reasonable limits) using this methodology.

16.5 Standardized Open-Hole Tension Test Method

Rather than utilize the PSC and ASC developed by Whitney and Nuismer (1974), as discussed in the previous sections of this chapter, it has become common simply to test a specimen of one specific geometry and dimensions and then use the measured strength as basis for comparison. That is, if a common specimen configuration is used, open-hole tensile strength results for different materials can be directly compared. The specimen most commonly used is 305 mm long and 38 mm wide and containing a centrally located 6.4-mm-diameter hole at the midlength of the specimen. This configuration was developed by the Boeing Company as Boeing Specification Support Standard BSS 7260 (1988) and has now been standardized by both SACMA (Suppliers of Advanced Composite Materials Association) as Recommended Method SRM 5R-94 (1994) and as ASTM (American Society for Testing and Materials) D5766 (2011).

The laminate configuration and specimen thickness are somewhat arbitrary, but of course the results obtained will be dependent on these parameters as well as the type of material tested. Although any laminate configuration can be used, most commonly a $[0/\pm45/90]_{ns}$ (quasi-isotropic) laminate is selected. Laminates with higher or lower percentages of 0° plies are tested when of specific interest to the intended design application.

Lower-strength laminate orientations such as the quasi-isotropic layup can normally be tested without specimen tabs. When laminates with higher percentages of 0° plies are to be tested, tabs may be necessary. However, because a strength-reducing hole is present, it may be possible to successfully test an open-hole tension specimen without tabs.

The open-hole tensile strength is calculated from:

$$\sigma_N = P_{max}/A \tag{16.16}$$

fracture load, and A is the cross-sectional area of the specimen ($A = wh$), where w and h are the specimen width and thickness, respectively.

16.6 Standardized Open-Hole Compression Test Methods

There are two standardized open-hole compression test methods in use; Boeing Specification Support Standard BSS 7260 (1988) and the Northrop Open-Hole Compression Test Method NAI-1504C (1988). The Boeing method was recently standardized as ASTM D6484-09 (2009).

16.6.1 Boeing Open-Hole Compression Test Method

The specimen is 305 mm long and 38 mm wide and contains a 6.4-mm-diameter hole centered at the midlength of the specimen. Again, the laminate configuration and specimen thickness are somewhat arbitrary, and the results obtained are dependent on these parameters as well as the material being tested.

A special fixture has been designed to load the specimen in compression while preventing gross (Euler) buckling, as shown in Figure 16.11. The 305-mm-long specimen is installed such that its ends are flush with the outer ends of the fixture halves. Thus, essentially the entire length of the specimen is supported against buckling, with only a small gap existing between the fixture halves themselves so that they do not come into contact when the compressive loading is applied. The standards originally specified that the ends of the fixture be clamped in hydraulic grips, the load being introduced into the specimen by shear. Because of the assembly thickness (approximately 35 mm with a specimen installed) and width (76 mm) of the fixture, this requires large hydraulic wedge grips of

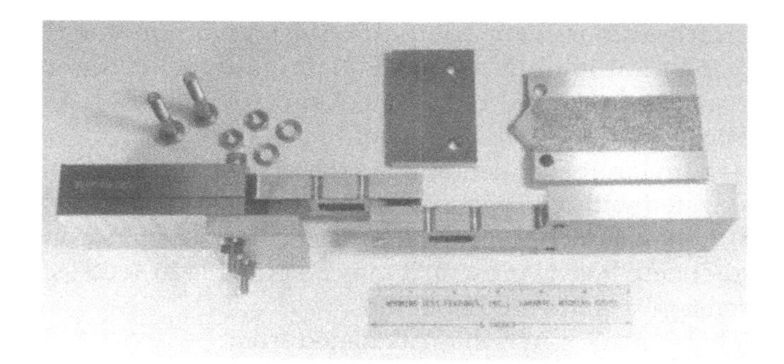

FIGURE 16.11
Boeing open-hole compression test fixture (ASTM D 6484). (Photograph courtesy of Wyoming Test Fixtures, Inc.)

250-kN or greater. There is no need for grips of this loading capacity in many testing laboratories. These large grips are relatively expensive and massive and thus difficult to handle during installation and removal from the testing machine. For example, a 250-kN hydraulic grip weighs about 125 kg.

Because of this, the fixture of Figure 16.11 is often loaded directly on its ends between compression platens. The faces of the fixture in contact with the specimen are thermal sprayed with tungsten carbide particles, as discussed in Chapters 8 and 9. Thus, if the fixture is clamped tightly to the specimen, the loading will become a combined end loading and shear loading, just as for the Wyoming combined loading compression test fixture described in Chapter 9, and end crushing of the specimen should be avoided. This success is particularly likely because, as noted, the strength of an open-hole compression specimen is typically not very high. The option of end loading has now been added to ASTM D6484 (2009). However, if direct end loading is to be used, care must be taken to secure the fixture so that it does not slip out between the platens during testing which may cause injury to nearby personnel. For example, restraint boxes attached directly to the testing machine can be used instead of platens (Coguill and Adams 2000); the ends of the fixture slip into these restraint boxes as shown in Figure 16.12.

Although strain gages can be used, as specified in Boeing Specification Support Standard BSS 7260 (1988) and SACMA Recommended Method SRM 3R-94 (1994), the ASTM standard eliminated their use as not providing useful information. As for the open-hole tension testing, the cross-sectional compressive strength is calculated using Equation (16.16).

16.6.2 Northrop Open-Hole Compression Test Method

The Northrop open-hole compression test method, Northrop Specification NAI-1504C (1988), is shown in Figure 16.13. The specimen contains a

FIGURE 16.12
Restraint boxes for use with the Boeing open-hole compression test fixture when applying loading directly through the specimen ends (lower half of fixture shown in a box; both boxes to be attached directly to the testing machine). (Photograph courtesy of Wyoming Test Fixtures, Inc.)

6.4-mm-diameter hole centered at its midlength. The Northrop open-hole compression test specimen is only 76 mm long and 25 mm wide, that is, only one-fourth as long and two-thirds as wide as the Boeing specimen. However, as discussed at the beginning of this chapter, it may be desirable to use a wider specimen to reduce edge effects. Thus, a modification of the Northrop fixture to accommodate 38-mm-wide specimens has also been proposed (Coguill and Adams 2000). It has been shown to produce compressive strengths comparable to the standard Boeing test while reducing the volume of specimen material required to one-fourth—still a significant savings.

FIGURE 16.13
Northrop open-hole compression test fixture (Northrop Specification NAI-1504). (Photograph courtesy of Wyoming Test Fixtures, Inc.)

The Northrop test fixture shown in Figure 16.13 is designed such that the untabbed specimen is installed flush with the ends of the fixture and is directly end loaded. Although a moderate clamping force is exerted on the specimen when the fixture screws are tightened to the recommended 6.8 N·m torque, the faces of the fixture are smooth and not intended to transfer a shear loading to the specimen. That is, the force applied to the specimen is essentially all end loading. Note that the cover plates, which serve to hold the fixture clamps in alignment, are bolted together with wing nuts, which are to be tightened only "finger tight," the Northrop standard specifying a torque of 1.1 N·m. This very light clamping reduces the possibility of the cover plates creating a redundant load path. That is, due to friction, a portion of the applied end load could pass through the plates rather than all of the load passing into the specimen. Actually, the cover plates could be eliminated, the specimen itself being stiff enough to keep the assembly in alignment.

As for the Boeing fixture, the specimen is supported over almost its full length, although a small gap is maintained between the fixture halves to prevent them from coming into contact with each other when the compressive loading is applied. Because the fixture is compact, with a base comparable in dimensions to its height, there is little danger of it being ejected from the test machine when loaded, and thus it need not be constrained. The fixture has recesses machined into it to permit the use of strain gages, but as for the Boeing fixture, strain gages are not normally used. Compressive strength is calculated using Equation (16.16).

16.6.3 Comparison of the Boeing and Northrop Open-Hole Compression Test Methods

The performance of the Boeing and Northrop open-hole compression test methods and corresponding fixtures have been compared (Coguill and Adams 2000). In addition to 25-mm-wide specimens tested in the standard Northrop fixture, 38-mm-wide specimens were tested in a special fixture fabricated with this increased width capability. Although there were some minor differences in the results, there were no distinct trends to report. This included whether the Boeing specimen was shear loaded or end loaded and whether the Northrop specimen was 25 mm or 38 mm wide.

16.6.4 Filled-Hole Tension and Compression Test Methods

The discussion in this chapter addresses the influence of open (unfilled) holes in composite laminates. An unfilled hole will deform under loading. However, often a hole is created in a laminate to accommodate a fastener of some type (e.g., a bolt, pin, or rivet). The presence of a close-fitting fastener will restrict the deformation of the (filled) hole, thus changing the state of stress in the laminate and possibly the failure stress. ASTM D6742 (2012) describes the tensile and compressive testing of laminates with filled holes. The test fixtures and procedures are similar to those outlined in the previous sections for unfilled holes. However, the laminate failure modes may change because of the fastener interference. The differences in failure mode are discussed by Sawicki and Minguet (1999).

17

Characterization of Delamination Failure

17.1 Introduction

The interlaminar mode of fracture (delamination) has aroused considerable attention since the pioneering work by Pipes and Pagano (1970). With the introduction of laminated composites into structures subjected to service loads, it has become apparent that the delamination failure mode has the potential for being the major life-limiting failure process. These delaminations are typically induced in composite laminates during service. However, delaminations may also be introduced during processing of the layup (e.g., as a result of contamination of the prepreg), leading to locally poor ply adhesion, or they may form locally in regions of high void content. Delamination may also be introduced during postfabrication handling and hard object impact loading of the structure.

It is recognized that a delamination represents a crack-like discontinuity between the plies, and that it may propagate during application of mechanical or thermal loads or both. It thus seems appropriate to approach the delamination using fracture mechanics (Section 2.7), which indeed has evolved as a fruitful approach for material selection and assessment of structural integrity. Fracture mechanics of delaminations is commonly based on the strain energy release rate, and fracture toughness is expressed as the work of fracture. Consequently, many new fracture tests have been devised for measuring the static interlaminar fracture toughness, as well as the crack propagation rate during cyclic loading. Most such tests and standard test procedures are limited to unidirectional $[0]_n$ laminates in which a delamination propagates between the plies along the fiber direction. In laminates with multidirectional plies, the crack may have a tendency to branch through the neighboring plies, invalidating the coplanar assumption in fracture analysis (Nicholls and Galagher 1983, Robinson and Song 1992, Shi et al. 1993). Composites with tough resin films (called interleaves) between the plies may experience peculiar delamination resistance behavior depending on crack path selection (i.e., if the crack propagates cohesively in the tough interlayer or adhesively at the film–composite interface) (Ozdil and Carlsson 1992). In woven fabric composites, a delamination crack will interact with matrix regions and interlacing yarns during its propagation and as a result will experience varying growth resistance (Alif et al. 1997). Composites with through-thickness reinforcement may experience large extended regions where the reinforcements bridge the crack (bridging zones), which invalidates data-reduction schemes based on linear elastic fracture mechanics (Cox et al. 1997). Although fiber bridging is common in unidirectional (all 0° plies) composites, characterization of the delamination resistance of such composites tends to be associated with fewer complications. Consequently, we here limit attention to unidirectional composites.

Fracture mechanics analysis, preparation of test specimens, testing, and data reduction are described for some contemporary interlaminar fracture test specimens, namely, the double-cantilever beam (DCB) specimen (mode I), end-notched flexure (ENF) specimen (mode II), the mixed-mode bending (MMB) specimen, and the edge crack torsion (ECT) specimen (mode III). The various fracture modes are defined in Figure 2.9.

17.2 Double-Cantilever Beam Test (Mode I)

The DCB specimen for mode I fracture testing and the test principle are shown in Figure 17.1. This is the standard specimen for ASTM D5528 (2007). The purpose of the test is to determine the opening mode interlaminar fracture toughness G_{IC} of continuous fiber composite materials with a polymer matrix. First developed in a tapered form by Bascom et al. (1980), the straight-sided geometry proposed by Wilkins et al. (1982), shown in Figure 17.1, has become standard. Although data reduction does not rely on the classical

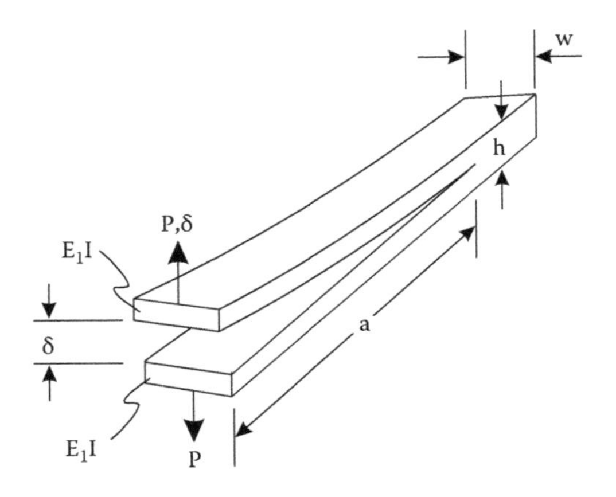

FIGURE 17.1
DCB specimen geometry.

beam theory approach used by Wilkins et al., the simplicity of this theory makes it easy to examine some features of the DCB specimen.

From classical beam theory, the load-point compliance, $C = \delta/P$, of the DCB specimen becomes

$$C = \frac{2a^3}{3E_1 I} \tag{17.1}$$

where P is the load applied, δ is the crack opening, a is the crack length, and $E_1 I$ is the flexural rigidity of each leg of the specimen, with E_1 being the modulus of the composite in the fiber direction and I the moment of inertia (Figure 17.1). The strain energy release rate, $G = G_I$, is obtained from Equation (2.59):

$$G = \frac{P^2}{2b}\frac{dC}{da} \tag{17.2}$$

in which b is the specimen width. Equations (17.1) and (17.2) give

$$G = \frac{P^2 a^2}{bE_1 I} \tag{17.3}$$

If G_{IC} is a true material constant, stable crack growth requires (see Section 2.7)

$$dG/da \leq 0 \tag{17.4}$$

For the DCB specimen under fixed-load conditions, dG/da is obtained from Equation (17.3) as

$$\frac{dG}{da} = \frac{2P^2 a}{bE_1 I} \tag{17.5}$$

This quantity is always positive; thus, the crack growth is unstable under load-controlled testing conditions.

For fixed-grip conditions, dG/da may be obtained by substitution of $P = \delta/C$ in Equation (17.2) and differentiation:

$$\frac{dG}{da} = \frac{-9\delta^2 E_1 I}{ba^3} \tag{17.6}$$

This quantity is always negative; thus, the crack growth should be stable. Experimentally, most testing is performed under fixed-grip conditions (displacement control), which should promote stable crack growth.

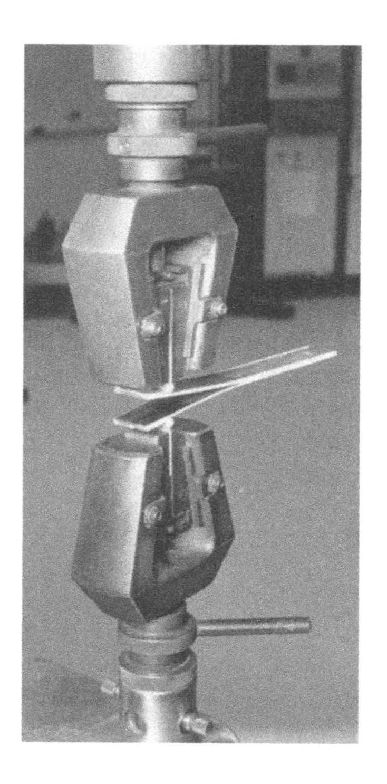

FIGURE 17.2
DCB test setup.

17.2.1 DCB Specimen Preparation and Test Procedure

The DCB specimen should be at least 125 mm long and between 20 and 25 mm wide. The number of plies, dimensions, and preparation of the panel are outlined in Appendix B. An even number of plies should be employed to achieve a thickness (h in Figure 17.1) between 3 and 5 mm. Variations in thickness should be less than 0.1 mm. Tough composites may require thicker specimens to avoid large displacements and nonlinear response. Figures 17.2 and 17.3 show the DCB specimen with hinge-loading tabs prepared and bonded as described in Chapter 4. The precrack is defined by inserting a thin film (<13 μm) at the midplane of the panel (see Appendix B). Crack length a is defined as the distance from the line of load application to the crack tip (Figure 17.3). The length of the film insert should be adjusted to obtain a pre-crack length a_0 of approximately 50 mm (see Appendix B).

Measure thickness and width of the specimen close to each end and at the center and calculate averages. Paint the specimen edges with a thin, white, brittle coating, such as typewriter correction fluid. To aid in recording crack length, mark the first 5 mm from the insert with thin vertical lines every 1 mm. Mark the remaining 20 mm every 5 mm.

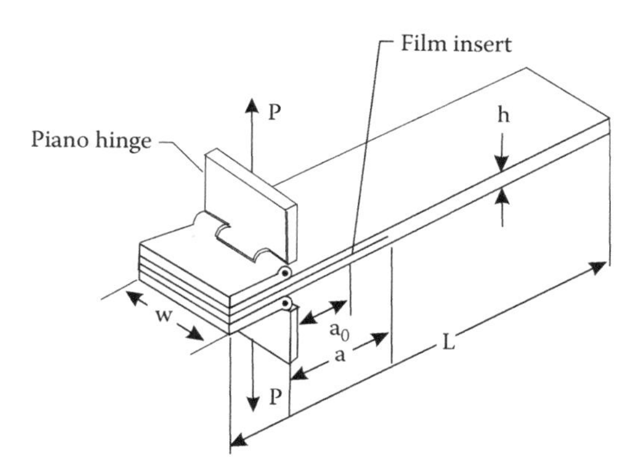

FIGURE 17.3
Hinge-loading tab arrangement for the DCB specimen.

The specimen should be mounted in the grips of a properly calibrated test machine with a sufficiently sensitive load cell. A traveling optical microscope with approximately ×10 magnification and a crosshair can be positioned on one side of the specimen to enable monitoring of the delamination crack tip and its extension during the fracture test within ±0.5 mm. Locate the crosshair at the delamination front without applying load to the specimen to obtain a record of the precrack length a_0 (Figure 17.3). Load the specimen at a crosshead rate at 0.5 mm/min and plot load versus crosshead displacement for real-time visual inspection of the load-displacement response. Displacement of the loaded ends (δ in Figure 17.1) can be taken as the crosshead travel, provided the machine and load cell are stiff enough not to deform more than 2% of the total opening displacement.

Observe the delamination front as the specimen is loaded. When the delamination begins to grow from the end of the insert, mark this incident as a_0 on the chart recording as indicated in Figure 17.4. Continue to observe the front of the growing crack and mark the chart accordingly. For the first 5 mm of crack growth, each 1 mm increment should be marked. After 5 mm of crack extension, the crosshead rate may be increased. Mark every 5 mm of crack length on the graph. Observe the opposite edge to monitor deviations from uniform crack extension across the beam width. The difference in crack length between the two edges should be less than 2 mm for a valid test. When the delamination has extended about 25 mm, the specimen may be unloaded while the unloading load-displacement response (see Figure 17.4) is recorded. A common occurrence in testing unidirectional DCB specimens is fiber bridging, which refers to debonded fibers bridging the fracture surfaces, as illustrated in Figure 17.5. The fiber bridging elevates the fracture resistance as a result of the closure tractions that develop in the fibers that bridge the crack faces behind the crack tip, and the energy consumed as the bridged fibers debond from the matrix (Suo et al. 1992).

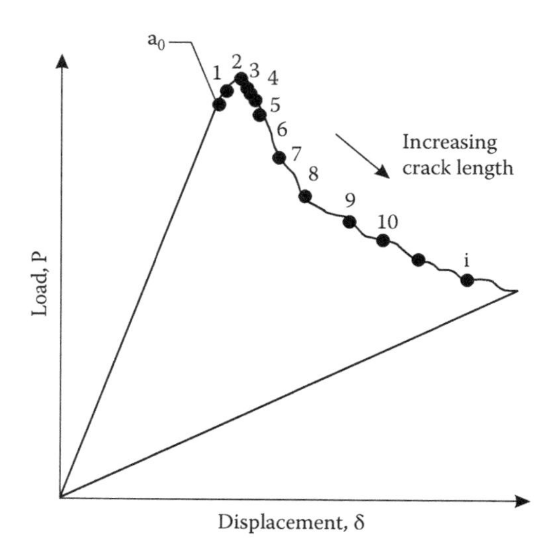

FIGURE 17.4
Schematic load-displacement record during crack growth for a DCB test.

It is common to display the fracture toughness measured at various crack lengths as a resistance curve (R-curve). As discussed by Suo et al. (1992), such R-curves do not represent true material behavior because they depend on specimen thickness. Fiber bridging is less likely to occur in multidirectional laminates used in composite structures because less opportunity exists for fiber wash (i.e., intermingling of wavy fibers between adjacent plies). Fiber bridging is thus likely to lead to nonconservative estimates of the actual delamination toughness. It is argued that the most meaningful, and also conservative, estimate of fracture toughness is the initiation toughness G_{IC} (init.) associated with the initial crack propagation from the Teflon insert because this value is not influenced by fiber bridging. Further discussion follows.

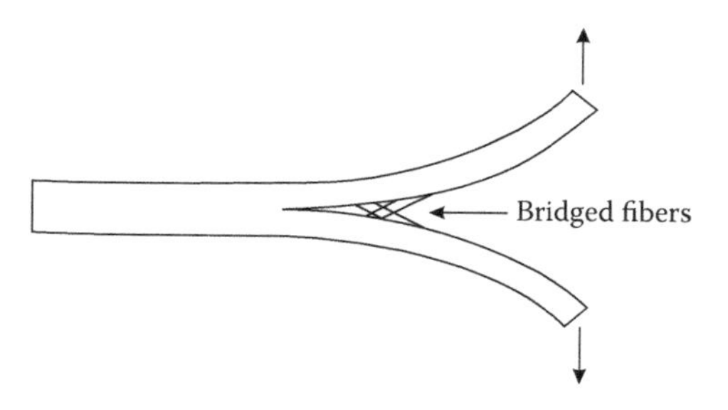

FIGURE 17.5
Fiber bridging in DCB testing.

17.2.2 DCB Data Reduction

Several data reduction methods for evaluating the mode I fracture toughness G_{IC} have been proposed (O'Brien and Martin 1993). A simple, yet accurate, method is the empirical compliance method suggested originally by Berry (1963); the beam compliance, $C = \delta/P$, is expressed as a power function of crack length:

$$C = \frac{a^n}{H} \tag{17.7}$$

where a is the crack length, and n and H are parameters determined experimentally. If classical beam theory and the assumption of fixed ends are valid, $n = 3$ and $H = 3E_1I/2$. In reality, the legs of the DCB specimen are elastically built into the uncracked portion of the specimen rather than being rigidly fixed. This will cause deviations from classical beam theory.

To establish the actual values of the empirical parameters in Equation (17.7), measured load and displacement data at each crack length are evaluated from the load-displacement graph (Figure 17.4), and the stiffness, that is, the inverse of the compliance $(1/C = P_c/\delta_c)$, is plotted versus crack length (a) in a double-logarithmic graph as shown in Figure 17.6. By fitting a straight line to the data, it is possible to establish the exponent n in Equation (17.7). Substitution of Equation (17.7) into (17.2) yields at fracture

$$G_{IC} = \frac{nP_c\delta_c}{2ba} \tag{17.8}$$

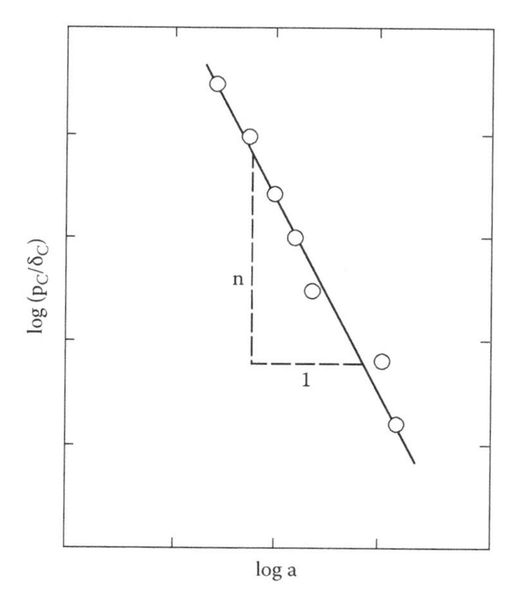

FIGURE 17.6
Log-log plot of DCB specimen stiffness versus crack length.

in which P_C and δ_C are the critical load and displacement associated with each crack length a.

Three toughness values corresponding to crack growth from the insert may be defined. G_{IC} (NL) refers to the critical load and displacement associated with the deviation from linear response (Figure 17.4). The second definition G_{IC} (vis.) refers to the visual observance of crack growth measured with the traveling microscope. The third definition G_{IC} (5%) uses the load and displacement at a 5% increase in compliance. G_{IC} (NL) is typically the most conservative estimate of the fracture toughness and is recommended as a measure of mode I delamination toughness. For subsequent crack growth, G_{IC} is calculated from Equation (17.8) using the recorded loads and crack lengths (Figure 17.4).

A crack growth resistance curve (R-curve) displaying G_{IC} versus crack extension can be constructed from the fracture toughness G_{IC} and crack length a data. Figure 17.7 shows an example of an R-curve for a carbon–polyetheretherketone (PEEK) composite. At the first loading increment, the delamination grows from the tip of the thin-film insert starter crack without any influence from fiber bridging. The corresponding three initiation fracture toughness values, G_{IC} (NL), G_{IC} (vis.), and G_{IC} (5%), are indicated in Figure 17.7. As the crack grows, the crack surfaces become more and more separated, and bridged fibers may fracture or become pulled out from the matrix, which causes the apparent fracture toughness to increase. With further crack extension, a steady-state toughness G_{IC} (prop.) is usually reached, corresponding to an equilibrium number of bridged fibers per unit crack area. As mentioned, the initial value associated with propagation of the crack from the film insert constitutes a well-defined measure of

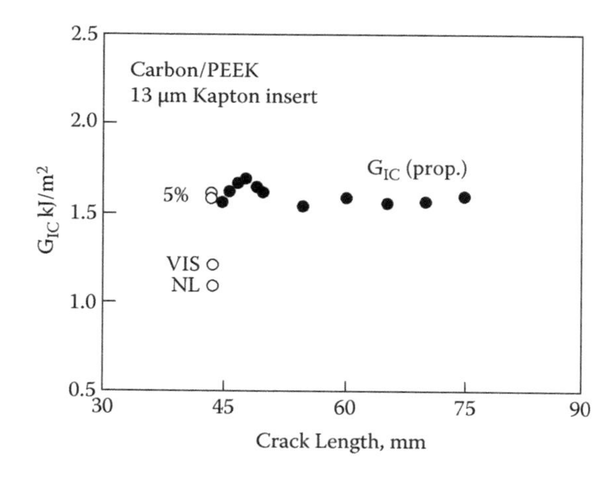

FIGURE 17.7
R-curve describing mode I interlaminar fracture resistance of carbon–PEEK with a 13-μm insert.

fracture toughness because it is unaffected by the subsequent fiber bridging that occurs with crack extension (Suo et al. 1992, O'Brien and Martin 1993).

17.3 End-Notched Flexure Test (Mode II)

The ENF specimen (Figure 17.8) was introduced as a pure mode II delamination specimen for testing of composites by Russell and Street (1982, 1985). The purpose of the ENF specimen is to determine the critical strain energy release rate in pure mode II loading of unidirectional composites. The ENF specimen produces shear loading at the crack tip without introducing excessive friction between the crack surfaces (Carlsson et al. 1986a, Gillespie et al. 1986). The ENF specimen is standardized in Europe, AECMA (Association Europeene de Constructeurs de Materiel Aerospatial; 1995), and Japan (Japan Industrial Standards [JIS] 1993) and has been studied extensively in the United States by the ASTM (American Society for Testing and Materials) D-30 Committee as a candidate for ASTM standardization. As will be discussed, however, several factors, such as the unstable crack growth under displacement control, has slowed acceptance of this specimen as a standard fracture test.

Assuming that classical beam theory is valid, an expression for the strain energy release rate G can be derived (Russell and Street 1982):

$$G = \frac{9P^2Ca^2}{2b(2L^3 + 3a^3)} \tag{17.9}$$

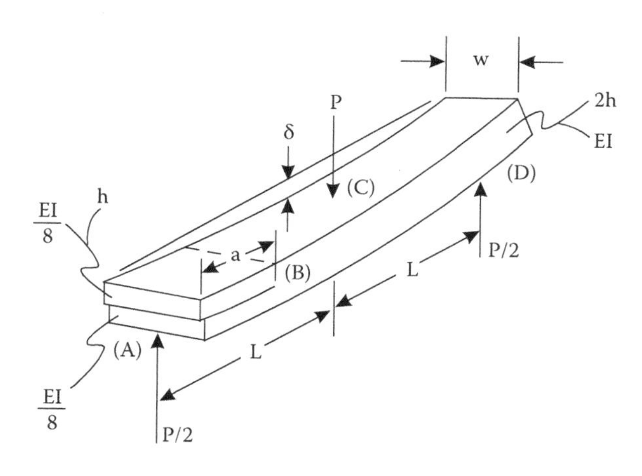

FIGURE 17.8
ENF specimen.

where P is the applied load, C is the compliance, a is the crack length, b is the specimen width, and L is the span between the central loading cylinders and the outer support cylinders (Figure 17.8). The specimen compliance as given by beam theory is

$$C = \frac{2L^3 + 3a^3}{8E_1 bh^3} \tag{17.10}$$

where E_1 is the flexural modulus, and h is one-half the total thickness of the beam (i.e., the thickness of each subbeam of the delaminated region).

The stability of crack growth may be judged from the sign of dG/da. For fixed-load conditions, Equations (17.9) and (17.10) give

$$\frac{dG}{da} = \frac{9aP^2}{8Eb^2h^3} \tag{17.11}$$

This quantity is positive; hence, the crack growth is unstable.

For fixed-grip conditions, Equations (17.9) and (17.10) give

$$\frac{dG}{da} = \frac{9\delta^2 a}{8E_1 b^2 h^3 C^2}\left[1 - \frac{9a^3}{2L^3 + 3a^3}\right] \tag{17.12}$$

Stable crack growth requires dG/da to be less than or equal to zero. This gives

$$a \geq L/\sqrt[3]{3} \approx 0.7L \tag{17.13}$$

Consequently, for the commonly used crack length $a = L/2$, the crack growth is unstable also under fixed-grip conditions. This has the consequence that only one measurement of the fracture toughness is obtained for each specimen.

17.3.1 ENF Specimen Preparation and Test Procedure

The ENF specimen is typically 120 mm long and 20 to 25 mm wide. Specimen thicknesses for unidirectional carbon- and glass-fiber composites are typically 3 and 5 mm (60% fiber volume fraction), respectively. The specimen is loaded in a three-point bend fixture (Figure 17.9) with a distance between the supports $2L$ of 100 mm. The loading and support cylinders should be about 5 mm in diameter. The crack-length-to-half-span ratio a/L should be 0.5. Panels should be prepared with a nonadhesive Teflon or Kapton film of thickness less than 13 μm placed at the midplane to define a starter crack. Further details of specimen preparation are presented in Chapter 4 and Appendix B. After specimens have been cut from the panel, the width and thickness at the center and 1 cm from each end should be measured for all

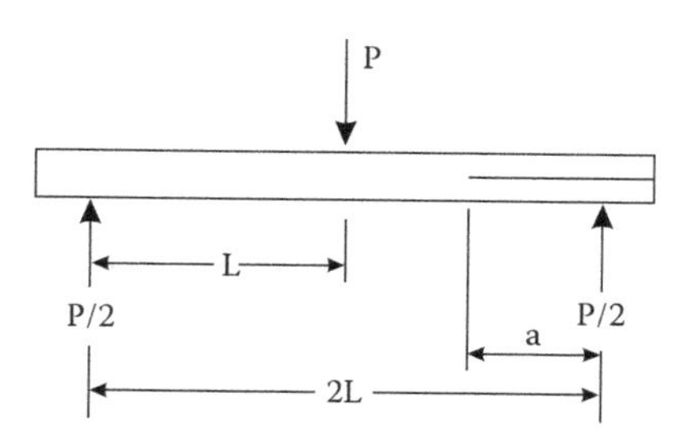

FIGURE 17.9
ENF specimen geometry parameters.

specimens. The thickness variations should not exceed 0.1 mm. Prior to testing, a brittle white coating should be applied to the specimen edges as described in Section 17.2.1.

The issue of whether precracking of the ENF specimen should be performed has long been discussed. Precracking in mode I is likely to create the fiber bridging discussed in Section 17.2 and is not recommended (O'Brien et al. 1989). A shear precrack may be achieved by loading the specimen in the stable crack length regime, $a > 0.7L$, according to Equation (17.13), until a short extension of the crack occurs. Unfortunately, however, it is difficult to detect the exact position and shape of the shear precrack after completion of the fracture test, and it is also difficult to obtain a straight and uniform crack front. For reasons of simplicity and consistency with the DCB procedure (Section 17.2), crack propagation from specimens with thin insert films, but without additional extension of the precrack, is advocated.

The ENF specimen is placed in a standard three-point bend fixture (ASTM D790 2010) so that a crack length a of 25 mm is achieved (Figures 17.9 and 17.10). To facilitate appropriate positioning of the crack tip, a low-magnification (×10) traveling microscope is useful. Mark the support location on the specimen edge for subsequent measurement of crack length. Measure the center beam deflection (load-point displacement) δ with a linear variable differential transformer (LVDT) or from the crosshead displacement corrected for the machine compliance. Use a crosshead rate in the range of 0.5 to 1 mm/min and monitor the load-displacement response. Both loading and unloading paths should be recorded. Observe the crack tip during loading (a traveling microscope is recommended) to detect any slow, stable crack propagation prior to fast fracture. Slow crack propagation preceding fast fracture is commonly observed in ductile matrix composites (Carlsson et al. 1986b) and leads to a nonlinear load-displacement curve (Figure 17.11). Indicate this event on the load-deflection curve. An example of a load-deflection curve for a brittle carbon–epoxy composite is shown in Figure 17.12. For this composite, fast fracture occurred without noticeable stable crack extension, and the response curve is essentially linear up to fracture.

FIGURE 17.10
ENF test setup.

17.3.2 ENF Data Reduction

Evaluation of the mode II fracture toughness G_{IC} requires a record of the load-displacement response (e.g., Figures 17.11 and 17.12). Toughness values G_{IC} (NL), G_{IC} (vis.), and G_{IC} (max.), referring to the loads at the onset of nonlinearity, visual stable crack extension, and maximum load, respectively (as illustrated

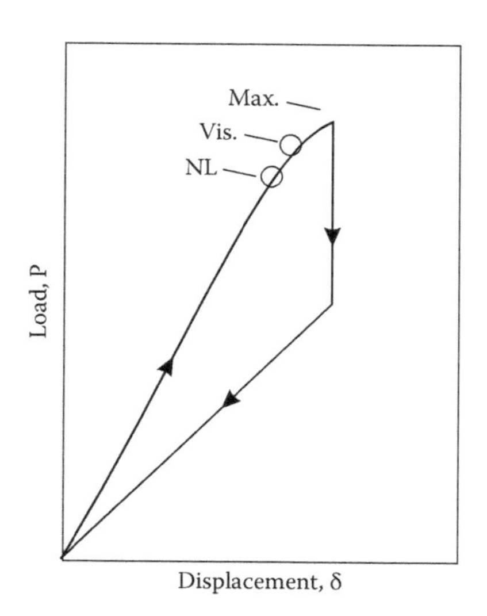

FIGURE 17.11
Schematic load-displacement curve for ENF fracture test of a ductile matrix composite. P(NL), P(vis.), and P(max.) denote loads at onset of nonlinearity, onset of visible stable crack growth, and onset of fast fracture, respectively.

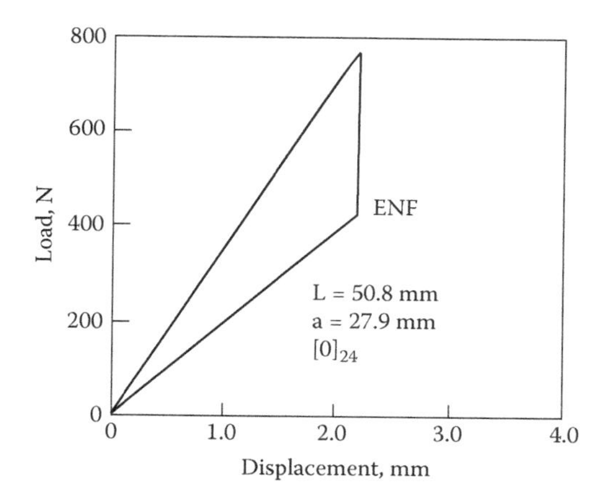

FIGURE 17.12
Load-deflection curve for a carbon–epoxy (AS4/3501-6) ENF specimen. $L = 50.8$ mm, $w = 25.4$ mm, and $a = 27.9$ mm.

in Figure 17.11) can be determined. For determination of G_{IIC}, the initial crack length is required. The initial crack length can be measured by cracking the failed specimen into two parts and measuring the distance between the support cylinders (marked on the specimen edge) and the initial crack front at three locations (each edge and center of the beam width). Commonly, the support cylinders leave imprints on the specimen surface that can be used to further verify the crack length measurements after the fracture test.

If the flexural modulus E_1 of the specimen is not known, the fracture toughness G_{IIC} may be calculated from the following beam theory expression using the measured compliance C:

$$G_{IIC} = \frac{9a^2P^2(C - C_{SH})}{4bL^3[1 + 1.5(a/L)^3]} \tag{17.14}$$

where C_{SH} is a compliance correction factor arising from interlaminar shear deformation calculated from

$$C_{SH} = \frac{6L + 3a - l^3/a^2}{20bhG_{13}} \tag{17.15}$$

In the calculation of C_{SH}, the interlaminar shear modulus G_{13} is required. If G_{13} is unknown, the in-plane shear modulus G_{12} (Chapter 10) can be used as an approximation to G_{13} for unidirectional composites. If the flexural modulus E_1 of the ENF specimen is known, it is most straightforward to determine G_{IIC} from a beam theory expression (Carlsson et al. 1986a):

$$G_{IIC} = \frac{9a^2P^2}{16b^2h^3E_1}\left[1+0.2\left(\frac{h}{a}\right)^2\frac{E_1}{G_{13}}\right] \tag{17.16}$$

To determine G_{IIC} (NL), G_{IIC} (vis.), and G_{IIC} (max.), the loads P(NL), P(vis.), and P(max.), defined in Figure 17.11, and the initial crack length are substituted in Equations (17.14) and (17.16). Consider, as an example, the load-displacement record shown in Figure 17.12 for a carbon–epoxy ENF specimen of dimensions L = 50.8 mm, a = 27.9 mm, $2h$ = 3.5 mm, b = 25.3 mm, and $G_{13} = G_{12} = 5\ GPa$. The critical load was 762 N, and the specimen compliance was 2.3 μm/N. Substituting these data in Equations (17.14) and (17.15) gives $G_{IIC} = 533 J/m^2$.

Note that the experimental compliance calibration method may be used for determination of the fracture toughness of the ENF specimen (O'Brien et al. 1989, Carlsson et al. 1986b). This method requires a long ENF specimen with a long precrack. By sliding of such a specimen across the test fixture, it is possible to achieve several specific crack lengths. Compliance is measured at constant span length at each crack length by loading the specimen at loads small enough not to promote crack extension. A set of compliance values over a range of discrete crack lengths a is obtained, and the data set is fitted by a third-order polynomial in crack length:

$$C = C_0 + C_3a^3 \tag{17.17}$$

Differentiation of this equation with respect to crack length and substitution into Equation (2.59) yields

$$G = \frac{3P^2C_3a^2}{2b} \tag{17.18}$$

Substitution of the corresponding critical loads (Figure 17.11) into this equation yields G_{IIC} (NL), G_{IIC} (vis.), and G_{IIC} (max.).

Unfortunately, this method tends to yield scattered G_{IIC} data for the ENF test. Davies et al. (1999) found that the coefficient of variation for G_{IIC} as determined for a carbon–epoxy composite using Equation (17.18) is 21%, whereas the corresponding value for the beam analysis method, Equation (17.16), is 14%. The reasons for the low precision are that the rate of change in the ENF specimen compliance with crack length is relatively small, and the experimental compliance method requires several accurate measurements of crack length.

17.4 Mixed-Mode Bending Test (Mixed-Mode I and II)

In most practical situations, delaminations in composite laminates tend to grow in mixed-mode stress fields (i.e., tension and shear stresses are acting ahead of the crack front). Previous works (e.g., Johnson and Mangalgiri 1987, Hashemi et al. 1990a,b) have shown that the resistance to delamination growth increases as the amount of shear loading (mode II) increases. Consequently, delamination characterization requires mixed-mode fracture testing. Several mixed-mode fracture tests exist in which various combinations of mode I and mode II can be generated. Most such methods, however, suffer from complicated test fixturing, a small range of mode mixities (G_{II}/G_I), and varying mode mixity as the crack grows.

The most promising test principle for mixed-mode delamination toughness testing is the MMB test proposed by Reeder and Crews (1990, 1991, 1992) (Figure 17.13). The MMB test is a superposition of the DCB and ENF tests discussed previously. The MMB method has been adopted as ASTM Standard D6671 (2006) because of simplicity of testing and the wide range of mode mixities possible.

Figure 17.14 depicts the geometry parameters and test principle of the MMB specimen. The loading lever adds an opening load at the cracked end and a compressive load at the midspan of the test specimen. The distance c between the point of load application and the midspan determines the ratio of the downward force P_d to upward force P_u and hence the mode mixity. Pure mode II loading is achieved when $c = 0$. The ratio G_{II}/G_I decreases with increasing distance c.

A vertical distance of 15 mm between the point of load application and the specimen midplane (Figure 17.14) has been found to minimize geometrical nonlinearity effects (Reeder and Crews 1991, 1992). Loading supports should be between 5 and 15 mm in diameter mounted on roller bearings. Detailed drawings are provided in ASTM Standard D6671. Figure 17.15 shows a photograph of the MMB test setup. The loading lever is a low-weight aluminum I-beam that is several orders of magnitude stiffer than the test specimen. The lever load, the midspan load, and the left support reaction are applied through bearing-mounted rollers to reduce frictional forces. The cracked end

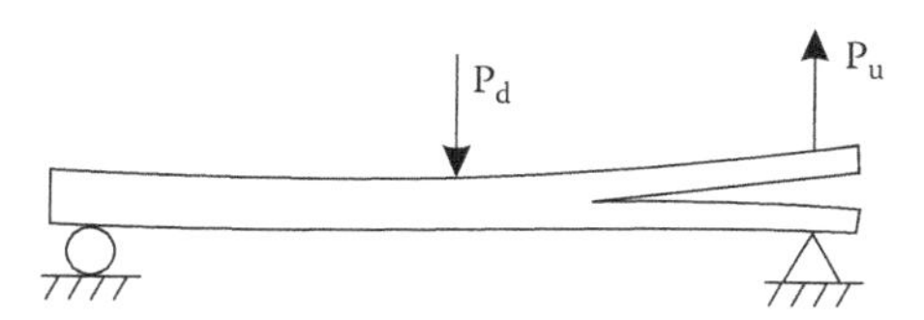

FIGURE 17.13
Principle of MMB test.

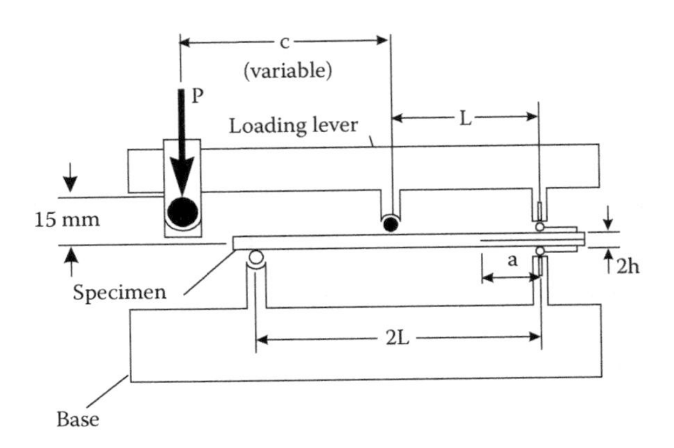

FIGURE 17.14
Definition of geometry parameters for the MMB specimen.

of the specimen is loaded through high-quality hinges bonded to the specimen arms. The apparatus rests on a thick steel base.

17.4.1 MMB Test Procedure

The MMB test specimen is 165 mm long, prepared with hinge-loading tabs as the DCB specimen discussed in Section 17.2 (see also Appendix B).

FIGURE 17.15
MMB test setup. (Courtesy of J. R. Reeder, NASA Langley Research Center.)

A specimen of 25 mm width and thickness between 3 and 4.4 mm should be appropriate for unidirectional carbon-epoxy. For fracture testing the crack length (a) should be 25 mm. The width and thickness of each specimen are measured to the nearest 0.025 mm at the midpoint and at 1 cm from both ends. Three thickness measurements are made at each of these positions, with one measurement close to each edge and one at the center. Variations in thickness should not exceed 0.1 mm. Average values of the width and thickness measurements shall be recorded.

Adjust the loading lever distance *c* (Figure 17.14) to achieve the following mode ratios: $G_{II}/G_I = 0.25$, 1, and 4 [using Equation (17.23) of the next section]. Prepare a minimum of three replicate specimens for each mode ratio.

Use a crosshead rate of 0.5 mm/min for consistency with the mode I and mode II tests discussed. Record the load-displacement response on an *x-y* recorder while monitoring the crack tip with a low-magnification traveling microscope. If slow, stable crack growth occurs, mark this event on the load-displacement curve. Figure 17.16 shows a load-displacement record for a carbon–PEEK composite. It is observed that the load-displacement record is similar to that of the ENF specimen (Figure 17.11), which allows evaluation of G_C (NL), G_C (vis.), and G_C (max.).

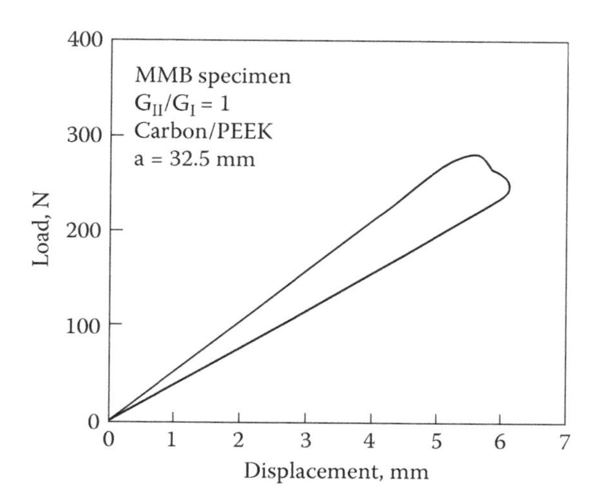

FIGURE 17.16
Load-displacement record for a carbon–PEEK MMB specimen, (After Reeder, J.R., and Crews, J.H., Jr., 1991, Redesign of the mixed mode bending test for delamination toughness, *Proceedings of the 8th International Conference on Composite Materials,* Honolulu, HI, July, Society for the Advancement of Materials and Process Engineering (SAMPE), Covina, CA.)

17.4.2 MMB Data Reduction

The following expressions to determine the mode I and mode II components of the energy release rate were derived by Hashemi et al. (1990a) and Kinloch et al. (1993):

$$G_I = \frac{12P_I^2(a+xh)^2}{b^2h^3E_1} \tag{17.19a}$$

$$G_{II} = \frac{9P_{II}^2(a+0.42xh)^2}{16b^2h^3E_1} \tag{17.19b}$$

where $G = G_I + G_{II}$, and P_I and P_{II} are the opening and shearing components of the applied load, Reeder and Crews (1990):

$$P_I = P\left(\frac{3c-L}{4L}\right) \tag{17.20a}$$

$$P_{II} = \frac{P(c+L)}{L} \tag{17.20b}$$

The correction term x in Equations (17.19) was obtained by curve fitting these equations to numerical (finite-element) data (Hashemi et al. 1990a, Kinloch et al. 1993):

$$x = \left[\frac{E_1}{11G_{13}}\left(3-2\left(\frac{\Gamma}{\Gamma+1}\right)^2\right)\right]^{1/2} \tag{17.21}$$

with

$$\Gamma = 1.18\frac{\sqrt{E_1E_2}}{G_{13}} \tag{17.22}$$

It may be verified that the ratio between the fracture modes (e.g., G_{II}/G_I), as given by Equations (17.19), is only weakly dependent on crack length. An approximate equation for the mode mixity is obtained from the asymptotic beam analysis presented by Reeder and Crews (1990):

$$\frac{G_{II}}{G_I} = \frac{3}{4}\left(\frac{c+L}{3c-L}\right)^2, \quad c \geq L/3 \tag{17.23}$$

For $c < L/3$, crack face contact may occur that corresponds to $G_1 = 0$ and invalidates the analysis given. Equation (17.23) can be used for initial (approximate) calculation of the mode ratio, which more accurately is calculated using Equations (17.19).

After testing is complete, split open the specimen and measure the crack length (the distance from the center of the hinge pin to the end of the delamination starter film). Measure the crack length at the edges and center of the specimen and obtain a mean value.

Calculations of G_I and G_{II} using Equations (17.19) require the critical load and several of the material properties (i.e., E_1, E_2, and G_{13}). The moduli E_1, E_2, and G_{13} (approximately equal to G_{12}) have to be known from previous tests (Chapters 8 and 10). The (flexural) modulus E_1 may also be calculated from the MMB compliance (ASTM D6671 2006):

$$E_1 = \frac{8(3c - L)^2(a + xh)^3 + (c + L)^2[4L^3 + 6(a + 0.42xh)^3]}{16CL^2bh^3} \tag{17.24}$$

where C is the specimen compliance corrected for the load cell compliance and (lever-length-dependent) fixture compliance.

As an alternative, simpler, and more accurate procedure to determine E_1, the uncracked portion of the beam may be tested in three-point bending as specified in Chapter 11.

The mode I and mode II components G_I and G_{II} of the mixed-mode fracture toughness are calculated from the measured critical load using Equations (17.19).

It has become customary to represent mixed-mode fracture toughness data in terms of the mode II fraction G_{II}/G, where $G = G_I + G_{II}$. Benzeggagh and Kenane (1996) proposed the following equation for description of the relation G_C versus G_{II}/G:

$$G_C = G_{IC} + (G_{IIC} - G_{IC})(G_{II}/G)^{\beta} \tag{17.25}$$

where β is an empirical factor determined from a fit of the experimentally determined G_C versus G_{II}/G data.

Figure 17.17 shows the relation between G_C and G_{II}/G filled to fracture toughness data for several of the carbon fiber, polymer matrix composites. It is observed that the fracture toughness of AS4/PEEK remains fairly independent of mode ratio, while the fracture toughness of the more brittle thermoset-matrix composites shows a quite large sensitivity to the mode ratio.

17.5 Edge-Cracked Torsion Test (Mode III)

The ECT test was introduced by Lee (1993) as a test method to determine the mode III delamination toughness of composites. Figure 17.18 shows the ECT test specimen and test fixture. The test specimen is a rectangular

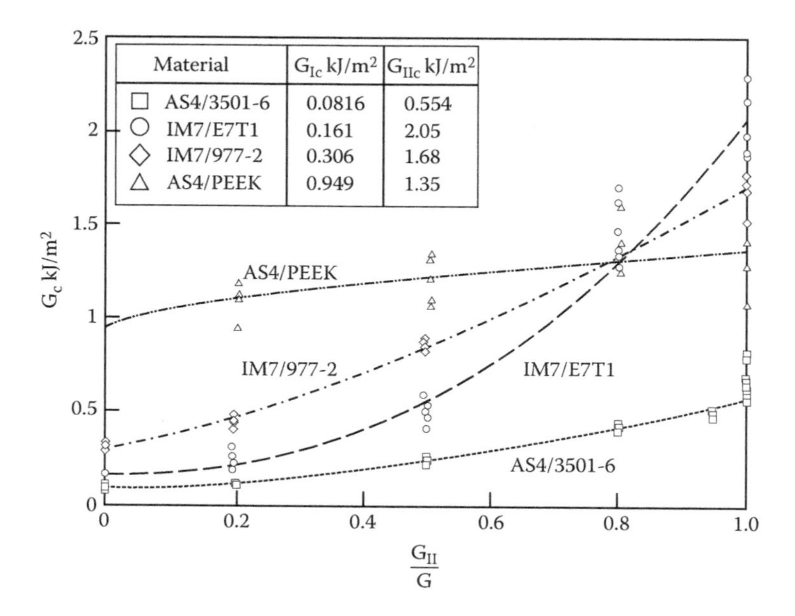

FIGURE 17.17
Mixed-mode interlaminar fracture toughness for a variety of carbon fiber, polymer matrix composites. The parameter β [Equation (17.25)] varies from 0.63 (AS4/PEEK) to 2.35 (IM7/977-2).

composite plate containing an edge delamination at the midplane. The long direction of the specimen defines the 0° ply direction. Layups such as $[90/(\pm 45)_n/(\mp 45)_n/90]_s$ are recommended. The integer n is 2 or 3, corresponding to a total of 20 or 28 unidirectional plies, respectively. A precrack is defined by inserting a strip of film of thickness less than 13 μm between

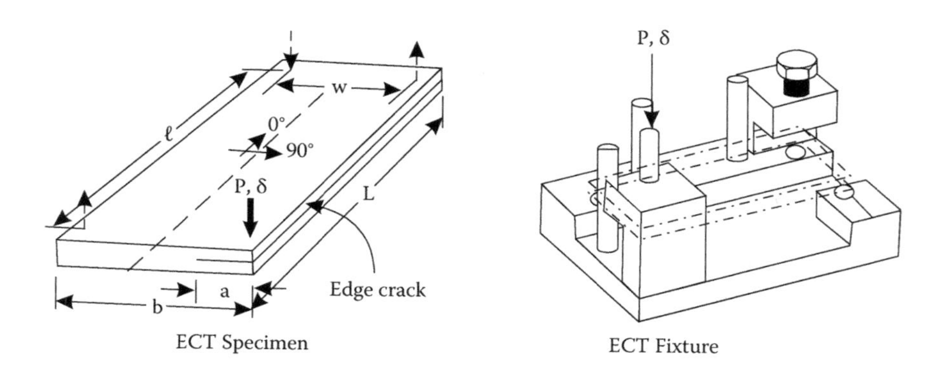

FIGURE 17.18
ECT specimen and test fixture.

the 90° plies at the midplane of the panel to define an edge crack of length a (Figure 17.18). By this configuration, the crack front is perpendicular to the fibers on both sides of the interface, consistent with the mode I and mode II and mixed-mode I and II tests discussed. For panel design, see Appendix B.

The test fixture (Figure 17.18) is designed so that three corners of the panel are supported, while one corner on the cracked side is displaced normal to the plane of the panel. This loading produces a pair of couples of equal magnitude but of opposite sign that induce twisting of the plate and the characteristic mode III deformation illustrated in Figure 2.9(c). Crack propagation should ideally occur uniformly in a direction perpendicular to the crack front at the midplane (i.e., parallel to the 0° fibers).

17.5.1 ECT Specimen Preparation

The ECT specimen (Figure 17.19) is a flat, rectangular plate, 83 mm long and 38 mm wide. Lay ups are $[90/(\pm 45)_n/(\mp 45)_n/90]_s$, where typically n = 2 or 3 for unidirectional carbon–epoxy and n = 3 for unidirectional glass–epoxy. Corresponding laminate thicknesses for carbon–epoxy are about 2.5 and 3.6 mm. An edge crack is defined by inserting a thin strip (<13 μm) of nonstick film such as Teflon, Kapton, or polypropylene. The film is inserted between the 90° plies at the midplane to define a straight precrack of the desired length. Panel design is outlined in Appendix B. Precrack lengths of 0, 8, 11, 15, 19, and 23 mm are recommended. Although testing of the uncracked specimen (a = 0) does not yield any toughness data, it provides a reference point for subsequent data reduction using the compliance calibration method. After the specimens are cut from the panel, measure the width b and length L to the nearest 0.1 mm and thickness $2h$ to the nearest 0.01 mm

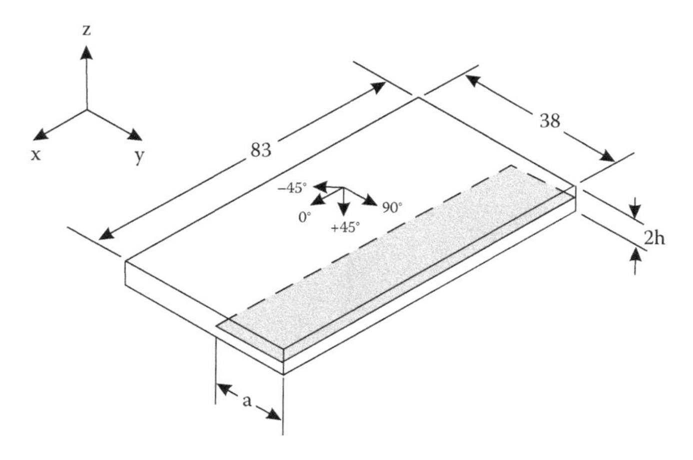

FIGURE 17.19
ECT specimen geometry and dimensions.

of each specimen. Measure the width and length near the corners and at the midlength of each side. Measure thickness at the center and near each corner. Thickness should not vary more than 0.1 mm. In a manner similar to that for the other fracture specimens, the short edges may be coated with a brittle white coating to aid in visual detection of crack extension.

17.5.2 ECT Test Fixture

The schematic in Figure 17.18 shows that the ECT specimen is constrained against lateral displacement at three corners and loaded by a concentrated normal force at the fourth corner. The distance w between the support-loading pins along the short edge is 31.8 mm. The distance l between the support-loading pins along the crack front is 76.2 mm.

17.5.3 ECT Test Procedure

Place the ECT test specimen in the test fixture. Adjust the threaded support pin (Figure 17.18) so that all four support-loading pins contact the specimen. Place the fixture in a properly calibrated load frame. Set the crosshead rate at 1.3 mm/min and load the specimen while recording the load P versus displacement δ response on an x-y recorder. Observe the crack front and P–δ record for indications of propagation of the crack.

Figure 17.20 shows schematic load-displacement records that are typically observed for the ECT test. The curve in Figure 17.20a indicates stable crack propagation under increasing load, whereas the curve in Figure 17.20b indicates some extent of unstable growth and a clearly defined maximum load P_C. For the curve in Figure 17.20a, the critical load for crack propagation P_C is determined by the 5% offset method. A straight line offset by a 5% increase in

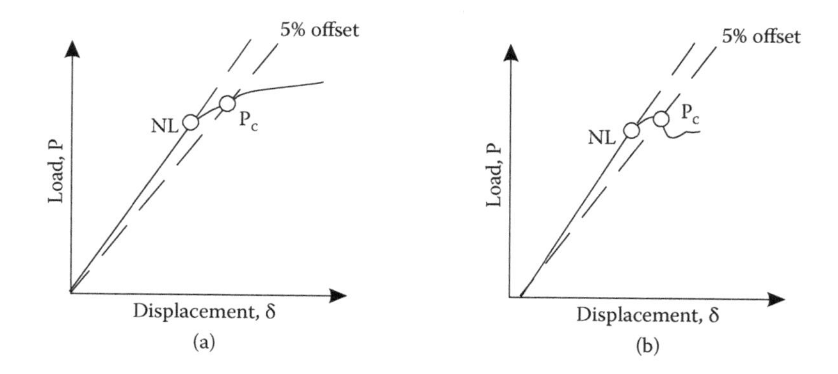

FIGURE 17.20
Schematic illustrations of load-displacement records and determination of critical load P_c for crack propagation in the ECT specimen: (a) stable growth and (b) initial unstable growth.

compliance is drawn as shown in Figure 17.20a, and P_C is defined as the load value where this line intersects the recorded P–δ curve. Notice that if the 5% offset line intersects the P–δ curve after the maximum load is reached, as in Figure 17.20b, P_C is defined as the maximum load.

After completion of the fracture test, unload the specimen and remove it from the fixture. Separate the fracture specimen into two halves. This enables accurate measurements of the precrack length (Figure 17.19) at the edges and midlength of the crack front. Although the final crack length is not used in the determination of G_{IIIC}, the crack front shape may be determined from crack length measurements at six or more equally spaced locations along the crack front.

17.5.4 ECT Data Reduction

Evaluation of the mode III delamination fracture toughness G_{IIIC} of the ECT specimen is based on the experimental compliance calibration method. Compliance C is determined from the linear slope of the load-versus-displacement record (Figure 17.20), $C = \delta/P$. After correction for machine and fixture compliance, the stiffness P/δ (i.e., the inverse of the compliance) is plotted versus the average initial crack length a normalized by edge length b for all the specimens tested (Figure 17.21). Analysis of the ECT test by Lee (1993) indicates a linear dependence of specimen stiffness on crack length. Such a relation is assumed when reducing the experimental data:

$$\frac{1}{C} = A\left[1 - m\left(\frac{a}{b}\right)\right] \tag{17.26}$$

where A is the intercept of the line at the $1/C$ axis, and Am is the magnitude of the slope of the line. Differentiation of Equation (17.26) with respect to

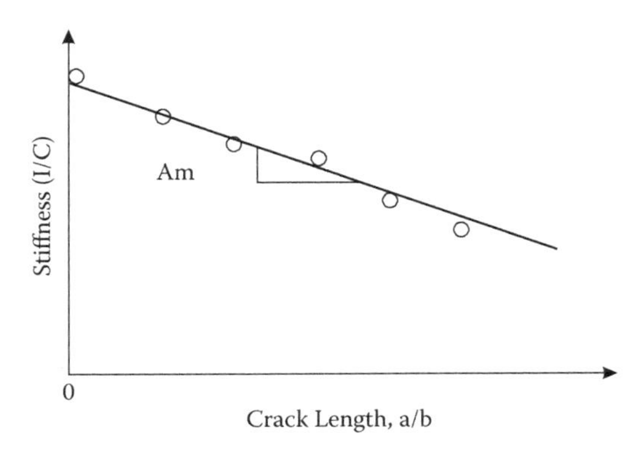

FIGURE 17.21
Stiffness (1/C) of ECT specimen plotted versus normalized crack length (a/b) for experimental determination of mode III toughness.

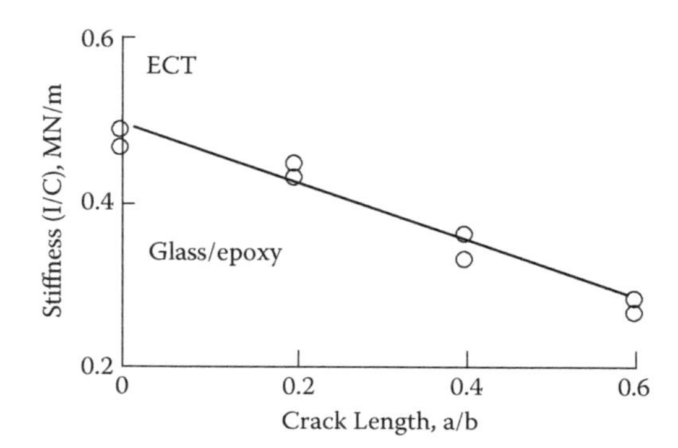

FIGURE 17.22
Stiffness versus crack length data for glass–epoxy ECT specimens. (From Li, X., Davies, P., and Carlsson, L.A., 2004, *Compos. Sci. Tech.*, 64, 1279–1286.)

crack length a yields, in conjunction with Equation (17.2), the strain energy release rate for the ECT specimen:

$$G = \frac{mCP^2}{2Lb(1 - m(a/b))} \tag{17.27}$$

where L is the total length of the crack along the ECT specimen (see Figure 17.18). Substitution of the critical load P_C into Equation (17.27) yields the mode III delamination toughness G_{IIIC}. Similar to the other delamination tests, G_{IIIC} (NL) and G_{IIIC} (max.) may be determined on the basis of the load-displacement record (Figure 17.20). Figure 17.22 shows an example of experimental stiffness data for glass–epoxy ECT specimens plotted versus crack length. For the glass–epoxy composite: G_{IIIC} (NL) = 1.23 ± 0.09 kJ/m^2 and G_{IIIC} (max.) = 1.48 ± 0.18 kJ/m^2.

18

Damage Tolerance Testing

18.1 Introduction

It can be important to test composite laminates' resistance to hard objects impacting at low velocity. As documented in numerous investigations (see, e.g., Cantwell and Morton 1991 and Abrate 1998), impact damage may not be visible to the naked eye, although its effect on the residual strength and reliability is important. Several failure modes have been identified, such as matrix permanent deformation (indentation), matrix cracking and delamination, and in severe cases fiber failures. When a laminate contains interlaminar damage, it is compressive strength that can be expected to be degraded because individual plies or groups of plies are free to exhibit local instabilities. Therefore, to simulate the action of lateral impact of foreign objects and subsequent delaminations, it has been common practice to conduct drop-weight testing of composite panels and subsequently conduct compression-after-impact (CAI) testing on the impacted panel to determine the damage tolerance of the material. Both these tests are American Society for Testing and Materials (ASTM) standards. The falling-weight impact test is designated ASTM D7136 (2012), and the CAI test is ASTM D7137 (2012). This chapter describes these two tests and provides examples of test results.

18.2 Falling Weight Impact Testing

18.2.1 Drop-Weight Impact Test Machine

The drop-weight impact test is designed for multidirectional composite laminates reinforced with continuous fibers in a polymer matrix. A flat rectangular composite plate is supported by a rigid fixture with a rectangular cutout of size smaller than the test panel to maintain support of the edges of the panel during the impact event. The impactor has a smooth hemispherical striker tip. The impact test is conducted by releasing the falling weight from

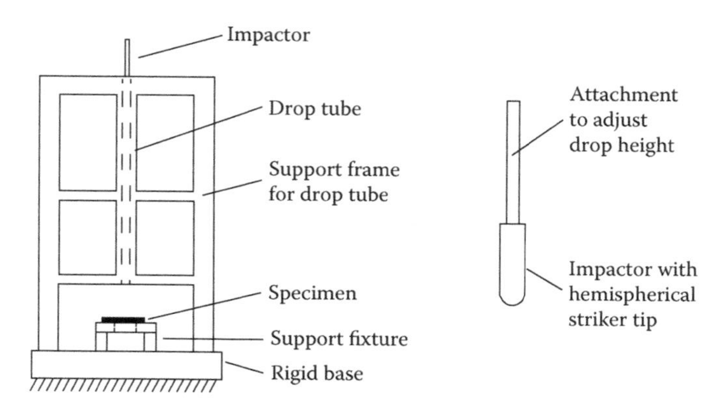

FIGURE 18.1
Falling-weight impact tester with drop tube impactor guide mechanism.

a certain height to impart impact loading on the panel. The tester may be instrumented with (optional) devices to measure velocity of the impactor prior to impact and contact force during the impact event.

Two basic types of impact testers are described in the ASTM D7136 standard. The simplest design employs a vertical cylindrical tube impactor guide (see Figure 18.1). In this device, a cylindrical impactor travels through the vertically arranged cylindrical drop tube. The impactor has a hemispherical striker tip attached to a rod or cord to support the impactor prior to impact testing at a given, well-defined drop height. The height of the tube should be sufficient to allow impact testing at the desired energy level. The clearance between the impactor and tube inner diameter should be less than 1 mm, and friction between the impactor and tube should be negligible. If friction is a factor, the velocity v of the impactor should be measured on exit from the tube. For such a case, the impact energy is calculated based on the kinetic energy K rather than the potential energy H:

$$K = m\frac{v^2}{2} \tag{18.1}$$

$$H = mgh \tag{18.2}$$

where m is the mass of the impactor (striker and rod or cord), v is the velocity of the impactor immediately before impact, g is the acceleration of gravity (9.81 m/s^2), and h is the drop height.

The other type of impact tester, schematically illustrated in Figure 18.2, employs a falling weight guided by double vertical columns. This design is most common for commercial impact testers. ASTM D7136 does not provide details on the design of the guide mechanism. As for the drop tube impactor discussed, it is essential that the falling weight travels with negligible

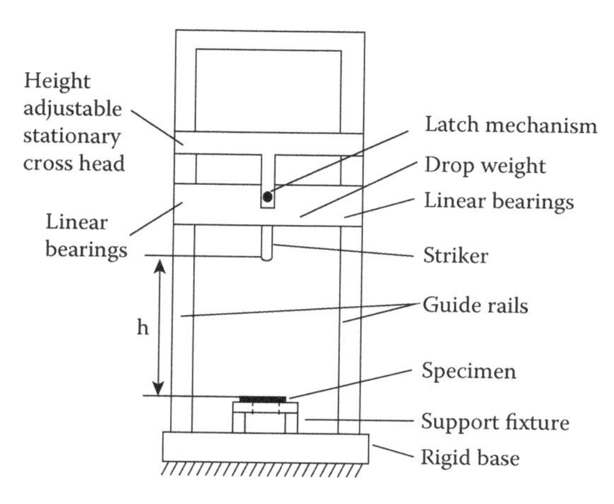

FIGURE 18.2
Schematic illustration of falling-weight impact tester.

friction, and that the striker hits the specimen at the intended site under well-defined conditions. Hence, the two vertical columns must be parallel, and smooth, and the moving crosshead should be supported by linear bearings. Furthermore, to minimize horizontal forces on the bearings and vertical columns during impact, the impactor should be mounted as close as possible to a vertical line through the center of gravity of the falling weight.

The impact tester should rest on a rigid base and be equipped with a rebound catcher. The rebound catcher is typically a mechanism that is triggered by the initial impact and stops the impactor from the next impact after it rebounds. The rebound catcher should not affect the initial impact event and should not engage until the striker head has lost contact with the specimen after the initial impact event. If special equipment to stop multiple impacts is not available, further impacts may be prevented by quickly sliding a piece of wood, plastic, or metal between the upward-moving impactor and the specimen.

18.2.2 Support Fixture and Impactor

An important component of the impact test set-up is the specimen support fixture. A schematic of the specimen support fixture is shown in Figure 18.3. The support fixture should accommodate the rectangular test specimen placed symmetrically over the rectangular hole of dimensions $125 \pm 1 \times 75 \pm 1$ (mm) in the top plate of the support fixture. The thickness of the top plate should be at least 2 cm, and it should be made from aluminum or steel. The top surface should be flat to within 0.1 mm to provide uniform support of the specimen during the impact event. Guiding pins may be used and positioned so that the 10×15 (cm) specimen becomes centered over the hole. The specimen is secured at the intended position

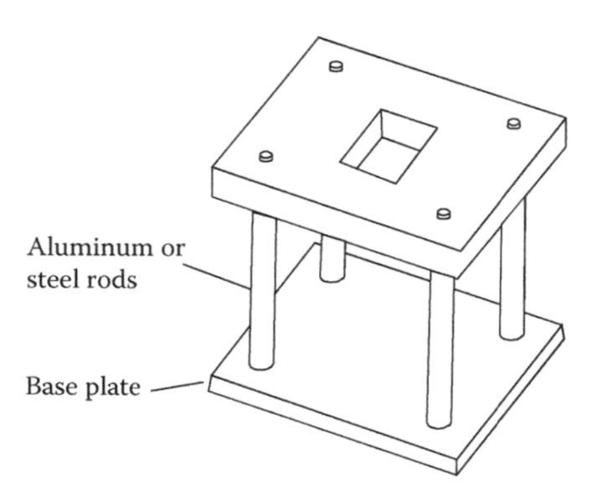

FIGURE 18.3
Schematic illustration of specimen support fixture.

during impact loading using four rubber-tip clamps (Figure 18.4). The four clamps should allow application of a clamping force of at least 1.1 kN to each corner of the specimen.

The top plate should be connected to a bottom (base) plate with solid, cylindrical aluminum or steel rods attached near the corners of the top support plate (Figure 18.3). The diameter of the columns should be 3.8 cm, and the length should be 30 cm. The base plate should be made from aluminum or steel 1.3 cm thick. The planar dimensions of the top supporting and base plates are 30 × 30 (cm). The top support and bottom base plates should be attached to the columns using recessed 1-cm diameter bolts. Similarly, the columns should be bolted to the base of the impact tester using 1-cm diameter bolts.

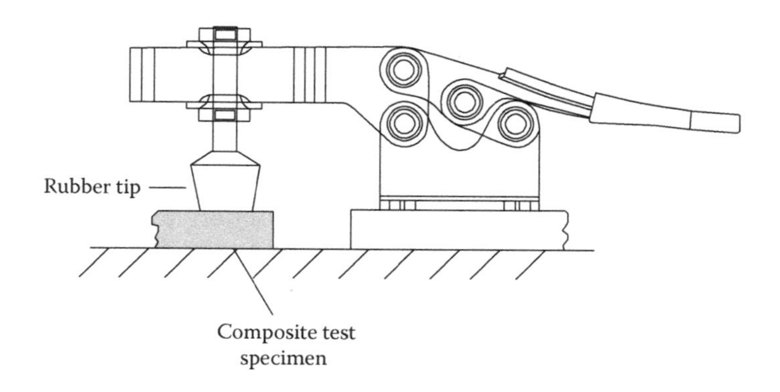

FIGURE 18.4
Rubber-tip clamps to secure specimen during impact loading. (From Griff Machine Products, Pittsburg, PA.)

The impactor should have a mass of 5.5 ± 0.25 kg. The striker tip should be hemispherical with a diameter of 16 ± 0.1 mm of hardness 60–62 HRC.

18.2.3 Impact Test Specimen

The ASTM D7136 standard stipulates testing of five or more replicate test specimens unless valid results can be achieved with fewer specimens. The standard specimen thickness is between 4 and 6 mm, with a target thickness of 5 mm.

The standard is limited to continuous fiber composites in the form of laminates made from unidirectional plies or woven fabric plies. For unidirectional tape laminates, the stacking sequence is $[45/0/-45/90]_{NS}$, where N is an integer selected to achieve a total cured thickness of 4–6 mm. For woven fabric composites, again the targeted thickness after cure is 5 mm. The fabric designation is 0° and 90° for the warp and weft fibers. The stacking sequence for the specimen should be $[(45/-45)/(0/90)]_{NS}$, where (45/-45) and (0/90) represent one layer of the woven composite. Laminates with other layups may be tested.

The test specimen dimensions are 15 x 10 (cm), although testing of larger specimens is allowed. Specimens should have uniform thickness and be without thickness taper greater than 0.08 mm along the length and width of the test panel. The coefficient of variation of the thickness should not exceed 2%. The specimens should be machined using water-lubricated precision sawing, milling, or grinding. It is important for subsequent compression testing (Section 18.3) that the edges are flat and parallel. ASTM D7136 specifies a tolerance of ±0.02 mm for the length and width dimensions. For measuring the dimensions, the accuracy should be within 1% for measuring length, width, and thickness. Typically, thickness should be measured within ±0.0025 mm, while length and width measurements of the panel should be within ±0.025 mm. If the specimens are not to be subsequently tested in compression, a caliper should be suitable for measuring the length and width dimensions.

18.2.4 Instrumentation

ASTM D7136 does not require data acquisition. If, however, the impact event is monitored, basic equipment consists of a load cell or accelerometer for measuring impact force and an instrument for measuring the velocity of the impactor prior to impact. Measurements of the contact force have been discussed by Corum et al. (1995). The contact force can be measured directly or indirectly. Direct methods measure the response of the impactor. Tan and Sun (1985) mounted a piezoelectric force transducer (PCB Piezoelectronics, Inc.) directly on the impactor. This transducer operates in the thickness compression mode. An amplifier was used to reduce the high impedance and amplify the output voltage. Lal (1983) mounted an accelerometer on the impactor. By recording the acceleration during the impact event, the contact force F is obtained from Newton's law:

$$F = ma \tag{18.3}$$

where m is the mass of the impactor, and a is the acceleration (retardation). Tan and Sun (1985) used an accelerometer to calibrate their force transducer under dynamic loading. In principle, the direct method of measuring contact force is simple and straightforward, but there is need for signal processing, as discussed further in this chapter.

The indirect method measures the response of the target (specimen) using a strain gage or strain gages. This was perused by Tan and Sun (1985). To determine the contact force history exerted on the specimen, structural analysis of the specimen subject to impact loading is necessary. This requires advanced dynamic analysis, which becomes a daunting task for impacts that inflict appreciable damage and nonlinear behavior.

For recording of the force or acceleration signal, a sampling rate of 100 kHz is recommended. It is well known that the contact-force-versus-time response curve contains high-frequency oscillations. As discussed in ASTM D7136, these oscillations arise due to two primary sources; the first is from the natural frequency (or frequencies) of the impactor. Such oscillations are often referred to as "impactor ringing." These oscillations are superposed onto the low-frequency oscillation caused by the flexible vibrations of the test specimen on impact.

The impact tester may be equipped with a velocity indicator placed close to the top surface of the specimen. Velocity measurements are necessary if friction or other factors make the free-fall formula inaccurate:

$$v = \sqrt{2gh} \tag{18.4}$$

where v is the velocity of the striker just prior to impact, and h is the drop height.

A common setup uses two light beams between a photodiode emitter and a detector. The beams are typically separated by a distance of 3–10 mm. The lower beam should be located 3–6 mm above the top surface of the test specimen. The impact velocity is calculated using the time difference between the first and second crossing of the light beam by the falling weight, and the distance between the parallel beams. The velocity measurements should be accurate to within 5 mm/s.

18.2.5 Impact Testing

ASTM D7136 specifies nondestructive inspection (NDI) of the test specimen prior to testing. If the specimens are conditioned in a certain environment, NDI should be performed prior to conditioning. Several more or less-sophisticated NDI methods have been developed for detecting surface flaws and interior defects. Visual inspection may identify surface flaws, while detection of interior flaws generally requires more sophisticated methods, such as x-ray radiography or ultrasonic techniques (see, for example, *Comprehensive Composite Materials*, 2002). After NDI, conditioning the specimens is required. Measure the specimen width and length at two locations and the thickness at four locations. All measurements should be accurate within ±1% of the dimension.

If possible, the impact testing should be conducted under the same environmental conditions as those during specimen conditioning. If this is not possible, the specimens should be tested quickly after removal from the environmental test chamber. Adjust the drop height h (Figure 18.2) to achieve the required energy level (H = mgh). The standard test utilizes a certain impact energy normalized by the specimen thickness t:

$$\frac{H}{t} = 6.7x10^{-3}(N) \tag{18.5}$$

where the potential energy H is measured in joules, and t is the specimen thickness (measured in meters). Make sure all components are tightly bolted down to minimize vibrations during the impact event.

Place the specimen on the support fixture and center it properly over the rectangular opening. Normally, the tool side of the specimen is impacted. Use the four rubber-tip clamps to hold the specimen in place during impact. By clamping the specimen, rebounding and vibrations during the impact event should be minimized. The clamp tips should be placed approximately 2.5 cm from the edges of the specimen. The impact test can then be performed by dropping the weight and preventing rebound impact loading as discussed in Section 18.2.1. If possible, the rebound height h_r should be estimated because the potential energy associated with the rebound mgh_r represents elastically stored energy in the specimen. Hence, the energy transferred to the specimen becomes

$$H_t = mg(h - h_r) \tag{18.6}$$

In this expression, the additional potential energy due to the deflection δ of the impacted panel has been neglected. This is justified if $\delta \ll h$.

18.2.6 Example of Impact Test Results

A drop weight impact tester was built at Florida Atlantic University. The design of the impact tester closely follows the ASTM D7136 recommendations (Dale et al. 2012a). The impact tester is instrumented with a Dytran accelerometer mounted directly on the falling crosshead above the striker. The

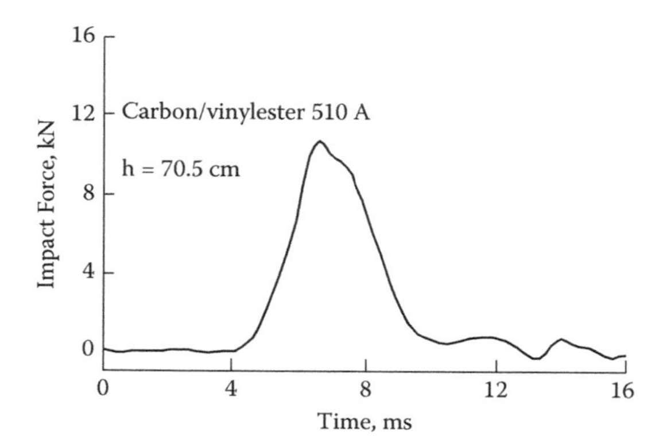

FIGURE 18.5
Force trace for an impact test on a carbon/vinylester panel. (From Dale, unpublished data.)

accelerometer is connected to a computer using a Measurement Computing breakout board that allows sampling up to a rate of 100 kHz. The acceleration-versus-time data acquisition is accomplished using LabView. Unfiltered acceleration data display many high-frequency oscillations (ringing), as discussed in Section 18.2.4. This component of the signal had a frequency content in the range of 30–35 kHz. A 10th-order low-pass Butterworth filter was used to filter the signal.

Figure 18.5 shows an example of the filtered impact force signal versus time for a 5.3-mm thick carbon–vinylester 510 woven fabric panel impacted at a height of 70.5 cm. The force was calculated from the mass of the falling weight (m = 7 kg) and the acceleration using Equation (18.3). The impact energy, Equation (18.2), was H = 48.4 J. It may be possible to identify damage initiation from a discontinuity in slope of the force trace (Figure 18.5). Identification of the maximum contact force F_{max} and the duration of the impact should be straightforward. For the contact force curve shown in Figure 18.5, no discontinuity in the force trace was observed. The peak contact force F_{max} was 10.6 kN, and the duration of the impact was about 6 ms.

18.2.7 Postimpact Damage Inspection

As discussed in the beginning of this chapter, composite materials subject to low-velocity impact loading by hard objects typically suffer from a host of damage mechanisms. Detection and quantification of damage can be accomplished by NDI and destructive techniques. If the impacted test panel is to be subsequently tested in, for example, compression (Section 18.3), only NDI inspection is possible. One such quantity is the indentation depth, which is the residual depth of the depression of the specimen surface formed by plastic deformation or damage of the material in the contact region by the hard impactor. The indentation depth can be measured with a depth gage micrometer. The probe tip should be hemispherical with a radius of 8 mm, and the accuracy should be ±25 μm. The depth should be measured immediately after the impact loading. The base of the depth gage should be at least 50 mm to span beyond the impacted region. Rotate the gage 90° to obtain two measurements.

Besides the dent created by the impactor, damage may possibly be detected by visual inspections or by examination in a microscope. Such forms of damage are matrix cracks in the surface layers of unidirectional tape composite laminates and delamination of the surface plies. Detection of internal damage using NDI is more cumbersome because this generally requires sophisticated equipment, such as ultrasonics, x-ray radiography, thermography, and eddy-current test methods (see Hellier 2001). If the panel will not be subsequently tested, destructive techniques involving cutting sections of the damage region, polishing, and microscopy can reveal much detail of the extent and type of damage inflicted on the composite.

To examine damage of the impacted carbon–vinylester 510 panel discussed previously (Figure 18.5), it was cut into four pieces along two symmetry lines centered through the impact site. The exposed cross sections near the damage site were sanded and polished (see Section 3.5.2) to enhance identification of damage. Figure 18.6 shows a photomicrograph of a cross section from the upper left quadrant of the sectioned panel, approximately 19 mm from

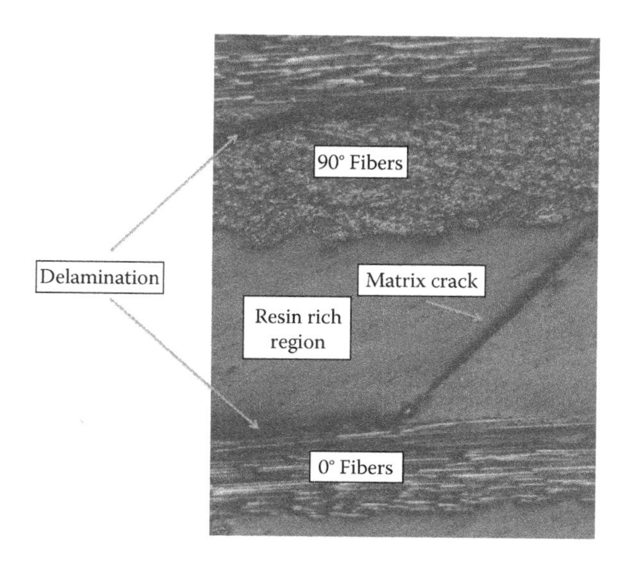

FIGURE 18.6
Photomicrograph of a cross section near the impact site of an impacted carbon–vinylester 510 panel. (From Dale et al., 2012 b.)

the center of the impact site, along the longitudinal direction. The region shown in Figure 18.6 is located about 3 mm below the impacted top surface of the 16-ply panel.

This photomicrograph reveals delamination between the 0° and 90° bundles as well as a crack in a resin-rich region between bundles. By examining this and similar photomicrographs, it was found that the extent of damage was different in the longitudinal and transverse directions of the specimen. On the long side, the damage extended approximately 25 mm on both sides of the impact site, while the transverse damage extended approximately 13 mm (on both sides). Hence, the damaged region can be approximated by an ellipse with a 50-mm longitudinal axis and 26-mm transverse axis.

18.3 Compression-after-Impact (CAI) Testing

The ASTM D7137 (2007) standard for CAI testing is titled, *Standard Test Method for Compressive Residual Strength Properties of Damaged Polymer Matrix Composite Plates*. The objective of this test method is to quantify the compression strength of composite plates that previously have been subject to the drop-weight impact test discussed in Section 18.2 or loaded statically. This test is limited to continuous-fiber polymer composite laminates with multidirectional fiber orientations. The laminate stacking sequence should be symmetric and balanced. The same specification of plate thicknesses and layup sequence as for the impact test specimens (Section 18.2.3) apply to this test, and the specimen dimensions are the same. The laminates should not have several adjacent plies with the same fiber orientation since such ply groups tend to weaken the material and result in complex damage mechanisms. When this test is used to characterize the compressive strength of plates that have been subject to drop-weight impact loading, it is commonly called the CAI method. This section focuses on this use of the method. In principle, this test can be used to characterize the compressive strength of undamaged panels, but most commonly such panels do not fail in pure compression. Typically, end crushing or instability failures tend to occur prior to failure of the material under pure compression. The IITRI (Illinois Institute of Technology Research Institute) and CLC (combined loading compression) test methods described in Chapter 9 are recommended for measuring the compressive strength of the undamaged material.

Figure 18.7 shows a schematic illustration of the CAI test fixture and test specimen. The CAI test fixture includes a base plate, two base side plates,

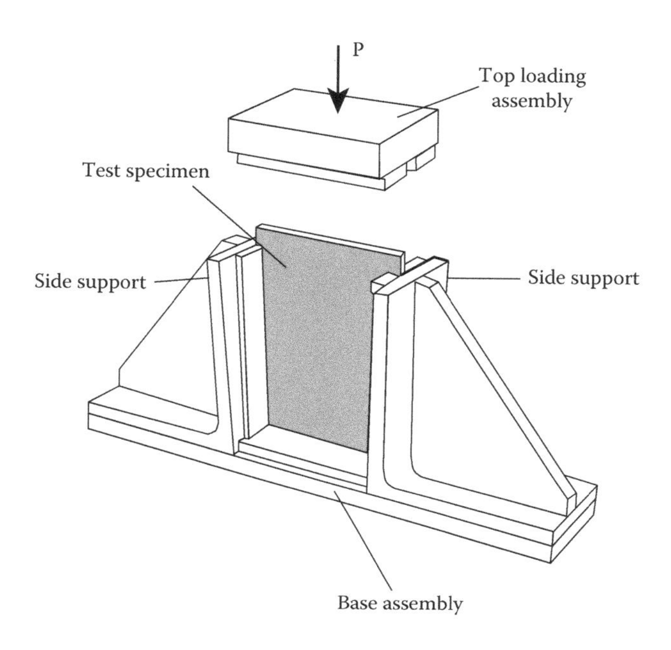

FIGURE 18.7
CAI test fixture and specimen.

two side supports, and a top assembly with two side supports. ASTM D7137 provides detailed drawings for the support fixture and top assembly.

18.3.1 CAI Test Fixture

It is well known that the edge support conditions of compression-loaded panels greatly affect buckling mode shapes, buckling loads, and the collapse load of the panel. The ASTM D7137 fixture has side supports in the form of knife-edges. Such supports constrain the vertical edges from out-of-plane displacements but do not impose rotational constraints. The horizontal top and bottom edges of the panel are in contact with flat steel platens on the top and bottom of the fixture. Out-of-plane displacements of these edges are prevented by side plates with 90° planar end surfaces put in contact with the test panel and bolted to the top and bottom (base) assemblies. These support conditions provide some degree of restraint of rotation of the horizontal edges, which thus are not strictly simply supported. Proper determination of the compression strength requires that the edge supports are in contact with the specimen without providing clamp-up. There should be no gaps between the side plates and the specimen. Such gaps diminish the edge support and may influence the failure mode of the test specimen. The fixture allows for

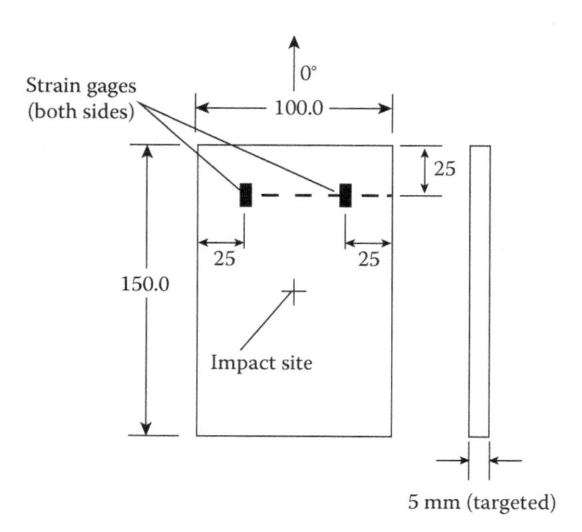

FIGURE 18.8
CAI test specimen. All dimensions are in millimeters.

small variations in specimen dimensions by adjusting the top, bottom, and side supports. The small gap between the ends of the side plates and the top plate (Figure 18.7) allows for application of compressive displacement to the specimen during the test.

18.3.2 CAI Test Specimen

The CAI specimen is shown in Figure 18.8. It has the same geometry and dimensions as the falling-weight impact specimen discussed in Section 18.2.3 and the same stacking sequences for unidirectional tape and woven fabric plies. Laminates with other layups may be tested, but ASTM stipulates a minimum of 5% of the fibers in each of the four basic directions (0°, 90°, and ±45°). Such laminates promote acceptable failure modes (discussed in the next section). Laminates with less than 5% of the fibers in any of these four directions may fail in complex failure modes. 0° is defined as the direction along the compressive loading direction.

The specimen has a relatively small width (10 cm). It is recommended that the size of the damaged central area be limited to half the unsupported width (4 cm) to achieve results representative for a large impact-damaged panel. ASTM D3137 stipulates testing of at least five replicate specimens.

The specimen may be instrumented with longitudinal (0°) strain gages on the front and back sides as indicated in Figure 18.8. Strain measurement is not required but is recommended to verify uniform loading and indicate buckling of the panel. If the test specimens are not instrumented by strain gages, ASTM recommends the use of an "alignment plate," which is an undamaged specimen instrumented with strain gages. It could also be a metal plate of similar dimensions as the composite test specimen instrumented with strain gages.

18.3.3 CAI Test Procedure and Evaluation of Test Results

Specimen dimensions should be measured prior to the impact test (see Section 18.2.3. The specimen should be installed between the vertical knife-edge side supports in the properly aligned test fixture (Figure 18.7). The center of the impact site on the specimen should coincide with the vertical line of load application. Adjust the position of the side supports to achieve a 0.8–1.5 mm gap between each support and the vertical edges of the test specimen. Tighten the bolts holding the side supports. Center the specimen by sliding the adjustable plates attached to the side supports and base plate. Hand tighten the attachment screws for the adjustable plates and place the top loading plate on the upper edge of the specimen. Hand tighten the screws for the side plates on the top loading plate. Adjust the gaps between the vertical edges of the specimen and the side plates using a feeler gage to ensure the gaps do not exceed 0.05 mm. Torque the fixture screws by approximately 7 Nm. The fixture/specimen assembly is placed between well-aligned, fixed flat plates of a general-purpose test machine. The plates should be parallel within 0.025 mm across the (100-mm) length of the top loading plate. Apply a preload of approximately 400–500 N to ensure that the loading platen is in contact with the upper edge of the test specimen and check the output from the strain gages to verify alignment of the specimen. Reduce the load to about 150 N and zero all data recording devices. Load the specimen at a crosshead rate of 1–2 mm/min to about 10% of the anticipated failure load and unload again to about 150 N. Examine the response from the strain gages. If load is applied uniformly, the response curves for the four gages should be similar. If the strain response curves are dissimilar, this would indicate that the load is applied nonuniformly or the specimen is undergoing buckling out of the plane. If this occurs, the fixture should be inspected for the presence of gaps, loose fasteners, and misalignment. If the test specimens are not instrumented with strain gages, the specially prepared alignment plates should be used to verify alignment and proper specimen support.

After the test fixture has been aligned and prepared according to the procedure outlined, the actual test specimen should be installed into the test fixture and preloaded to 450 N to ensure proper engagement and alignment and unloaded to a load of approximately 150 N. Apply compressive displacement at a rate of 1–2 mm/min while recording load, crosshead displacement, and strain data until the specimen fails in compression.

Preferred failure modes are those that are triggered by the impact damage at the center of the test panel. Sometimes, however, if the extent of impact

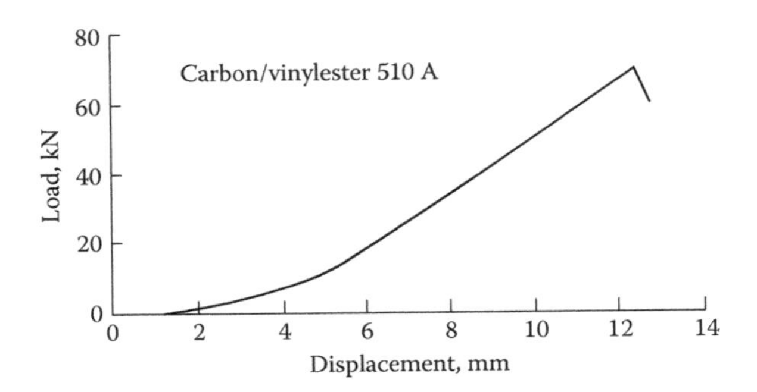

FIGURE 18.9
Load-versus-displacement response in compression for a carbon–vinylester 510 panel impacted at an energy of 48.4 J. (From Dale et al., 2012b.)

damage is small, failures are observed outside the central damaged region. Unacceptable failure modes are those related to load introduction and support conditions (end crushing) and buckling failure of the specimen. The CAI strength is calculated using

$$\sigma_{CAI} = \frac{P_{max}}{A} \tag{18.7}$$

where P_{max} is the magnitude of the compression failure load, and A is the gross section cross section area of the test panel.

18.3.4 Example of CAI Test Results

The CAI test was conducted on a carbon–vinylester panel previously impacted at a height of 70.5 cm (H = 48.4 J). The actual test panel dimensions were 10.1 × 15.3 × .529 (cm). The force trace and damage of a similar panel are shown in Figures 18.5 and 18.6. The CAI test was performed in a universal testing machine at a crosshead rate of 1.27 mm/min. The impacted panel was inspected visually, and the only visually apparent indications of damage were a circular indentation at the impact site and some longitudinal and transverse matrix splits radiating out from the impact site. Figure 18.9 shows a representative load-displacement curve for an impacted specimen. The CAI strength, Equation (18.7), is $\sigma_{CAI} = 129MPa$. As a reference, the compression strength of the undamaged material, measured using the CLC test (Section 9.4) is $X_c = 263MPa$. Hence, the reduction of compressive strength due to impact loading is 134 MPa (51% loss of strength), which is significant.

19

Sandwich Beam and Core Testing

19.1 Introduction

A structural sandwich (Figure 19.1) consists of two thin face sheets made from a dense and stiff material bonded to a low-density core. This structure provides very high bending and torsional stiffnesses at very low weight and has become popular for use in weight-critical structures such as parts of airplanes, space structures, ship hulls, sporting goods, and the internal structure of the blades for wind-power generation. As discussed in several classical and recent texts on sandwich structures (Allen 1969, Plantema 1966, Whitney 1987, Vinson 1999, Zenkert 1997, Carlsson and Kardomateas 2011), the primary function of the core, similar to the web in an I-beam, is to separate and support the face sheets against local buckling (wrinkling or face dimpling) and to transfer shear force between the faces without failure or excessive deformation. Also, the bond between the face and core is a critical component of this structure that is required to maintain the high stiffness of the sandwich panel.

Core materials (Figure 19.2) are classified within two broad categories: "cellular" or "structural." Cellular implies that the material consists of "cells" containing open space enclosed by walls in a repetitive manner so that space filling is achieved. Cellular foams (e.g., polymer or metal foams) and honeycomb and balsa wood cores are common in structural applications. Structural core, also called "web core," consists of a continuous web made from a solid material formed in such a way that it separates the faces.

Because core materials typically are lightweight and of low extensional stiffness and strength (Gibson and Ashby 1997), the core is commonly the weak constituent of a sandwich. In several instances, the bond between face and core may be critical for the integrity of the sandwich (Carlsson and Kardomateas 2011). In some cases, the core shear strength can be critical, and in all cases, the shear stiffness of the core is important to sandwich performance.

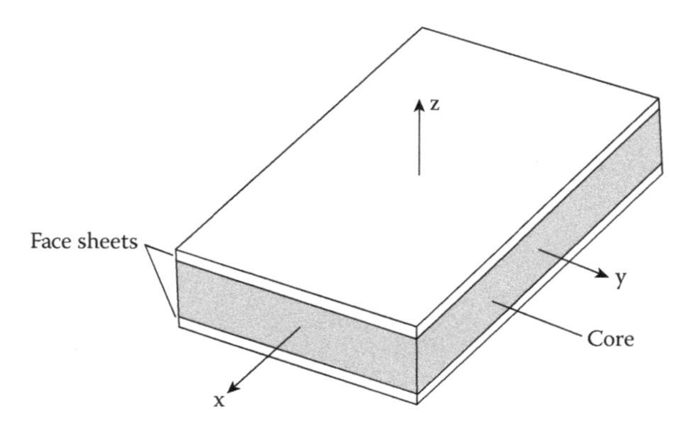

FIGURE 19.1
Sandwich panel consisting of two face sheets bonded to a low-density core.

19.2 Sandwich Beam Flexure

The American Society for Testing and Materials (ASTM) C393 (2011) standard test method covers determination of the core shear properties by flexural testing of sandwich beams. This standard applies to balsa and foam

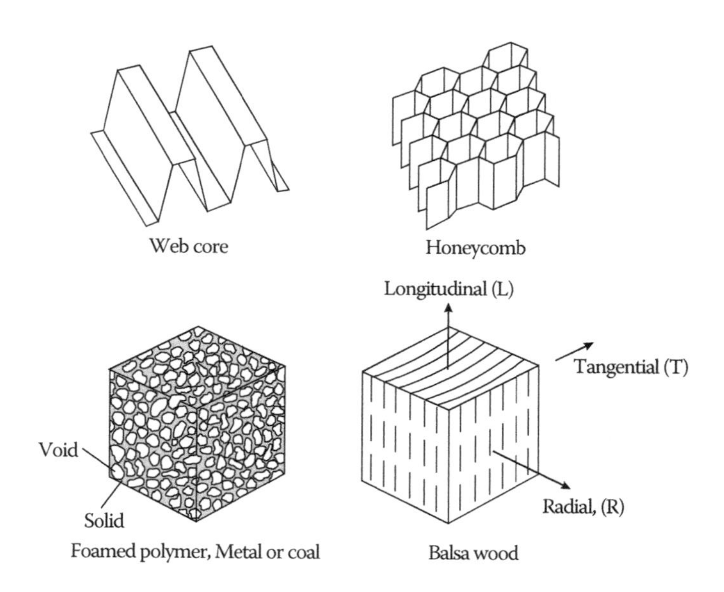

FIGURE 19.2
Core concepts in sandwich structures.

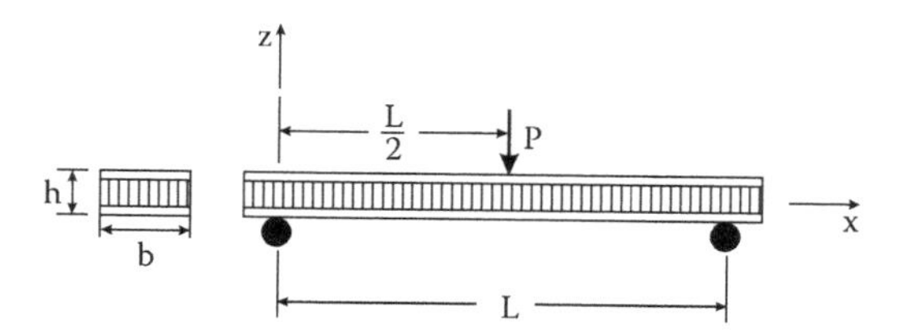

FIGURE 19.3
Three-point flexure loading of a sandwich beam.

cores as well as honeycomb cores, although outside the standard, this test can also be used to determine the effective shear modulus of web core sandwich (Nordstrand and Carlsson 1994). This section describes determination of both the core shear modulus and shear strength of the core and the flexural stiffness of a sandwich beam.

Consider a sandwich beam loaded in three-point flexure as shown in Figure 19.3. The deflection δ under the centrally applied load is given by the following equation (Carlsson and Kardomateas 2011):

$$\delta = \frac{PL^3}{48bD} + \frac{PL}{4h_c bG_c} \quad (19.1)$$

$$D = \frac{bh_f E_f}{2}\left[\frac{h_f^2}{3} + \left(h_c + h_f\right)^2\right] \quad (19.2)$$

where P is the central load applied, L is the span length, b is the beam width, h_c is the core thickness, D is the bending stiffness per unit width, E_f is the face elastic modulus, h_f is the face sheet thickness, and G_c is the core shear modulus.

This equation allows for determination of the bending stiffness of the sandwich beam and core shear modulus, as discussed in this section. Similar to the three-point flexure loaded composite beam discussed in Chapter 11, testing of long sandwich beams allows determination of the face sheet strength in tension or compression, while testing of short beams allows determination of the shear strength of the core.

The standard test specimen is a 20 cm long and 7.5 cm wide beam with rectangular cross section. For nonstandard specimens, the width should be within the range [2h, 6h], where h is the total beam thickness, to avoid anticlastic bending-induced lateral deformations. Furthermore, the width should exceed three times the cell size of the core and be less than half the span length. The total beam length should exceed the span length by 5 cm or

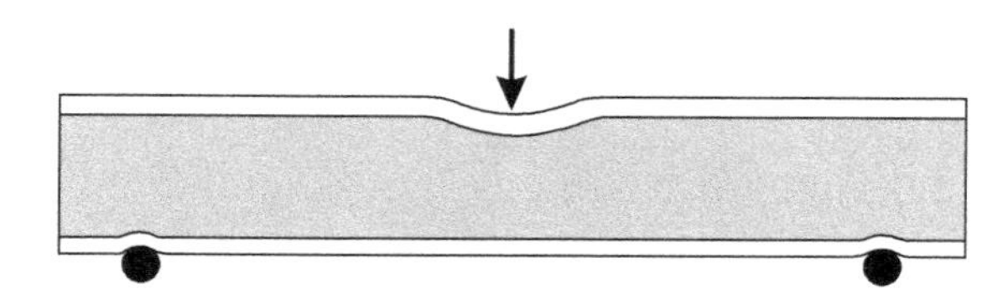

FIGURE 19.4
Indentation of sandwich beam with thin face sheets and soft core at load introduction and support regions.

half the total beam thickness h, whichever is greater. After the beam specimens are prepared, measure the cross-section dimensions at several locations along the span length. The accuracy of length measurement should be ±0.25 mm, and the cross-sectional measurements should be accurate within ±0.025 mm.

Three-point flexure loading of sandwich beams with thin face sheets and a soft core or web core is complicated by local out-of-plane deformation (indentation) of the face sheets at support and load introduction regions, as schematically illustrated in Figure 19.4. For web-core sandwich beams, indentation is accentuated if the loading and support locations coincide with unsupported regions of the web core.

Indentation at the central line of load introduction is expected to be more severe than at the supports because this load is twice as large as the support loads. If local deformation is expected or detected in trial tests, ASTM C393 recommends placing 25 mm wide and 3 mm thick rubber pressure pads between load introduction and support points and the sandwich to distribute the load over a large area. ASTM C393 also provides some general instructions for the selection of testing machines, fixturing, and instrumentation and some instructions on the flexure loading fixture. It is common procedure to use steel cylinders with a diameter of 25 mm at the load introduction and support regions. For measuring deflection δ of the beam under the central loading line, a displacement gage with an accuracy within ±0.025 mm should be used. A linear variable differential transformer (LVDT) (see Section 4.12) is often a good choice.

Testing of the sandwich beams should be performed at a constant crosshead speed of 6 mm/min. For determination of the bending stiffness and core shear modulus, the beam compliance C, defined as deflection under the load δ divided by the load P (i.e., $C = \delta/P$) should be measured at several span lengths. This requires application of a small load to avoid plastic deformation of the specimen. After completion of each loading-unloading cycle, the beam is unloaded, and the span length may be adjusted and the compliance measured again. This procedure is nondestructive and should provide a set of compliance values for the beam. Allen (1969) proposed the following data reduction methodology for the three-point flexure test based on the following compliance expression obtained from Equation (19.1):

$$C = \frac{L^3}{48bD} + \frac{L}{4h_c bG} \tag{19.3}$$

This equation may be expressed in two ways:

$$\frac{C}{L} = \frac{L^2}{48bD} + \frac{1}{4bh_c G_c} \tag{19.4a}$$

$$\frac{C}{L^3} = \frac{1}{48bD} + \frac{1}{4bh_c G_c}\frac{1}{L^2} \tag{19.4b}$$

Consequently, C/L is plotted versus L^2 and C/L^3 is plotted versus $1/L^2$ (see Figure 19.5). The bending and shear stiffnesses are determined from the slopes m_1 and m_2 of the graphs according to Equations (19.4):

$$m_1 = \frac{1}{48bD} \tag{19.5a}$$

$$m_2 = \frac{1}{4bh_c G_c} \tag{19.5b}$$

The slopes are best evaluated by fitting a linear equation to the data. D and G_c are easily calculated from Equations (19.5).

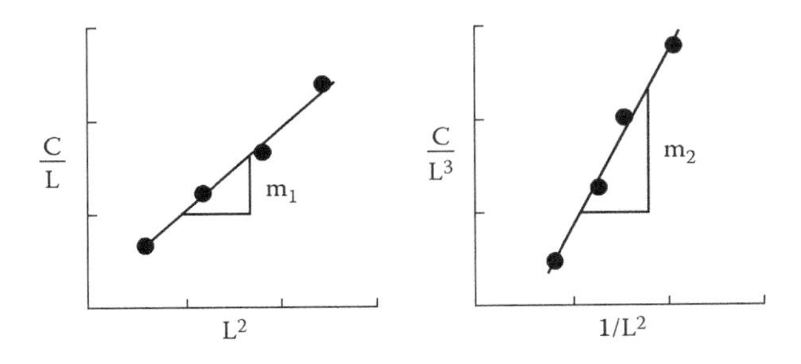

FIGURE 19.5
Plots of sandwich beam compliance to determine bending stiffness and core shear modulus.

TABLE 19.1

Span Lengths and Experimental Compliance Values for a Three-Point Flexure-Loaded Sandwich Beam

L (cm)	C (μm/N)
25.4	1.59
35.6	2.38
45.7	3.18
55.9	4.13

To demonstrate the procedure for determining bending stiffness and shear modulus, actual test data for a sandwich beam consisting of 3.02-mm thick aluminum face sheets and a 26-mm thick H80 polyvinyl chloride (PVC) foam core is examined. The total length of the beam was 61 cm (nonstandard), and the width b was 52 mm. Notice that the width is twice the beam thickness and the length about 24 times the beam thickness, which satisfies the ASTM C393 requirements. The beam was tested in three-point flexure at span lengths L of 25.4, 35.6, 45.7, and 55.9 cm. The deflection δ under the central load was measured with an LVDT. The load levels were adjusted to obtain a reasonable part of the initial linear response region without causing yielding of the face sheets and core. Span lengths and measured compliance values are listed in Table 19.1.

Figure 19.6 shows compliance data plotted according to Figure 19.5 and straight lines fitted to the data. The slopes of the lines are $m_1 = 4.35x10^{-6}(Nm^2)^{-1}$ and $m_2 = 5.933x10^{-6} N^{-1}$.

Solving for D and G_c using the slopes in Figure 19.6 and Equations (19.5) yields $D = 92.1$ kNm and $G_c = 31.2$ MPa. The core shear modulus is very close

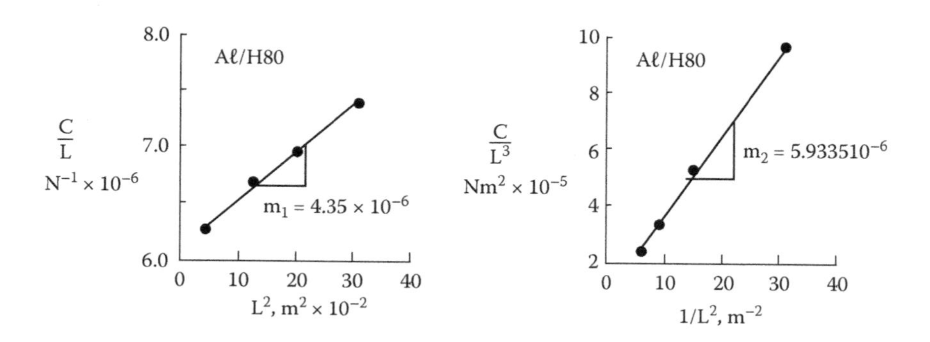

FIGURE 19.6
Compliance plots for sandwich beam with aluminum face sheets and H80 PVC foam core for determination of bending stiffness and core shear modulus.

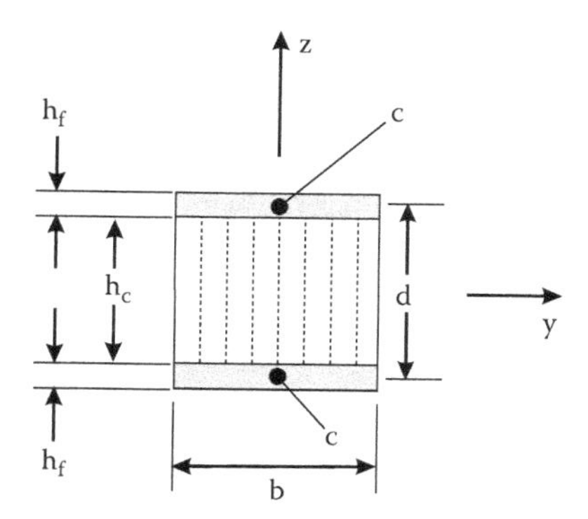

FIGURE 19.7
Cross-sectional view of a symmetric sandwich beam. c represents the centroid location for the face sheets, and y is the neutral axis of the beam.

to the value provided for the H80 foam by the foam manufacturer (DIAB): $G_c = 31\ MPa$ (http://www.diabgroup.com).

The bending stiffness, D = 92.1 kNm, can be used to calculate the elastic modulus of the face sheets. A symmetric sandwich beam with the cross section as indicated in Figure 19.7 is considered.

For a symmetric sandwich with thin faces, $h_c/h_f \geq 5.35$ (Carlsson and Kardomateas 2011), the bending stiffness is given by

$$D = \frac{E_f h_f d^2}{2} \tag{19.6}$$

where E_f is the elastic modulus of the face sheets, and h_f is the face thickness, h_c is the core thickness, and $d = h_c + h_f$. For the specific sandwich beam considered here, $h_c/h_f = 26/3.02 = 8.61$ (thin faces), and Equation (19.6) should apply. Calculation of the face modulus from the measured bending stiffness (D = 92.1 kNm) yields $E_f = 72.5\ GPa$, in good agreement with handbook values for aluminum alloys.

19.2.1 Evaluation of Core Shear and Face Tensile or Compressive Strengths

In addition to measurement of the bending stiffness and core shear modulus, the three-point flexure test may be used to determine the shear strength of the core and compression strength of the face sheets. Such testing requires specific design of the test since both failure modes will not occur simultaneously. Further, proper adhesion between the face sheets and core must be ensured. To achieve the desired failure mode of the beam, the stresses in the face sheets and core must be determined. We here examine a symmetric sandwich beam, that is, with identical face sheets, each of thickness h_f and elastic modulus E_f bonded to a low-modulus core of thickness h_c.

Analysis of the bending stress in the face sheets (Carlsson and Kardomateas 2011) reveals that the maximum bending stress (magnitude) occurs at mid-span and is given by

$$\sigma_x(\max) = \frac{PL}{4b(h_f + h_c)h_f} \tag{19.7}$$

Hence, if σ_x (max) exceeds the tensile or compressive strength of the face sheet, beam failure should occur. By observation of the failure mechanism, it should be possible to associate σ_x (max) with the tensile or compressive strength of the face sheet. Most composite materials are weaker in compression than in tension, and for metals, these two strengths are similar. Hence, beam failure occurs typically by face compression.

Shear failure of the core may occur if the core shear stress τ_{xz} exceeds the core shear strength. Analysis of the core shear stress (Carlsson and Kardomateas 2011) yields the following expression:

$$\tau_{xz} = \frac{P}{2bh_c} \tag{19.8}$$

The shear stress is taken as uniform through the thickness of the core and along the length of the beam (although a sign reversal occurs at $x = L/2$).

The ratio of maximum bending stress in the face sheets and shear stress in the core [Equations (19.7) and (19.8)] becomes

$$\frac{\sigma_x(\max)}{\tau_{xz}} = \frac{L}{2h_f(1 + h_f/h_c)} \tag{19.9}$$

This expression can be utilized for determining the geometry where a transition from core shear failure to face sheet failure occurs.

As an example, the sandwich beam examined in the example (aluminum faces and H80 PVC core) is considered. Equation (19.9) with the face and core thicknesses ($h_f = 3.02mm, h_c = 25mm$) substituted provides

$$\frac{\sigma_x(\max)}{\tau_{xz}} = \frac{L}{6.742} \tag{19.10}$$

where L is in units of millimeters. The yield stress of 6061-T6 aluminum is σ_{ys} = 270 MPa (Gere and Goodno 2013), and the shear strength of the

H80 PVC foam core is $\tau_c = 1.15\ MPa$ (http://www.diabgroup.com). With these data, Equation (19.10) yields L = 1.58 m. This indicates that beams with span lengths less than 1.58 m would fail by core shear failure. Thus, if the objective of the test is to determine strength of the face sheets in tension or compression, beams with very long span lengths should be tested, but such beams may undergo excessive deflection or may fail by other failure modes, such as face wrinkling (Carlsson and Kardomateas 2011). Overall, flexure tests are not desirable for materials testing.

19.3 Sandwich Core Shear Testing

The shear modulus and shear strength of the core are among the most important properties of a sandwich panel. As discussed, a major function of the core is to effectively transmit shear force between the top and bottom faces in a sandwich panel. Inadequate transfer of shear results in severely degraded performance of the sandwich structure. The shear response of honeycomb and web cores is typically measured in the two principal planes (yz and xz) as illustrated in Figure 19.8. Isotropic cores may be tested in any direction.

The ASTM standard C273 (2011) (i.e., the plate shear test; Figure 19.9) considers a sandwich or core specimen adhesively bonded to two steel blocks that are loaded in tension or compression.

The plate shear test does not produce a perfectly uniform state of shear in the entire test specimen. By using long specimens and adjusting the

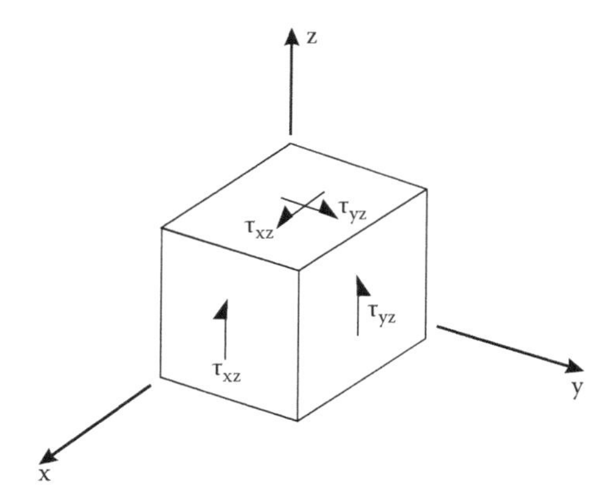

FIGURE 19.8
Core element under shear stresses τ_{xz} and τ_{yz}. The z-axis is the thickness coordinate.

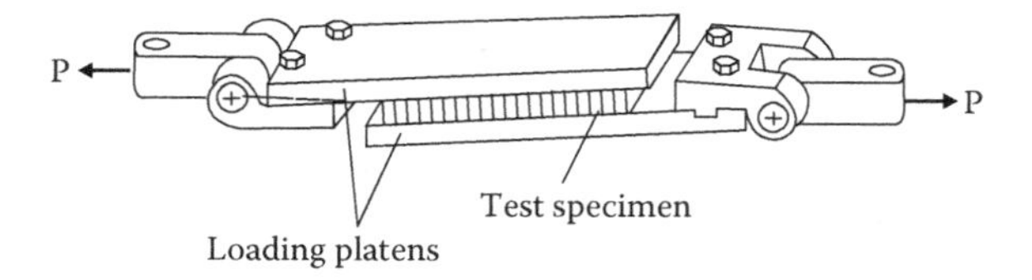

FIGURE 19.9
Plate shear specimen. A sandwich or core specimen is bonded to the plates.

specimen length so that the line of load passes through the diagonally opposite corners of the specimen (see Figure 19.10), it is possible to minimize the influence of secondary stresses on the response. Notice that the line of load action is defined by the centers of the pins at A and B in Figure 19.10.

The loading plates are typically made from steel. The thickness of the plates could be determined according to the ASTM C273 recommendation that their bending stiffness per unit with width EI/b should exceed $2.67x10^6$ Nmm per millimeter of core thickness. Typically, the plate thickness is about 1 cm. The test specimen should have a width of 50 mm or more and a length more than 12 times the thickness, while the loading line condition discussed (Figure 19.10) should be targeted as closely as possible. At least five replicate test specimens are required. The ASTM D5687 (2007) guide provides instructions for specimen preparation. Dimensions of specimens should be measured before bonding to the loading plates. Measurements of length and width should be within 1% of the dimension. The thickness should be measured within ±0.025 mm. To determine shear strength of strong core materials, a very strong aerospace-grade epoxy adhesive is required. Notice that a thick flexible bond layer will deform in shear and add to the measured compliance. Such effects will reduce the apparent shear modulus of the core in an uncontrolled manner. To promote adequate bonding of strong honeycomb cores, a deep adhesive fillet on the honeycomb cell walls may require two plies of film adhesive. Alignment of the specimen during bonding is important and difficult to achieve without proper care since the uncured, low-viscosity adhesive is slippery.

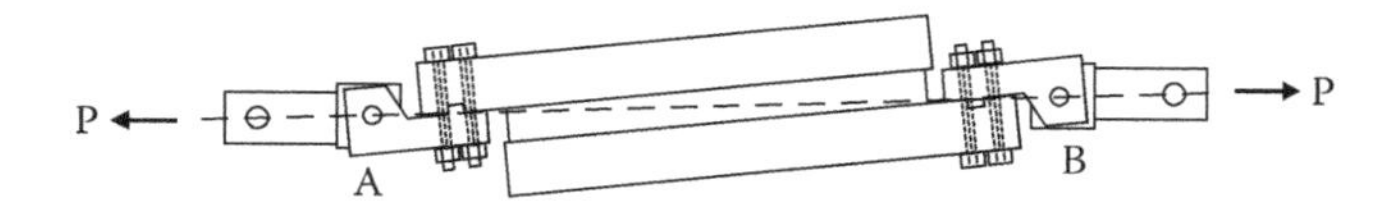

FIGURE 19.10
Specimen length is adjusted so that the line of load is passing through opposite corners of the specimen.

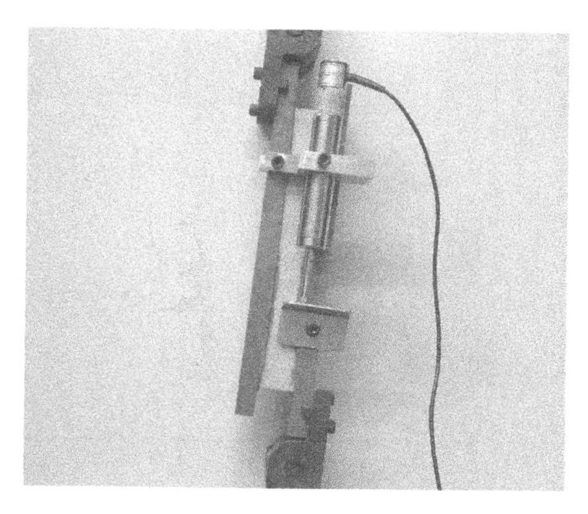

FIGURE 19.11
LVDT is attached to the plate shear test for measurement of shear displacement.

After mounting the fixture assembly in the test frame, a fixture that supports an LVDT or an extensometer may be attached to each of the steel plates to allow direct measurement of the shear displacement during the test. Figure 19.11 shows a photograph of such an arrangement. The specimen should be loaded at a crosshead rate of 0.5 mm/min while recoding load versus shear displacement until failure of the test specimen occurs. If the specimen fails at the specimen–steel plate interface, the test will not provide the shear strength of the core, while the shear modulus determined from the test at low stress and strain levels may still be valid.

The shear stress τ is determined from

$$\tau = \frac{P}{Lb} \tag{19.11}$$

where L and b are the length and width of the test specimen. The core shear modulus G_c is determined from

$$G_c = \frac{\tau}{\gamma} \tag{19.12}$$

where $\gamma = u/h_c$, with u being the relative displacement of the loading plates in the length dimension of the plates, and h_c is the core thickness. G_c is determined from the slope of the initial linear-elastic region of response

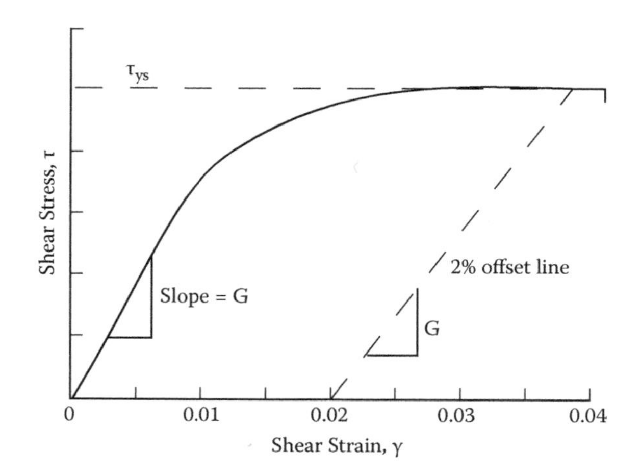

FIGURE 19.12
Schematic curve of shear stress versus shear strain for core material.

(see Figure 19.12). The shear strength is determined using the maximum load. Sometimes, a yield strength in shear τ_{ys} is evaluated using a 2% offset shear strain (ASTM C273 2011). Figure 19.12 illustrates schematically the shear response curve and the determination of the yield stress.

Figure 19.13 shows an example of a shear stress–shear strain curve for a F50 PES foam determined by the plate shear test. The foam specimen was 15.2 cm long, 5.08 cm wide, and 1.27 cm thick. The specimen was bonded to the steel plates using Hysol EA 9309.3NA epoxy adhesive, and displacement was measured using an LVDT as shown in Figure 19.11. Based on the

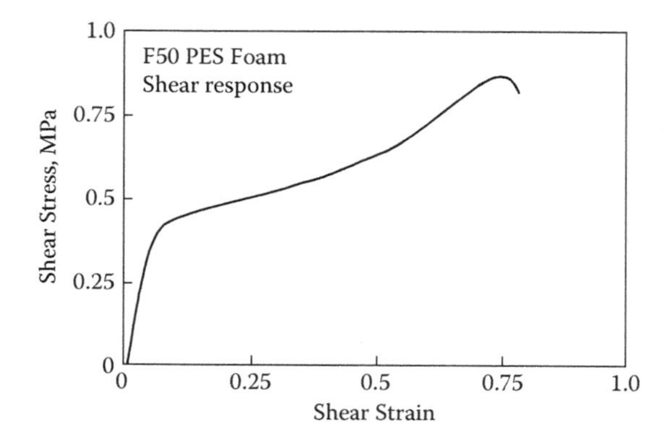

FIGURE 19.13
Shear-stress-versus-shear-strain response for a F50 PES foam. (After Saenz, 2012.)

stress–strain curve shown in Figure 19.13, the shear modulus, and shear strength τ_{ult} were determined as $G = 8.9$ MPa, and $\tau_{ult} = 0.87$ MPa.

19.4 Sandwich Through-Thickness Tensile Testing

The ASTM C297 standard (2010) is aimed at through-thickness tensile testing of a sandwich, or a core. Load is introduced in the sandwich through loading blocks adhesively bonded to the face sheets or to the core. If a sandwich specimen is tested, this method can be used to examine the integrity of the face–core bond. If a core material is tested, the test provides the through-thickness tensile strength of the core. The test specimens should have a square or circular cross section. The loading blocks should be self-aligning and thick enough to support the core specimen during loading. ASTM C297 suggests blocks 40–50 mm thick are adequate. Tolerance for the diameter or side length of the loading blocks is ±0.025 mm. Figure 19.14 shows the test configuration used by Viana and Carlsson (2002) for testing of PVC foam cores between adhesively bonded aluminum blocks of circular cross sections. The specimen gage length (L_G in Figure 19.14) is defined by the thickness of the foam panel. The minimum cross-sectional area is

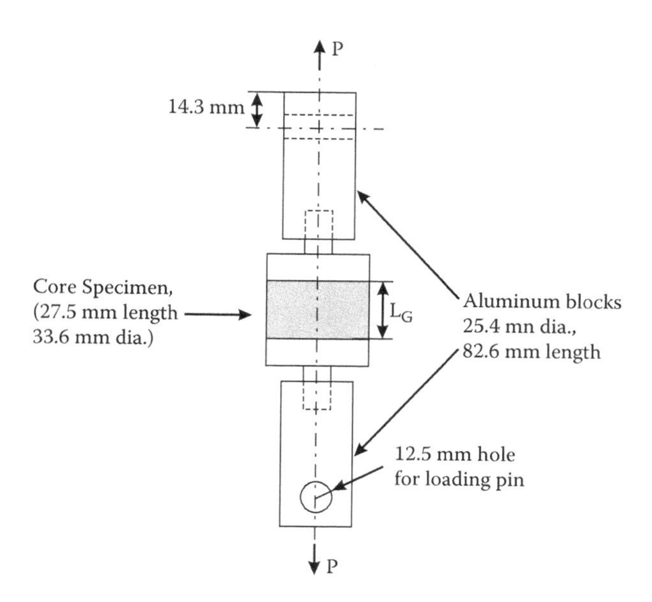

FIGURE 19.14
Through-thickness tension test setup.

625 mm² for foam and balsa wood cores and depends on the cell size for honeycomb cores. ASTM C297 recommends an area to accommodate at least 60 cells for honeycomb cores. The largest area recommended for this type of test is 5625 mm².

Specimen machining is important for this test. The specimens may be machined using water-lubricated precision sawing, milling, or grinding. Measure the specimen dimensions using an instrument with an accuracy of ±0.025 mm or more before bonding. The specimen diameter or side length should be within ±0.025 mm of the corresponding dimension of the loading block. The specimens should be bonded to the loading blocks using a strong aerospace-grade epoxy adhesive. The bonding surfaces should be prepared as outlined in section 4.6. At least five specimens should be prepared and tested.

Although the main focus of the ASTM C297 test is on tensile strength determination of the sandwich or core, it is possible to determine the out-of-plane modulus of the core E_z if an extensometer is fitted to the edge of the core specimen. It should, however, be pointed out that the end blocks constrain lateral deformation of the core, which tends to increase the apparent modulus. Mechanical properties should ideally be determined under uniaxial stress conditions using longer specimens.

After mounting the test specimen in the fixture, the test specimen is loaded to failure at a crosshead rate of 0.5 mm/min. Modulus E_z and tensile strength Z_t are determined from

$$E_z = \frac{\sigma_z}{\varepsilon_z} \tag{19.13}$$

$$Z_t = \frac{P_{ult}}{A} \tag{19.14}$$

where $\sigma_z = P/A$ (load/unit cross-sectional area), ε_z is the strain, and P_{ult} is the failure load of the specimen. The modulus E_z should be determined from the initial linear region of stress–strain response. Notice here that the only acceptable failure mode defining core strength is the one where the core fails. Such a failure may be difficult to achieve for honeycomb cores and end-grain balsa wood cores, for which the adhesive bond may be the weakest link.

Figure 19.15 shows an example of a through-thickness tensile stress–strain curve for an H100 PVC foam specimen (Viana and Carlsson 2002) using the test principle shown in Figure 19.14. The specimen was cut from a 27-mm thick foam panel using a circular saw and bonded to the loading blocks using epoxy adhesive. Strain was estimated from the crosshead displacement corrected for machine/fixture compliance. The gage length was 27 mm for this case.

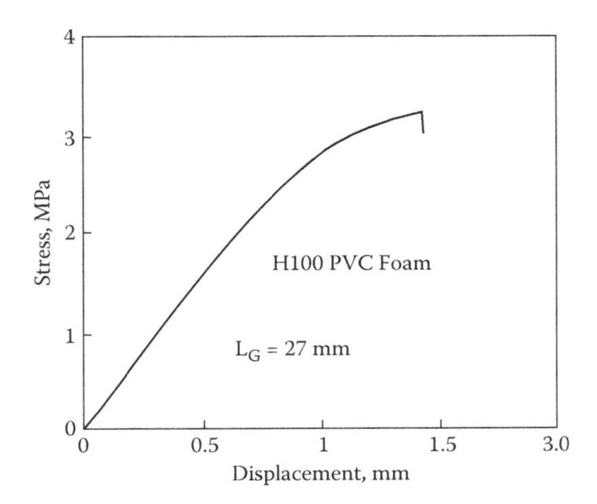

FIGURE 19.15
Tensile stress–strain response (through thickness) for H100 PVC foam. (After Viana and Carlsson, 2002, *J. Sandwich Struct. Mater.*, 4, 99–113.)

Based on the stress–strain curve shown in Figure 19.15, $E_z = 126MPa$ and $Z_t = 3.22MPa$. The modulus value should be considered as apparent due to the short gage length of the specimen.

19.5 Sandwich Core Compression Testing

The out-of-plane compression modulus and strength of the core are fundamental properties that govern wrinkling failure and resistance to localized loading. A compression test method consisting of compression of a short unsupported specimen between parallel metal platens is outlined in ASTM Standard C365 (2009). Such a test may be performed on a core or sandwich specimen. The test specimen should be machined from a core or sandwich panel and should have a circular or square cross section. The minimum cross-sectional area should be at least 625 mm^2 for foam and balsa cores. For honeycomb cores with cell size below 6 mm, the cross-sectional area should be 2500 mm^2. For cells greater than 6 mm, the area should be 5800 mm^2. To avoid crushing of the ends of honeycomb cores, the ends may be reinforced with a layer of resin, or thin faces may be bonded to the core. With reinforced ends, the test is called a "stabilized compression test;" tests on unreinforced honeycomb core are called "bare compression tests."

The specimens should have a square or circular cross section. The cross-sectional area should not exceed 100 cm^2. The flatwise compression test

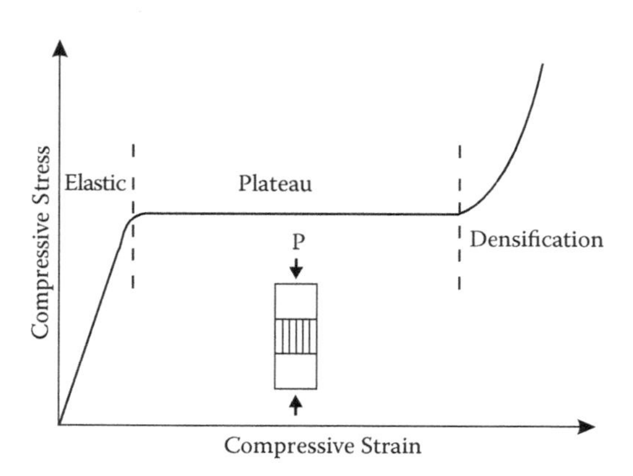

FIGURE 19.16
Schematic stress–strain curve for compression loading of cellular materials.

requires specimens of uniform thickness. Follow the machining recommendations outlined for the tension test in Section 19.4. For measuring the side length or diameter of the specimen, use an instrument with an accuracy of at least ±1%. The thickness should be measured with an accuracy of ±0.001 mm.

Test at least five specimens. The specimen should be compressed between two parallel platens at 0.5 mm/min. The loading platens should be self-aligned through a spherical ball joint to promote a uniform distribution of compressive stress over the end surfaces. Displacement of the specimen during loading may be monitored using the crosshead travel or, more accurately, by an extensometer fitted to the top and bottom ends of the gage section of the specimen.

As discussed by Gibson and Ashby (1997), compression loading of cellular materials tends to produce a stress–strain curve of the type shown in Figure 19.16, where the stress reaches a constant plateau level after the initial linear region of response, and a steeply increasing region at large strains.

The initial linear region corresponds to elastic extensional and bending deformation of the cell edges and walls, while the plateau region occurs due to the formation of a band where localized buckling and collapse of the cells occur. On further deformation, this band progresses through the gage region until all the empty space within the core is consumed, and the cell walls begin to impinge on each other and take the direct load in compression. At this point (or before), the compression test should be stopped to protect the test equipment. The modulus and compressive strength E_z and Z_c, respectively, are reduced from the compressive stress–strain curve (Figure 19.16).

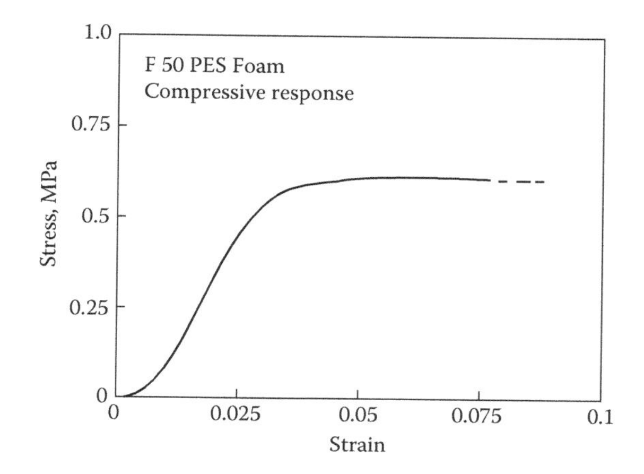

FIGURE 19.17
Stress–strain curve in compression (through thickness) for a F50 PES foam. (After Saenz, 2012).

An example of compressive stress–strain response for a F50 PES foam is shown in Figure 19.17. The specimen was 2.54 cm long, and the cross section dimensions were 5.08 x 5.08 (cm).

The strain is here defined as crosshead displacement over specimen length, which is not an accurate strain measure. The compressive strength, however, should be accurate, $Z_c = 0.32 MPa$.

19.6 Sandwich Edge Compression Test

The objective of the sandwich edge compression test, ASTM Standard C364 (2007), is to determine the compression strength of a sandwich panel loaded uniaxially in the plane of the sandwich panel. Figure 19.18 illustrates the test procedure schematically.

A relatively short specimen is centered and clamped in the test fixture and loaded until it fails in compression. This test is specifically used for a sandwich panel with a cellular foam core or balsa wood or honeycomb core. The edgewise compression strength provides a basis for determining the compressive strength of a sandwich panel and the "in situ" compression strength of the face sheets. As discussed in the ASTM standard and by several investigators (e.g., Allen 1969, Zenkert 1997), the face sheets may undergo face dimpling or wrinkling failure prior to material failure in compression (Chapter 9). Such instabilities are especially likely if the face sheets are thin.

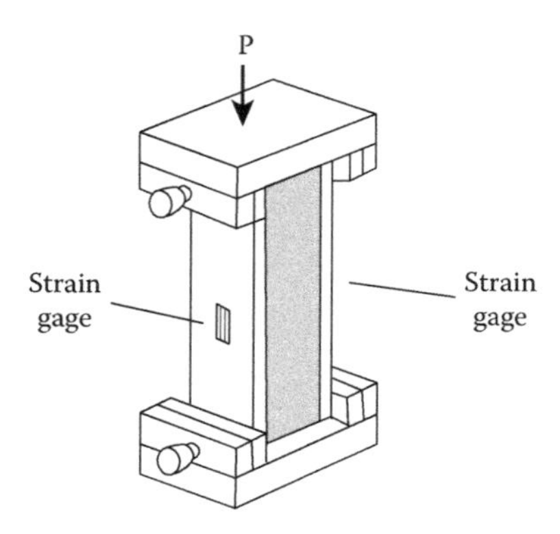

FIGURE 19.18
Sandwich edge compression test.

The sandwich column, even if short, commonly undergoes instability failure. Such failures are representative for an actual sandwich structure and are considered acceptable. Unacceptable failures are those occurring at or near the supported ends.

19.6.1 Test Specimen

The edgewise compression test specimen is a rectangular test piece cut from a larger sandwich panel using a fluid-cooled saw or water jet (see Chapter 4 and ASTM D5687 Guide 2007). Of particular importance for this end-loaded compression test is that the end surfaces are flat and perpendicular to the loading axis. Figure 19.19 shows the specimen geometry and dimensions. At least five specimens should be prepared and tested.

The dimension L in Figure 19.19 refers to the loading direction. ASTM C364 specifies that $L \leq 8h$, where h is the thickness of the sandwich panel

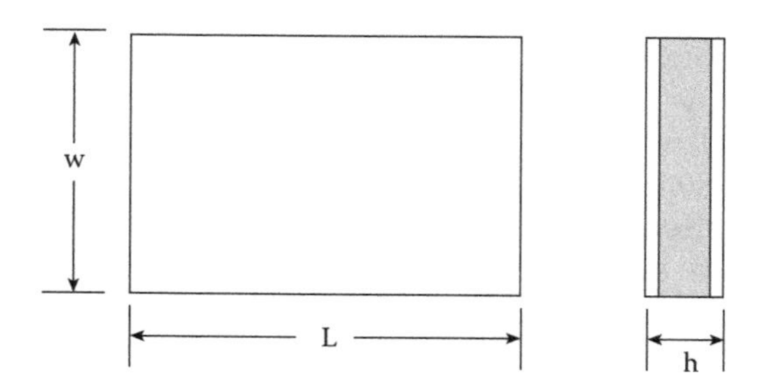

FIGURE 19.19
Edgewise compression test specimen geometry and dimensions.

(Figure 19.19). This restriction on the length is to avoid global buckling failure of the specimen. The width w should be greater than 5 cm and less than L, that is, $\{5cm \leq w \leq L\}$. Furthermore, $w \geq 2h$, and for honeycomb cores, $w \geq 4$ x cell size. Measure the panel dimensions at several locations. The length and width measurements should be within 1% of the dimension, and the thickness measurement should be within ±25 μm. The dimensional tolerances are ±0.25 mm.

The specimen is to be instrumented with at least two axial strain gages bonded to opposite faces. The gages should be bonded at the center of each face sheet (Figure 19.18).

19.6.2 Compression Test Fixture

The test specimen and fixture are illustrated in Figure 19.18. The clamps should be made from steel arranged so that centering and clamping of the specimen can be accomplished. The overwhelming majority of compressive load imposed on the specimen is due to end pressure. The clamps serve to achieve centering of the specimen and partial support against bending deformation to prevent buckling instability failure. The cross-sectional dimensions of the rectangular clamping bars in touch with the specimen surfaces should be at least 6 mm. ASTM C364 suggests alternate clamping arrangements, although these appear less practical and versatile than the adjustable steel clamps discussed.

19.6.3 Compression Test Procedure

Install the specimen and the test fixture in a universal test machine. An LVDT or noncontact laser displacement sensor can be useful for monitoring lateral bending deformation (buckling) of the specimen. The LVDT should be placed in contact with one of the face sheets at a point near the central gage section. After clamping the specimen and centering the fixture in the test machine, zero the load, strain gage, and LVDT signals. Set the crosshead rate at 0.5 mm/min and apply load until the specimen fails while recording the load, strains, and displacement signals. Observe how the specimen fails during the test.

19.6.4 Postfailure Observations and Data Reduction

Inspect the failed specimen and record the mode of failure and its location. Specimen failure by end crushing in the clamped region is not an acceptable failure mode, and the test results should be rejected. Global buckling of the panel manifested in out-of-plane bending is another unacceptable failure mode. Face sheet compression failure, face sheet wrinkling or dimpling failures, and debonding of one face sheet from the core are examples of acceptable failure modes.

TABLE 19.2

Layer Thicknesses and Material Properties of Sandwich Panel with Glass–Vinylester Face Sheets and H100 PVC Foam Core

h_f (mm)	h_c (mm)	E_f (GPa)	E_c (MPa)	G_c (MPa)	X_f^c (MPa)
3.6	50	21.3	111	40	365

If the specimen fails in an acceptable failure mode, the ultimate compression strength of the face sheets is calculated as follows:

$$\sigma_f^c = \frac{P_{\max}}{2wh_f} \tag{19.15}$$

where $P_{\max}$ is the failure load, w is the specimen width, and h_f is the thickness of each face sheet.

19.6.5 Examples of Edgewise Compression Testing

An edgewise compression test fixture was built according to the ASTM C 364 standard (see Figure 19.18). Compression tests were conducted on sandwich specimens machined from a sandwich panel with glass–vinylester face sheets and an H100 PVC foam core (Vadakke and Carlsson 2004). Table 19.2 lists the thicknesses and material properties of the constituent face and core materials.

Rectangular steel clamps were used to position the specimen at the center of the fixture and to clamp the sandwich specimen. The length of the clamped region was 25.4 mm. All specimens were nominally 37.5 mm wide. Load was applied to the ends of the sandwich faces through platens attached to the crosshead and base of the test machine. Both sides of the specimen were fitted with axially oriented strain gages to monitor the strain during loading. Notice that the specimen width w is 3.75 cm, somewhat less than the 5 cm specified as a minimum in the ASTM C364 standard. The total thickness h of the specimen is $h = 2h_f + h_c = 5.72$ cm (Table 19.2). Again, the width falls below the minimum specified in the ASTM C364 standard ($w \geq 2h$). More important is that the length of the tested specimen is not too long to avoid global buckling failure. In this study, specimens of length L = 7.58 cm and 12.6 cm were tested corresponding to unsupported lengths (gage lengths) of 2.5 and 7.5 cm. These lengths conform to the ASTM recommendation ($L \leq 8h$ = 45.8 cm) and exceed the width (w = 3.75 cm). As will be demonstrated, these specimens did not fail by global instability.

Compression testing was conducted on the sandwich specimens at a crosshead speed of 0.8 mm/min. Figure 19.20 shows the load-versus-strain results.

The response curves shown in Figure 19.20 reveal almost identical load–strain response for both face sheets. This is an indication that the load is distributed uniformly, and that global buckling did not occur. The short specimen (L = 7.58 cm) failed by "edge failure" inside the top grip region,

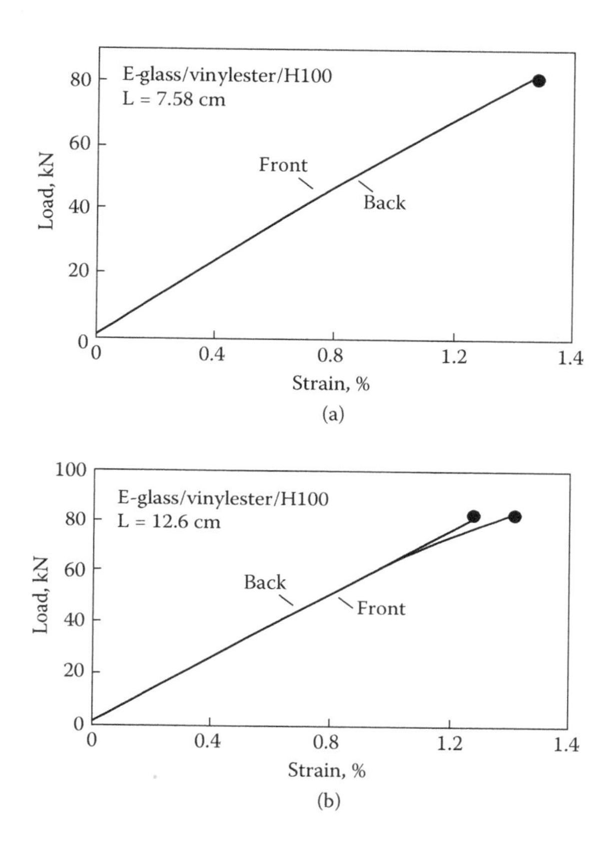

FIGURE 19.20
Load–strain response for edge compression loaded sandwich specimens: (a) $L = 7.58$ cm; (b) $L = 12.6$ cm.

which is considered unacceptable. The longer specimen ($L = 12.6$ cm), however, failed by local buckling or wrinkling of the face sheets in an antisymmetric mode (see Vadakke and Carlsson 2004). This failure mechanism is consistent with the slight divergence of the strain response observed near the failure load (Figure 19.20b). Such a failure mode is considered acceptable. The failure load is about 80 kN. Equation (19.15) provides a face sheet stress of $\sigma_f^c = 296 MPa$. This stress falls below the independently measured compression strength of the face sheets: $X_f^c = 365 MPa$ (Table 19.2). The face wrinkling stress may be estimated using the classical Hoff-Mautner (1945) equation:

$$\sigma_w = 0.5\sqrt[3]{E_f E_c G_c} \tag{19.16}$$

where the factor 0.5 is an empirical factor (see Zenkert 1997). With the data in Table 19.2, Equation (19.16) yields $\sigma_w = 228 MPa$, which agrees reasonably with the test results.

Appendix A: Compliance and Stiffness Transformations and Matrix Operations

Transformation of plane stress compliance (S_{ij}) and reduced stiffness (Q_{ij}) elements

$$
\begin{aligned}
\bar{S}_{11} &= m^4 S_{11} + m^2 n^2 (2S_{12} + S_{66}) + n^4 S_{22} \\
\bar{S}_{12} &= m^2 n^2 (S_{11} + S_{22} - S_{66}) + (m^4 + n^4) S_{12} \\
\bar{S}_{22} &= n^4 S_{11} + m^2 n^2 (2S_{12} + S_{66}) + m^4 S_{22} \\
\bar{S}_{16} &= 2m^3 n (S_{11} - S_{12}) + 2mn^3 (S_{12} - S_{22}) - mn(m^2 - n^2) S_{66} \\
\bar{S}_{26} &= 2mn^3 (S_{11} - S_{12}) + 2m^3 n (S_{12} - S_{22}) + mn(m^2 - n^2) S_{66} \\
\bar{S}_{66} &= 2m^2 n^2 (2S_{11} + 2S_{22} - 4S_{12} - S_{66}) + (m^4 + n^4) S_{66}
\end{aligned}
\tag{A.1}
$$

$$
\begin{aligned}
\bar{Q}_{11} &= m^4 Q_{11} + 2m^2 n^2 (Q_{12} + 2Q_{66}) + n^4 Q_{22} \\
\bar{Q}_{12} &= m^2 n^2 (Q_{11} + Q_{22} - 4Q_{66}) + (m^4 + n^4) Q_{12} \\
\bar{Q}_{22} &= n^4 Q_{11} + 2m^2 n^2 (Q_{12} + 2Q_{66}) + m^4 Q_{22} \\
\bar{Q}_{16} &= m^3 n (Q_{11} - Q_{12}) + mn^3 (Q_{12} - Q_{22}) - 2mn(m^2 - n^2) Q_{66} \\
\bar{Q}_{26} &= mn^3 (Q_{11} - Q_{12}) + m^3 n (Q_{12} - Q_{22}) + 2mn(m^2 - n^2) Q_{66} \\
\bar{Q}_{66} &= m^2 n^2 (Q_{11} + Q_{22} - 2Q_{12} - 2Q_{66}) + (m^4 + n^4) Q_{66}
\end{aligned}
\tag{A.2}
$$

The matrices [a], [b], [c], and [d] may be determined from

$$
\begin{aligned}
[a] &= [A^*] - [B^*][D^*]^{-1}[C^*] \\
[b] &= [B^*][D^*]^{-1} \\
[c] &= -[D^*]^{-1}[B][A]^{-1} \\
[d] &= [D^*]^{-1}
\end{aligned}
\tag{A.3}
$$

where

$$
\begin{aligned}
[A^*] &= [A]^{-1} \\
[B^*] &= -[A]^{-1}[B] \\
[C^*] &= [B][A]^{-1} \\
[D^*] &= [D] - [B][A]^{-1}[B]
\end{aligned}
\tag{A.4}
$$

For symmetric laminates, [B] = [0]. For that case,

$$
\begin{aligned}
[a] &= [A]^{-1} \\
[b] &= [c] = [0] \\
[d] &= [D]^{-1}
\end{aligned}
\tag{A.5}
$$

Appendix B: Preparation of Test Specimens and Panels

In this appendix, test specimen geometries for some of the more basic tests are summarized. In addition, suggested geometries and dimensions of panels are presented. Note that there are many ways in which the panels could be designed. The dimensions of the flat panels herein should be appropriate for a laboratory-size autoclave. Dimensions of each and every test specimen and panel are not specified. Rather, the examples provided can be adapted to, or easily modified for, preparation of those types of specimens that are not specifically included here.

The illustrations (Figures B1–B20) detail the specimen geometries and dimensions and show suggested panel layups and dimensions (in millimeters). The layups shown for the delamination beam specimens are all unidirectional $[0]_{24}$ laminates that should be appropriate for carbon/epoxy composites. Ductile-matrix composites or composites with lower-modulus fibers require thicker specimens (more plies). The delamination fracture specimens (Chapter 17) incorporate a thin film at the laminate midplane to define an initial delamination. Panels for delamination testing should therefore contain an even number of plies and be manufactured with a nonadhesive Teflon or Kapton film at the laminate midplane. The film thickness should not exceed 13 μm, and the film may be sprayed with a mold-release agent before it is inserted between the plies. The insert length should extend an appropriate distance from the front edge of the specimen to achieve the correct precrack length, as illustrated in the figures. It is difficult to detect the thin insert film when viewed from the edge of a cut specimen. Therefore, the area covered by the insert should carefully be marked on the panel before the specimens are cut.

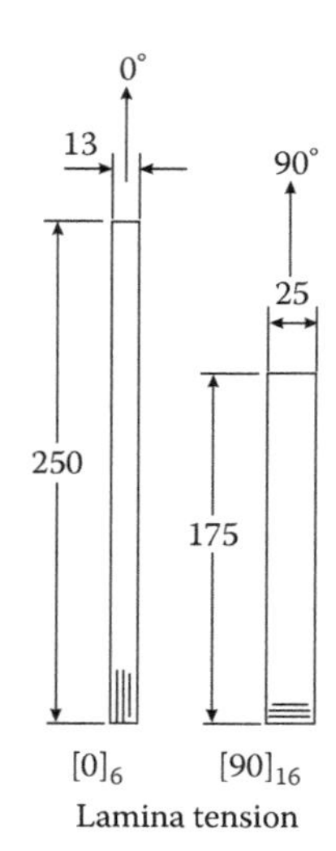

FIGURE B.1
Specimen geometries, dimensions (millimeters), and layups for lamina tension experiment.

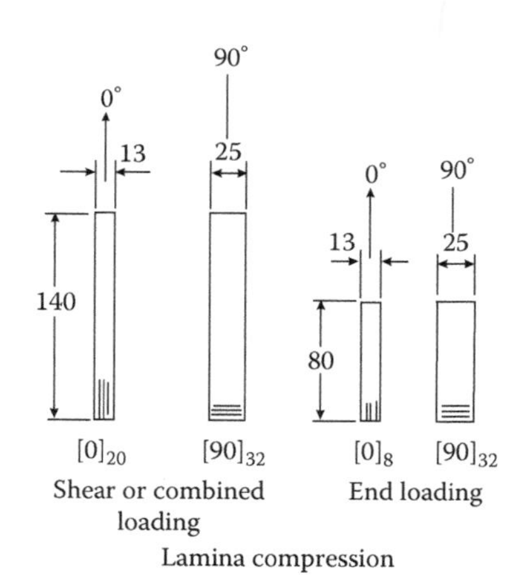

FIGURE B.2
Specimen geometries, dimensions (millimeters), and layups for lamina compression experiment.

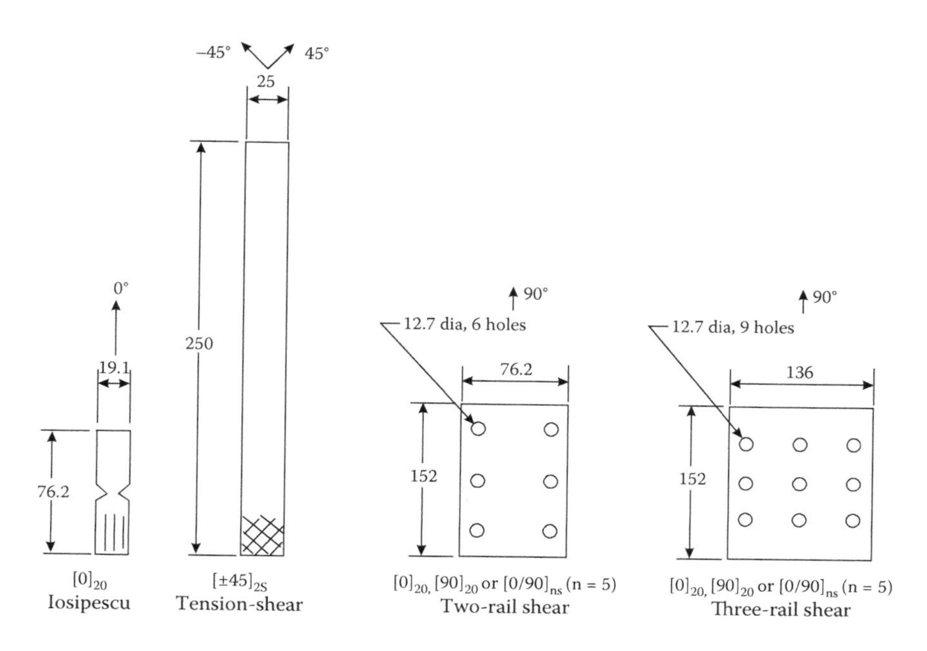

FIGURE B.3
Specimen geometries, dimensions (millimeters), and layups for lamina shear experiment.

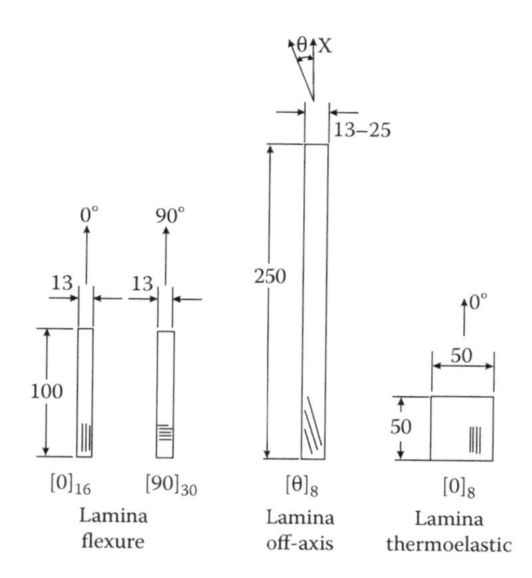

FIGURE B.4
Specimen geometries, dimensions (millimeters), and layups for lamina flexure, lamina off-axis, and lamina thermoelastic experiments.

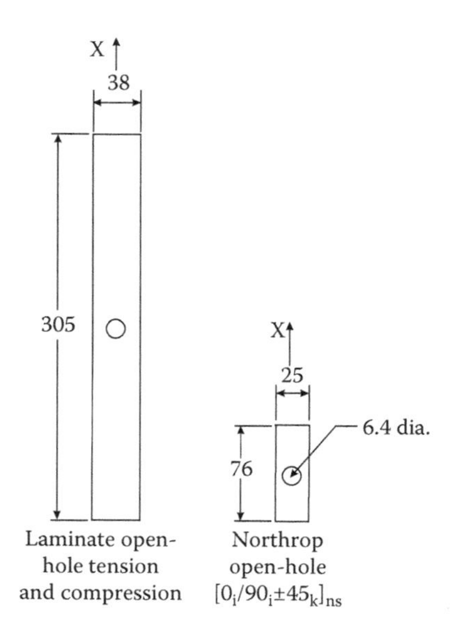

FIGURE B.5
Specimen geometries and dimensions (millimeters) for laminate open-hole tension and compression experiments.

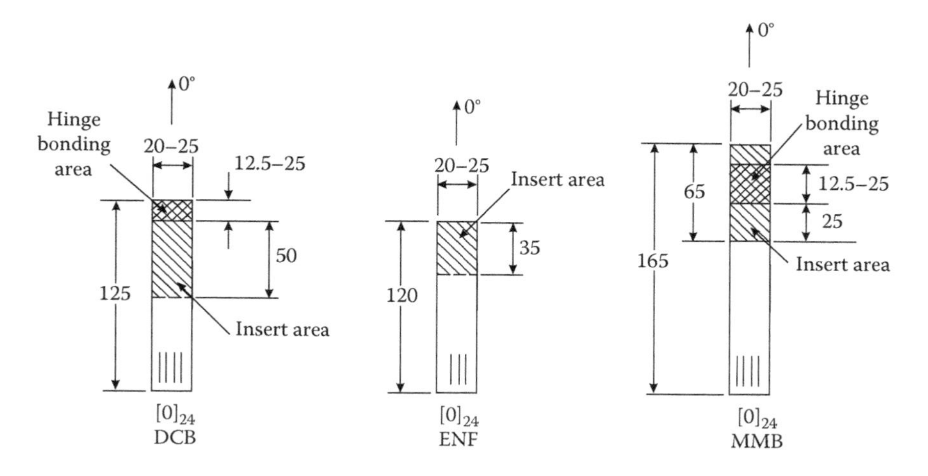

FIGURE B.6
Geometries, dimensions (millimeters), and layups for DCB, ENF, and MMB specimens.

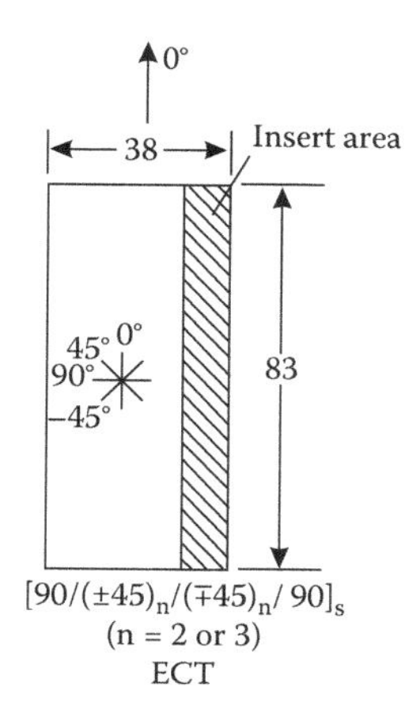

FIGURE B.7
Geometry, dimensions (millimeters), and layup for ECT specimen

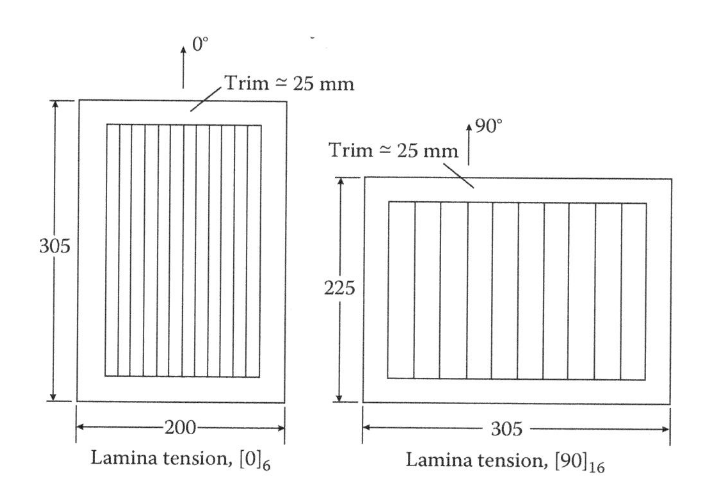

FIGURE B.8
Suggested panel dimensions (millimeters) for 0° and 90° lamina tension experiment.

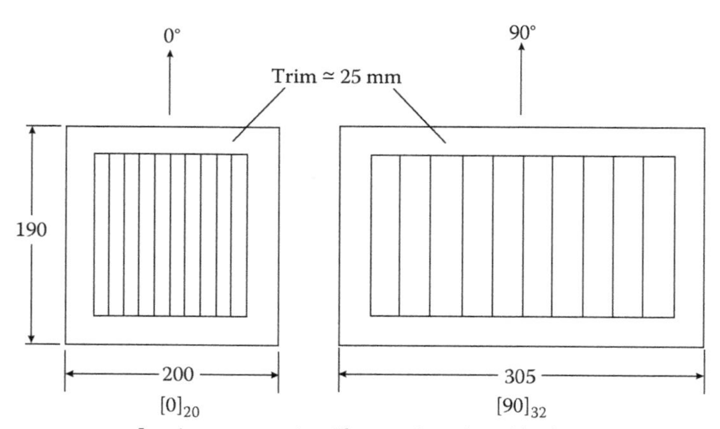

FIGURE B.9
Suggested panel dimensions (millimeters) for lamina compression experiment (shear and combined loading).

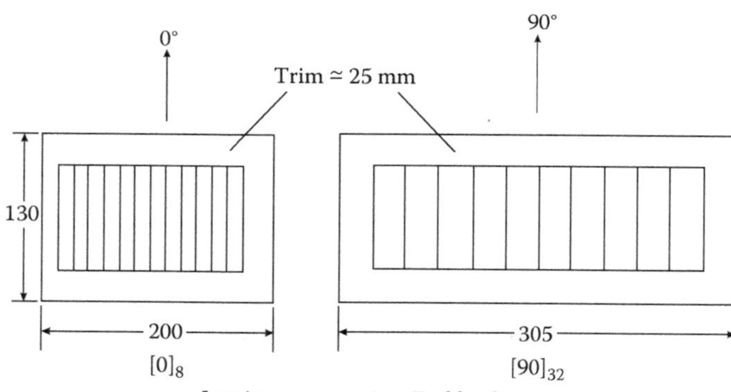

FIGURE B.10
Suggested panel dimensions (millimeters) for lamina compression experiment (end loading).

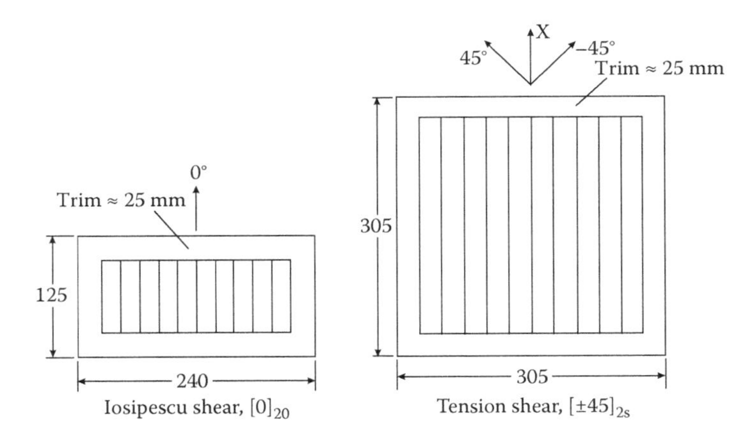

FIGURE B.11
Suggested panel dimensions (millimeters) for Iosipescu and tensile shear experiments.

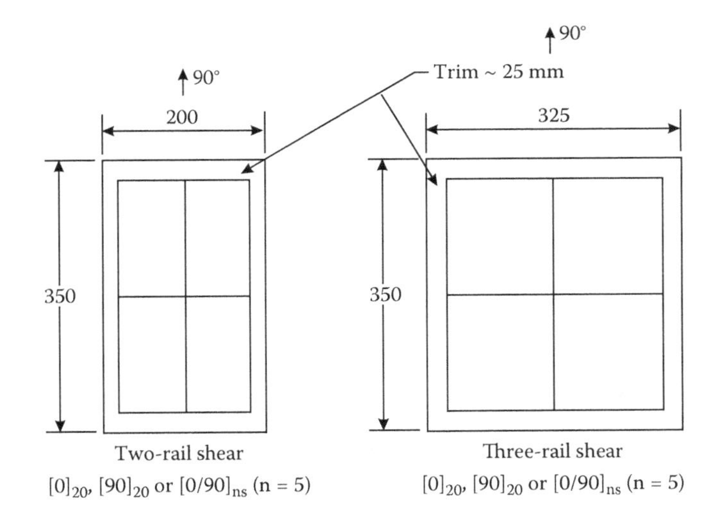

FIGURE B.12
Suggested panel dimensions (millimeters) for two-rail and three-rail shear experiments.

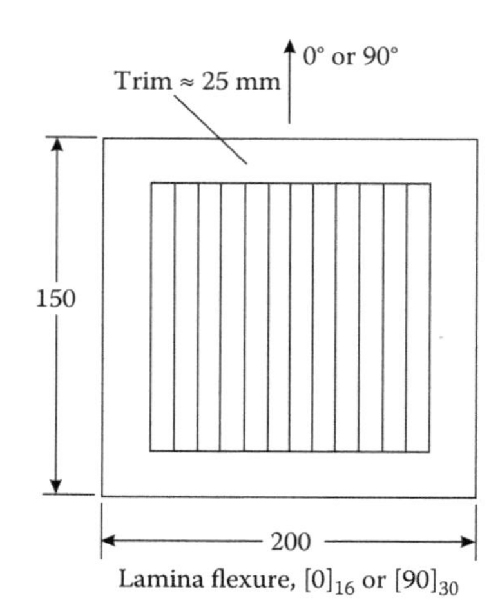

FIGURE B.13
Suggested panel dimensions (millimeters) for lamina flexure experiment.

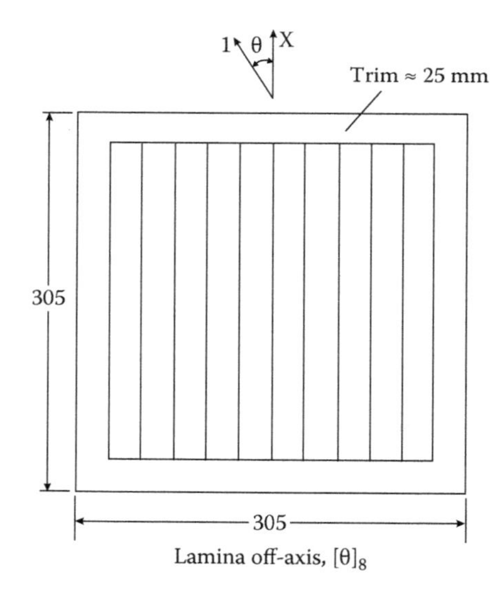

FIGURE B.14
Suggested panel dimensions (millimeters) for lamina off-axis experiment.

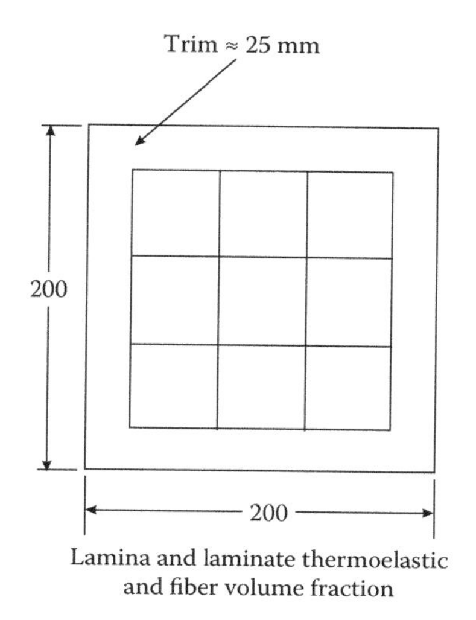

FIGURE B.15
Suggested panel dimensions (millimeters) for lamina and laminate thermoelastic and fiber volume fraction experiments.

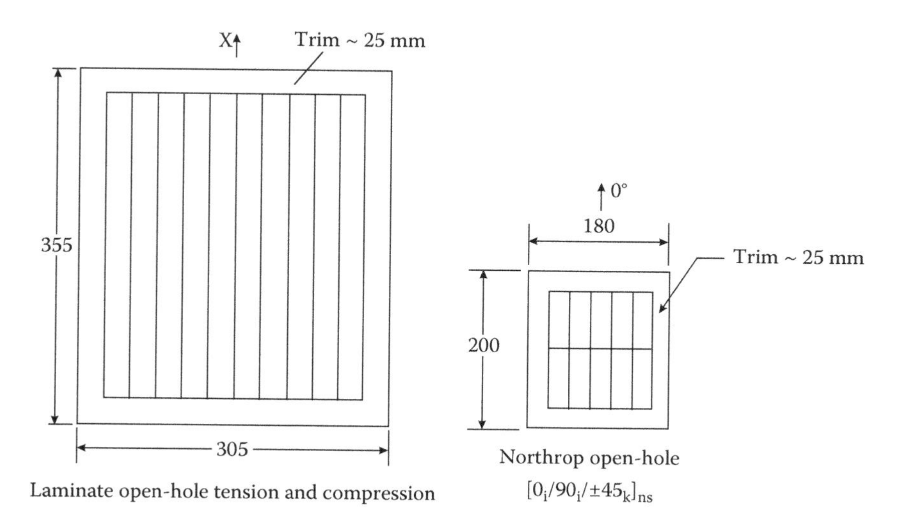

FIGURE B.16
Suggested panel dimensions (millimeters) for laminate and Northrop open-hole experiments.

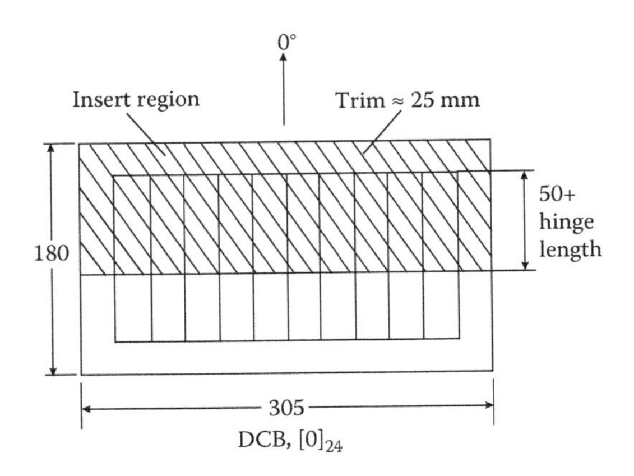

FIGURE B.17
Suggested panel dimensions (millimeters) for DCB specimen. Insert film should be placed at midplane.

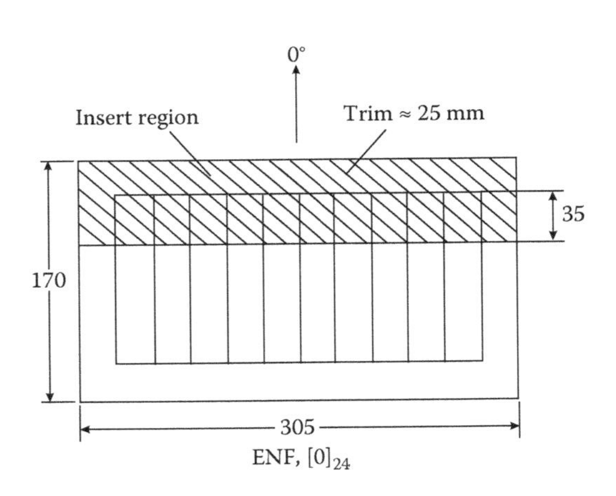

FIGURE B.18
Suggested panel dimensions (millimeters) for ENF specimen. Insert film should be placed at midplane.

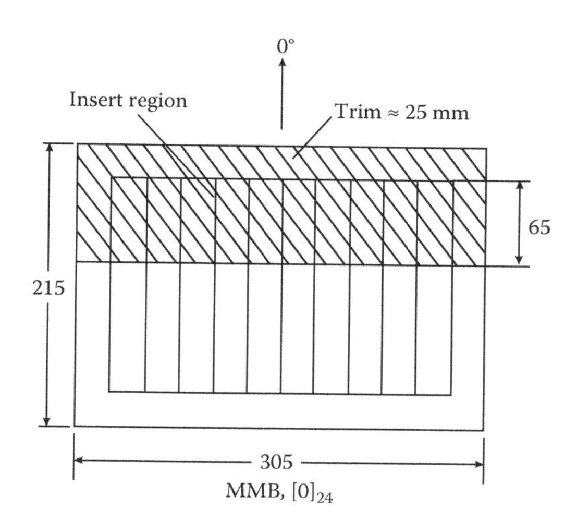

FIGURE B.19
Suggested panel dimensions (millimeters) for MMB specimen. Insert film should be placed at midplane.

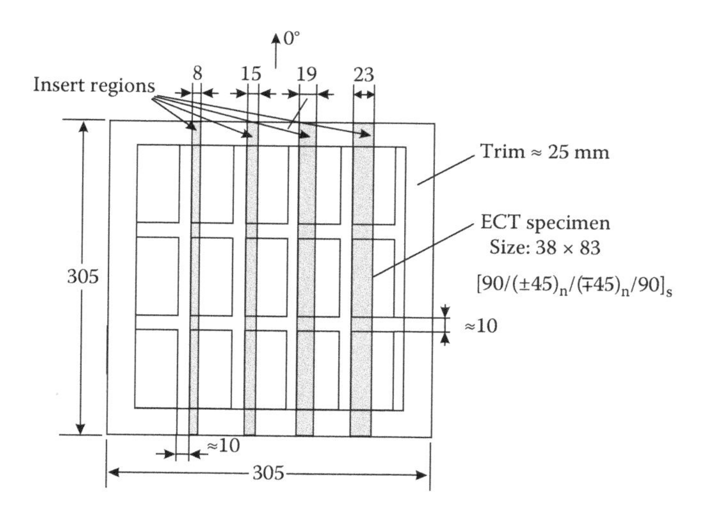

FIGURE B.20
Suggested panel dimensions (millimeters) for ECT specimen. Insert film should be placed at midplane.

Appendix C: Sample Laboratory Report

Lamina Tensile Response

The lamina tensile response of a carbon–fiber, epoxy–matrix composite was examined experimentally to establish the intrinsic mechanical properties. The test specimen geometries were chosen according to the outline presented in Chapter 8, in accordance with American Society for Testing and Materials (ASTM) standards. The specimens were loaded to failure in a tensile testing machine utilizing serrated wedge grips. Average test results and standard deviations were as follows

Elastic modulus in the fiber direction	$E_1 = 126 \pm 2$ GPa
Elastic modulus transverse to the fiber direction	$E_2 = 10.2 \pm 0.4$ GPa
Poisson's ratios: Major	$\nu_{12} = 0.30 \pm 0.01$
______________ Minor	$\nu_{21} = 0.024$
Ultimate tensile stress in the fiber direction	$X_1^T = 2040 \pm 85$ MPa
Ultimate tensile stress in the transverse direction	$X_2^T = 53 \pm 8$ MPa
Ultimate tensile strain in the fiber direction	$e_1^T = 0.015$
Ultimate tensile strain in the transverse direction	$e_2^T = 0.0057$

Procedure

The procedure for this experiment is detailed in Chapter 8. Briefly, unidirectional panels were configured for achieving test specimens with 0° and 90° orientation as shown in Appendix B. After the edges of the panels were trimmed, tabs made from a glass–fabric epoxy laminate were adhesively bonded to both surfaces at two opposite edges of the panels. Four specimens of each orientation were machined to the appropriate widths using procedures detailed in Chapter 4. The 0° specimens were nominally 12.7 mm wide, whereas the 90° specimens were 25.4 mm wide. The 0° specimens were 8 plies thick, whereas the 90° specimens were 16 plies thick. To establish the axial stiffness E_1, Poisson's ratio ν_{12}, and the overall stress–strain response of the 0° specimens, a bidirectional (0°/90°) strain gage rosette was bonded at the geometric center on one surface of each specimen. In addition, an axial gage was bonded on the opposite surface of the specimen. For the 90° specimens, a single-element strain gage oriented along the length of the specimen was bonded to each surface of the specimen in the gage section to determine the axial stress–strain response. No strain gages transverse to the specimen loading axis were used because the minor Poisson's ratio ν_{21} may be calculated from E_1, E_2, and ν_{12} (Chapter 2). Each specimen was tested in a general-purpose testing machine at a crosshead rate of 2 mm/min. Specimen load and strains were sampled throughout the test using a PC-driven data acquisition system. The specimens were loaded to failure.

Specimen Dimensions

Specimen cross-sectional dimensions were recorded as follows:

Specimen	Orientation (degrees)	Width w (mm)	Thickness t (mm)
1	0	12.78	1.067
2	0	12.78	1.067
3	0	12.65	1.067
4	0	12.75	1.067
5	90	25.40	2.184
6	90	25.35	2.185
7	90	25.45	2.134
8	90	25.53	2.236

Stress–Strain Data

The load readings were converted to axial stress readings using the cross-sectional dimensions reported. Examples of stress and strain data recorded using the data acquisition system are tabulated in the following:

Stress–Strain Data for Specimen 2 ($[0]_s$) (Reduced Set from Original Record)

σ_1 (MPa)	ε_1 (μstrain)	ε_1 (μstrain)	$-\varepsilon_2$ (μstrain)
0	0	10	0
36	310	320	120
72	590	600	200
108	860	870	280
144	1140	1160	340
180	1420	1440	420
252	2010	2000	570
395	3050	3030	880
647	4900	4850	1380
1006	7490	7430	2040
1294	9470	9420	2540
1617	11640	11590	3070
1977[a]	14060	13990	3610

The last two columns are strain readings from the (0°/90°) strain gage rosette.

[a] Ultimate stress.

Stress–Strain Data for Specimen 6 ($[90]_{16}$)
(Reduced Set from Original Record)

σ_2 (MPa)	ε_2 (μstrain)	ε_2 (μstrain)
0	0	0
1.77	180	200
3.54	350	380
5.31	520	550
10.6	1040	1120
17.7	1750	1860
23.0	2290	2410
30.1	2990	3130
35.4	3520	3690
40.7	4120	4330
49.6	5050	5280
60.2	6220	6510
63.4[a]	6580	6879

[a] Ultimate stress.

Test Results

Test results for three representative 0° test specimens are presented in graphical form in Figures C.1–C.3. The linear response region in the fiber direction is bounded by a strain of about 0.004. It is noteworthy that the stress–strain response exhibits strain-stiffening characteristics—a reflection

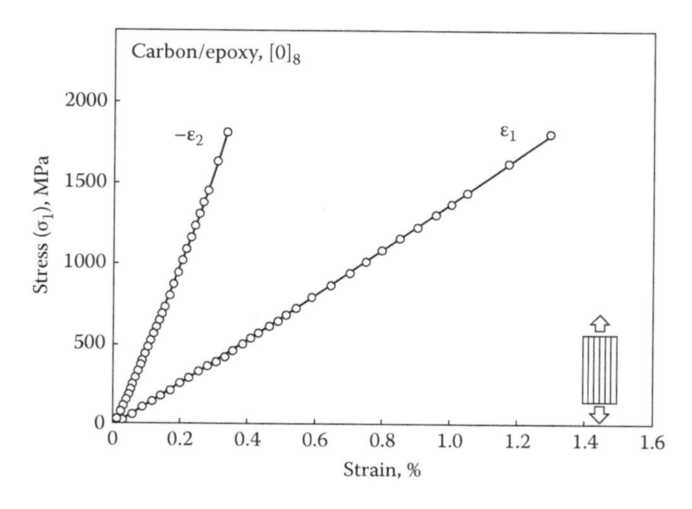

FIGURE C.1
Stress–strain results for specimen 1 (0°).

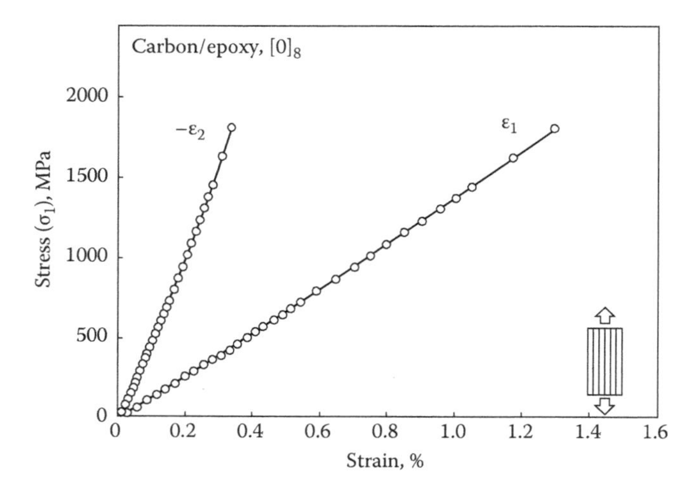

FIGURE C.2
Stress–strain results for specimen 2 (0°).

of the behavior of carbon fibers. Results for three representative 90° specimens are shown in Figures C.4–C.6. Here, only a modest nonlinearity in the stress–strain response is observed. The strain softening is attributed to the nonlinear response of the epoxy matrix.

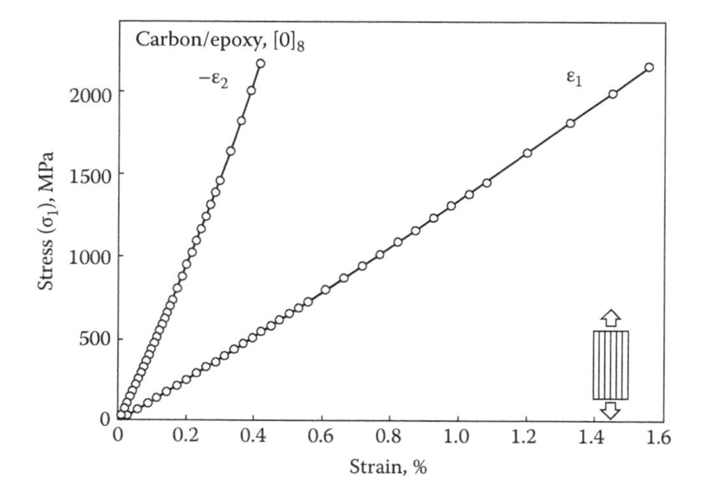

FIGURE C.3
Stress–strain results for specimen 3 (0°).

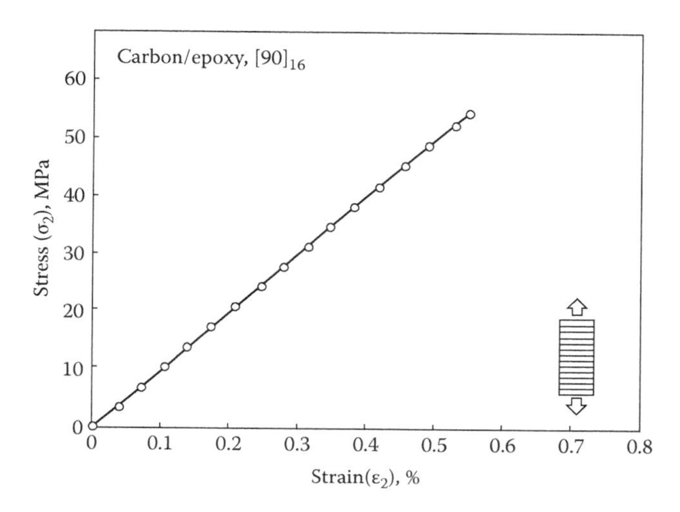

FIGURE C.4
Stress–strain results for specimen 5 (90°).

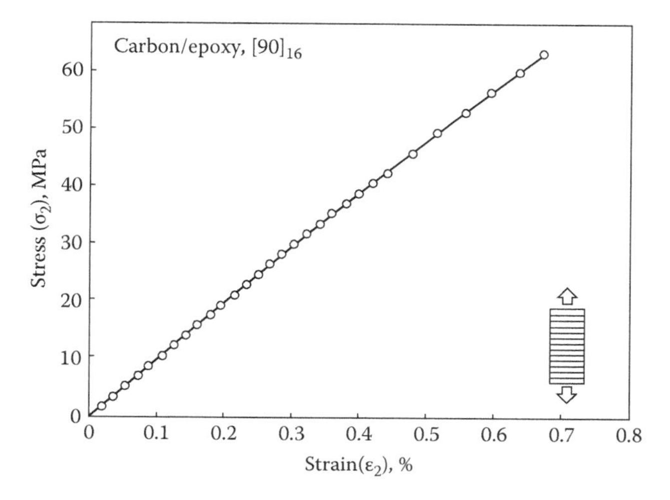

FIGURE C.5
Stress–strain results for specimen 6 (90°).

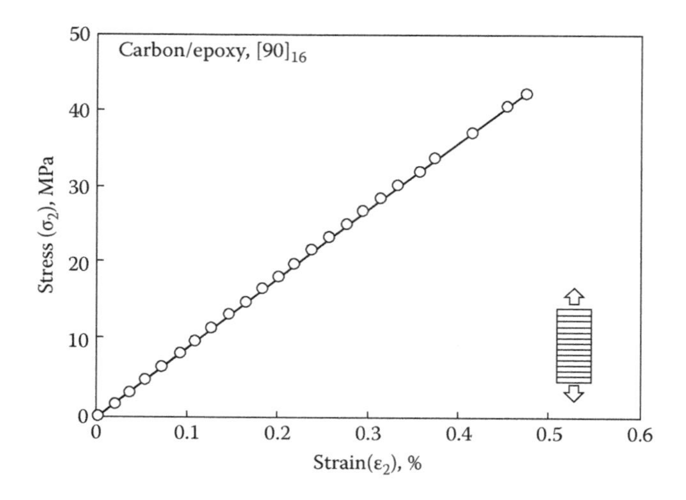

FIGURE C.6
Stress–strain results for specimen 7 (90°).

Reduced Data

The mechanical properties were reduced from the measured data using procedures and equations provided in Chapter 8. The following equations were employed (Figures C.1–C.6):

$$E_1 = \frac{\sigma_1}{\varepsilon_1} \tag{C.1}$$

$$\upsilon_{12} = \frac{-\varepsilon_2}{\varepsilon_1} \tag{C.2}$$

$$X_1^T = \sigma_1^{ult} \tag{C.3}$$

$$E_2 = \frac{\sigma_2}{\varepsilon_2} \tag{C.4}$$

$$X_2^T = \sigma_2^{ult} \tag{C.5}$$

where σ_1 and σ_2 refer to the load per unit cross-sectional area [$\sigma = P/(wt)$] for the 0° and 90° tests, respectively. Note that it was not possible to evaluate experimentally the minor Poisson's ratio ν_{21} because the 90° specimens were not instrumented with a transversely oriented strain gage. The reduced data are summarized as follows:

	E_1 (GPa)	ν_{12}	X_1^T (MPa)	e_1^T	E_2 (GPa)	X_2^T (MPa)	e_2^T
	128	0.295	2034	0.015	9.92	54.5	0.0056
	127	0.292	1800	0.013	9.79	63.4	0.0067
	124	0.299	2158	0.016	10.5	45.6	0.0046
	125	0.319	1979	0.014	10.4	48.3	0.0049
Average	126	0.301	1990	0.015	10.2	53.0	0.0055
STD[a]	2	0.012	149	0.001	0.4	7.9	0.0010

[a] STD = standard deviation.

Using the reciprocal relation between the elastic moduli and Poisson's ratios, Equation (2.8), the minor Poisson's ratio was determined as

$$\nu_{21} = \nu_{12}E_2/E_1 = 0.301 \times 10.2/126 = 0.024$$

Uncertainty Analysis

An uncertainty analysis was performed to estimate the possible scatter range in the mechanical properties as a result of uncertainties in the primary measurements of load, strain, and specimen dimensions. Procedures for such estimation are outlined in the text by Holman and Gajda (1993). Here, we perform a simple, conservative propagation of error analysis on the governing Equations (C.1)–(C.5) used for data reduction and property determination. Such an analysis yields

$$\Delta E_i = E_i\left[\frac{\Delta P}{P} + \frac{\Delta w}{w} + \frac{\Delta t}{t} + \frac{\Delta \varepsilon}{\varepsilon}\right] \quad i = 1,2 \tag{C.6}$$

$$\Delta \upsilon_{12} = \upsilon_{12}\left[\frac{\Delta \varepsilon_1}{\varepsilon_1} + \frac{\Delta \varepsilon_2}{\varepsilon_2}\right] \tag{C.7}$$

$$\Delta X_i^T = X_i^T\left[\frac{\Delta P}{P} + \frac{\Delta w}{w} + \frac{\Delta t}{t}\right] \quad i = 1,2 \tag{C.8}$$

Consider the uncertainties in measuring the load P, strain ε, and dimensions w and t:

$$\Delta P = \pm 10\ N$$

$$\Delta \varepsilon = \pm 5 \times 10^{-6}$$

$$\Delta w = \pm 0.025\ mm$$

$$\Delta t = \pm 0.025\ mm$$

With these uncertainties in the load and strain data and in the cross-sectional dimensions, load and strain values were inserted into Equations (C.6)–(C.8) to yield the uncertainties in the reduced mechanical properties. When considering uncertainties in the elastic moduli (E_i) and Poisson's ratio (ν_{12}), the load and strains in the middle of the linear response region (Figures C.1–C.6) were used. For uncertainty analysis of the strengths (X_i^T), the ultimate loads were used. The calculations yielded the following uncertainties:

$$\Delta E_1 = 3.2\ GPa$$

$$\Delta \nu_{12} = 0.002$$

$$\Delta X_1^T = 52\ MPa$$

$$\Delta E_2 = 0.3\ GPa$$

$$\Delta X_2^T = 1.0\ MPa$$

The uncertainties are all below 4% of the corresponding average values, which indicates that the measuring accuracy was reasonable. For several of the mechanical properties, the standard deviation exceeds the estimated uncertainties presented, which indicates that the variability of the material properties contributes to the scatter.

Micromechanics Predictions

It is useful to compare the measured properties to those predicted by the micromechanics analyses discussed in Chapter 2. Previous laboratory experiments using an AS4/3501-6 carbon–epoxy composite gave a fiber volume fraction of 0.55 (see Chapter 3). Application of the micromechanics relations for E_1, ν_{12}, and E_2 given in Chapter 2 [i.e., Equations (2.25) and (2.26)], together with the following data for AS4 carbon fibers and 3501-6 epoxy obtained from Daniel and Ishai (2006):

Fiber Data	**Matrix Data**
Axial modulus (E_1), 235 GPa	Young's modulus E, 4.28 GPa
Transverse modulus (E_T), 13.8 GPa	Poisson's ratio ν, 0.35
Axial Poisson's ratio (ν_{LT}), 0.20	

yields the following estimate of the mechanical properties of the composite:

$$E_1 = 131 \text{ GPa}$$

$$\nu_{12} = 0.27$$

$$E_2 = 8.3 \text{ GPa}$$

The estimated properties agree reasonably well with the measured data. The differences may be due to variations in fiber volume fraction (E_1 and ν_{12}) and the inaccuracies of the micromechanical models.

Appendix D: Unit Conversions

Quantity	SI to US	US to SI
Length	1 m = 39.37 in 1 cm = 0.394 in 1 mm = 0.0394 in	1 in = 0.0254 m 1 in = 2.54 cm 1 in = 25.4 mm
Force	1 N = 0.225 lb	1 lb = 4.448 N
Stress	1 MPa = 145 psi	1 psi = 6.895 kPa
Work	1 J = 1 Nm = 8.851 in-lb	1 in-lb = 0.1130 J
Surface energy	1 J/m^2 = 5.71 × 10^{-3} in-lb/in^2	1 in-lb/in^2= 175 J/m^2
Temperature	°F = 1.8°C + 32	°C = 0.556°F − 17.8
Coefficient of thermal expansion	$\alpha\left[\frac{1}{°F}\right] = 0.556\alpha\left[\frac{1}{°C}\right]$	$\alpha\left[\frac{1}{°C}\right] = 1.8\alpha\left[\frac{1}{°F}\right]$

References

Abrate, S., 1998. *Impact on Composite Structures*, Cambridge University Press, Cambridge, UK.

Abrate, S. and Walton, D.A., 1992. Machining of composite materials. Part I Traditional methods, *Compos. Manuf.*, 3, 75–83.

Adams, D.F., 1990, The Iosipescu shear test method as used for testing polymers and composites, *Polym. Compos.*, 11, 286–290.

Adams, D.F., 2000a. Dimensional stability (thermal), in *Comprehensive Composite Materials*, A. Kelly and C. Zweben, Eds., Technomic, Lancaster, PA.

Adams, D.F., 2000b. Test methods for mechanical properties, in *Comprehensive Composite Materials*, A. Kelly and C. Zweben, Eds., Elsevier Science, Oxford, UK, Vol. 5, 113–148.

Adams, D.F., 2010, January. Shear test methods: Iosipescu vs. v-notched rail, *High Performance Composites*, 13–14.

Adams, D.F. and Finley, G.A., 1996. Experimental study of thickness-tapered unidirectional composite compression specimens, *Exp. Mech.*, 36(4), 348–355.

Adams, D.F., King, T.R., and Blacketter, D.L., 1990. Evaluation of the transverse flexure test method for composite materials, *Compos. Sci. Technol.*, 39, 341–353.

Adams, D.F. and Lewis, E.Q., 1995. Experimental study of three- and four-point shear test specimens, *J. Compos. Technol. Res.*, 17, 341–349.

Adams, D.F. and Lewis, E.Q., 1997. Experimental assessment of four composite material shear test methods, *J. Test. Eval.*, 25, 174–181.

Adams, D.F. and Odom, E.M., 1991a. Influence of specimen tabs on the compressive strength of a unidirectional composite material, *J. Compos. Mater.*, 25, 774–786.

Adams, D.F. and Odom, E.W., 1991b. Influence of test fixture configuration on the measured compressive strength of a composite material, *J. Compos. Technol. Res.*, 13, 36–40.

Adams, D.F. and Walrath, D.E., 1982. *Iosipescu Shear Properties of SMC Composite Materials*, ASTM Spec. Tech. Publ. 787, American Society for Testing and Materials, West Conshohocken, PA, 19–33.

Adams, D.F. and Welsh, J.S., 1997. The Wyoming combined loading compression (CLC) test method, *J. Compos. Technol. Res.*, 19, 123–133.

Adams, D.O., Moriarty, J.M., Gallegos, A.M., and Adams, D.F., 2002. Development and evaluation of a V-notched rail shear test, *Proceedings of the 2002 SAMPE Technical Conference*, Baltimore, MD.

Adams, D.O., Moriarty, J.M., Gallegos, A.M., and Adams, D.F., 2003, *Development and Evaluation of the V-Notched Rail Shear Test for Composite Laminates*, Federal Aviation Administration Final Report DOT/FAA/AR-03/63, September 2003.

Adsit, N.R., 1983. Compression testing of graphite/epoxy, in *Compression Testing of Homogeneous Materials and Composites*, ASTM Spec. Tech. Publ. 808, American Society for Testing and Materials, West Conshohocken, PA, 175–196.

Advani, S.G., 1994. *Flow and Rheology in Polymer Composites Manufacturing*, in the series *Composite Materials*, R.B. Pipes, Ed., Elsevier, Amsterdam, 465–511.

AECMA Aerospace Series, 1995. *Carbon Fiber Reinforced Plastics: Determination of Interlaminar Fracture Toughness Energy in Mode I— G_{IC} (prEN6033) and Mode II—G_{IIC} (pr EN 6034),* Association Europeene de Constructeurs de Materiel Aerospatial, Paris.

Air Force Flight Dynamics Laboratory, 1973. *Advanced Composites Design Guide,* 3rd ed., R.M. Neff, project engineer, Air Force Flight Dynamics Laboratory, Wright Patterson Air Force Base, Dayton, OH.

Alif, N. and Carlsson, L.A., 1997. *Failure Mechanisms of Woven Carbon and Glass Composites,* ASTM Spec. Tech. Publ. 1285, American Society for Testing and Materials, West Conshohoken, PA, 471–493.

Alif, N., Carlsson, L.A., and Gillespie, J.W., Jr., 1997. *Mode I, Mode II, and Mixed Mode Interlaminar Fracture of Woven Fabric Carbon/Epoxy,* ASTM Spec. Tech. Publ. 1242, American Society for Testing and Materials, West Conshohoken, PA, 82–106.

Allen, H.G., 1969. *Analysis and Design of Structural Sandwich Panels,* Pergamon Press, Oxford, UK.

Allen, S.R., 1987. Tensile recoil measurement of compressive strength for polymeric high performance fibers, *J. Mater. Sci.,* 22, 853–859.

Anderson, T.L., 2005. *Fracture Mechanics: Fundamentals and Applications,* 3rd ed., CRC Press, Boca Raton, FL.

Aronsson, C.G., 1984. Tensile fracture of composite laminates with holes and cracks, PhD dissertation, Royal Institute of Technology, Stockholm, Sweden.

Arridge, R.G.C., Barham, P.I., Farell, C.J., and Keller, A., 1976. The importance of end effects in the measurement of moduli of highly anisotropic materials, *J. Mater. Sci.,* 11, 788–790.

ASTM C273, 2011. *Standard Test Method for Shear Properties of Sandwich Core Materials,* American Society for Testing and Materials, West Conshohocken, PA.

ASTM C297/C297M-04, 2010. *Tensile Strength of Flat Sandwich Constructions in Flatwise Plane,* American Society for Testing and Materials, West Conshohocken, PA.

ASTM C364/C364M-07, 2007. *Standard Test Method for Edgewise Compressive Strength of Sandwich Construction,* American Society for Testing and Materials, West Conshohocken, PA.

ASTM C365/C365M-05, 2009. *Flatwise Compressive Properties of Sandwich Cores,* American Society for Testing and Materials, West Conshohocken, PA.

ASTM C393, 2011. *Standard Test Method for Core Shear Properties of Sandwich Constructions by Beam Flexure,* American Society for Testing and Materials, West Conshohocken, PA.

ASTM D638-10, 2010. *Standard Test Method for Tensile Properties of Plastics,* American Society for Testing and Materials, West Conshohocken, PA.

ASTM D695-10, 2010. *Standard Test Method for Compressive Properties of Rigid Plastics,* American Society for Testing and Materials, West Conshohocken, PA.

ASTM D696-08, 2008. *Standard Test Method for Coefficient of Linear Thermal Expansion of Plastics between –30°C and 30°C with a Nitreous Silica Dilatometer,* American Society for Testing and Materials, West Conshohocken, PA.

ASTM D790-10, 2010. *Standard Test Methods for Flexural Properties of Unreinforced and Reinforced Plastics and Electrical Insulating Materials,* American Society for Testing and Materials, West Conshohocken, PA.

ASTM D1505-10, 2010. *Standard Test Method for Density of Plastics by the Density-Gradient Technique,* American Society for Testing and Materials, West Conshohocken, PA.

ASTM D2344/D2344M-00, 2006. *Standard Test Method for Short Beam Strength of Polymer Matrix Composite Materials and Their Laminates*, American Society for Testing and Materials, West Conshohocken, PA.

ASTM D2471-99, 2008. *Standard Test Method for Gel Time and Peak Exothermic Temperature of Reacting Thermosetting Resins*, American Society for Testing and Materials, West Conshohocken, PA (withdrawn in 2008).

ASTM D2566-86, 1986. *Standard Test Method for Linear Shrinkage of Cured Thermosetting Casting Resins during Cure*, American Society for Testing and Materials, West Conshohocken, PA (withdrawn in 1993).

ASTM D2584-11, 2011. *Standard Test Method for Ignition Loss of Cured Reinforced Resins*, American Society for Testing and Materials, West Conshohocken, PA.

ASTM D2719-07, 2007. *Standard Test Methods for Structural Panels in Shear Through-the-Thickness*, American Society for Testing and Materials, West Conshohocken, PA.

ASTM D2734-09, 2009. *Standard Test Methods for Void Content of Reinforced Plastics*, American Society for Testing and Materials, West Conshohocken, PA.

ASTM D3039/D3039M-098, 2008. *Standard Test Method for Tensile Properties of Polymer Matrix Composite Materials*, American Society for Testing and Materials, West Conshohocken, PA.

ASTM D3171-11, 2011. *Standard Test Method for Constituent Content of Composite Materials*, American Society for Testing and Materials, West Conshohocken, PA.

ASTM D3379-75, 1989. *Standard Test Method for Tensile Strength and Young's Modulus for High-Modulus Single Filament Materials*, American Society for Testing and Materials, West Consohocken, PA (withdrawn in 1998).

ASTM D3410/3410M, 2008. *Standard Test Method for Compressive Properties of Polymer Matrix Composite Materials with Unsupported Gage Section by Shear Loading*, American Society for Testing and Materials, West Conshohocken, PA.

ASTM D3518/3518M, 2007. *Standard Test Method for In-Plane Shear Response of Polymer Matrix Composite Materials by Tensile Test of a ±45° Laminate*, American Society for Testing and Materials, West Conshohocken, PA.

ASTM D3800M-11, 2011. *Standard Test Method for Density of High Modulus Fibers*, American Society for Testing and Materials, West Conshohocken, PA.

ASTM D4255/D4255M-01, 2007. *Standard Test Method for In-Plane Shear Properties of Polymer Matrix Composite Materials by the Rail Shear Method*, American Society for Testing and Materials, West Conshohocken, PA.

ASTM D4473-08, 2008. *Standard Test Method for Plastics: Dymanic Mechanical Properties: Cure Behavior*, American Society for Testing and Materials, West Conshohocken, PA.

ASTM D5379/5379M-05, 2005. *Standard Test Method for Shear Properties of Composite Materials by the V-Notched Beam Method*, American Society for Testing and Materials, West Conshohocken, PA.

ASTM D5467/5467M-97, 2010. *Standard Test Method for Compressive Properties of Unidirectional Polymer Matrix Composite Materials Using a Sandwich Beam*, American Society for Testing and Materials, West Conshohocken, PA.

ASTM D5528-01, 2007. *Standard Test Method for Mode I Interlaminar Fracture Toughness of Unidirectional Fiber-Reinforced Polymer Matrix Composites*, American Society for Testing and Materials, West Conshohocken, PA.

ASTM D5687/D5687M-95, 2007. *Standard Guide for Preparation of Flat Composite Panels with Processing Guidelines for Specimen Preparation*, American Society for Testing and Materials, West Conshohocken, PA.

ASTM D5766/D5766M-11, 2011. *Standard Test Method for Open-Hole Tensile Strength of Polymer Matrix Composite Laminates*, American Society for Testing and Materials, West Conshohocken, PA.

ASTM D6272-10, 2010. *Standard Test Method for Flexural Properties of Unreinforced and Reinforced Plastics and Electrical Insulating Materials by Four-Point Bending*, American Society for Testing and Materials, West Conshohocken, PA.

ASTM D6415/D 6415M-06a, 2006. *Standard Test Method for Measuring the Curved Beam Strength of a Fiber-Reinforced Polymer-Matrix Composite*, American Society for Testing Materials, West Conshohocken, PA.

ASTM D6484/D6484M-09, 2009. *Standard Test Method for Open-Hole Compressive Strength of Polymer Matrix Composite Laminates*, American Society for Testing and Materials, West Conshohocken, PA.

ASTM D6641/D6441M-09, 2009. *Standard Test Method for the Compressive Properties of Polymer Matrix Composite Materials Using a Combined Loading Compression (CLC) Test Fixture*, American Society for Testing and Materials, West Conshohocken, PA.

ASTM D6671/D6671M-06, 2006. *Standard Test Method for Mixed Mode I-Mode II Interlaminar Fracture Toughness of Unidirectional Fiber Reinforced Polymer Matrix Composites*, American Society for Testing and Materials, West Conshohocken, PA.

ASTM D6742/D6742M-12, 2012. *Standard Practice for Filled-Hole Tension and Compression Testing of Polymer Matrix Composite Laminates*, American Society for Testing and Materials, West Conshohocken, PA.

ASTM D7078/D7078M-05, 2005. *Standard Test Method for Shear Properties of Composite Materials by the V-Notched Rail Shear Method*, American Society for Testing and Materials, West Conshohocken, PA.

ASTM D7136/D7136M-12, 2012. *Standard Test Method for Measuring the Damage Resistance of Fiber-Reinforced Polymer Matrix Composites to a Drop Weight Impact Event*, American Society for Testing and Materials, West Conshohocken, PA.

ASTM D7137/D7137M-12, 2012. *Standard Test Method for Compressive Residual Strength Properties of Damaged Polymer Matrix Composite Plates*, American Society for Testing and Materials, West Conshohocken, PA.

ASTM D7264/7264M, 2007. *Standard Test Method for Flexural Properties of Polymer Matrix Composite Materials*, American Society for Testing and Materials, West Conshohocken, PA.

ASTM D7291/D7291M-07, 2007. *Standard Test Method for Through-Thickness "Flatwise" Tensile Strength and Elastic Modulus of a Fiber-Reinforced Polymer Matrix Composite Material*, American Society for Testing and Materials, West Conshohocken, PA.

ASTM E228-11, 2011. *Standard Test Method for Linear Thermal Expansion of Solid Materials with a Push-Rod Dilatometer*, American Society for Testing and Materials, West Conshohocken, PA.

ASTM E289-04, 2010. *Standard Test Method for Linear Thermal Expansion of Rigid Solids with Interferometry*, American Society for Testing and Materials, West Conshohocken, PA.

ASTM E831-12, 2012. *Standard Test Method for Linear Thermal Expansion of Solid Materials by Thermomechanical Analysis*, American Society for Testing and Materials, West Conshohocken, PA.

Aviles, F., Carlsson, L.A., Browning, G., and Millay, K., 2009, Investigation of the plate twist test, *Exp. Mech.*, 49, 813–822.

Awerbuch, J. and Madhukar, M.S., 1985. Notched strength of composite laminates, *J. Reinf. Plast. Compos.*, 4, 3–159.

Bascom, W.D., Bitner, R.J., Moulton, R.J., and Siebert, A.R., 1980. The interlaminar fracture of organic-matrix woven reinforced composites, *Composites*, 11, 9–18.

Bathe, K.-J., 1982. *Finite Element Procedures in Engineering Analysis*, Prentice-Hall, Englewood Cliffs, NJ.

Beer, F.P. and Johnston, E.R., Jr., 1992. *Mechanics of Materials*, 2nd ed., McGraw Hill, New York.

Benzeggagh, M.L. and Kenane, M., 1996. Measurement of mixed-mode delamination fracture toughness of unidirectional glass/epoxy composites with mixed-mode bending apparatus, *Compos. Sci. Technol.*, 56, 439–449.

Berg, C.A., Tirosh, J., and Israeli, M., 1972. *Analysis of Short Beam Bending of Fiber Reinforced Composites*, ASTM Spec. Tech. Publ. 497, American Society for Testing and Materials, West Conshohocken, PA, 206–218.

Berglund, L.A. and Kenny, J.M., 1991. Processing science for high performance thermoset composites, *SAMPE J.*, 27, 27–37.

Bergner, H.W., Jr., Davis, J.G., Jr., and Herakovich, C.T., 1977. *Analysis of Shear Test Methods for Composite Laminates*, Department of Engineering Science and Mechanics, Report VPI-E-77-14, Virginia Polytechnic Institute and State University, Blacksburg, VA.

Berry, J.P., 1963. Determination of fracture energies by the cleavage technique, *J. Appl. Phys.*, 34, 62–68.

Boeing Specification Support Standard BSS 7260, 1988, *Advanced Composite Compression Tests*, Boeing, Seattle, WA (originally issued February 1982).

Bogetti, T.A. and Gillespie, J.W., 1991. Two-dimensional cure simulation of thick thermosetting composites, *J. Compos. Mater.*, 25, 239–273.

Boller, K.H., 1969. *A Method to Measure Intralaminar Shear Properties of Composite Laminates*, Technical Report AFML-TR-69-311, Air Force Materials Laboratory, Wright-Patterson Air Force Base, Dayton, OH.

Broutman, L.J., 1969. *Measurement of the Fiber-Polymer Matrix Interfacial Strength*, ASTM Spec. Tech. Publ. 452, American Society for Testing and Materials, West Conshohocken, PA.

Broutman, L.J. and McGarry, F.J., 1963. Glass-resin joint strengths studies, *Proceedings of the 17th Annual Technical and Management Conference*, Society of Plastics Industry, 1–7.

Browning, C.E., Husman, G.E., and Whitney, J.M., 1977. *Moisture Effects in Epoxy Matrix Composites*, ASTM Spec. Tech. Publ. 617, American Society for Testing and Materials, West Conshohocken, PA, 481–496.

Byun, J.-H. and Chou, T.–W., 2000. Mechanics of textile composites, in *Comprehensive Composite Materials*, A. Kelly and C. Zweben, Eds., Elsevier, Oxford, UK, Vol. 1, 719–761.

Cantwell, W.J. and Morton, J., 1991. The impact resistance of composite materials—a review, *Composites*, 22, 347–362.

Carlsson, L.A., 1981. Out-of-plane hygroinstability of multi-ply paperboard, *Fibre Sci. Technol.*, 14, 201–212.

Carlsson, L.A., Gillespie, J.W., Jr., and Pipes, R.B., 1986a. On the analysis and design of the end notched flexure (ENF) specimen for mode II testing, *J. Compos. Mater.*, 20, 594–604.

Carlsson, L.A., Gillespie, J.W., Jr., and Trethewey, B.R., 1986b. Mode II interlaminar fracture of graphite/epoxy and graphite/PEEK, *J. Reinf. Plast. Compos.*, 5, 170–187.

Carlsson, L.A. and Kardomateas, G.K., 2011. *Structural and Failure Mechanics of Sandwich Composites*, Springer, Dordrecht, Netherlands.

Castro, J.M. and Macosko, C.W., 1982. Studies of mold-filling flows in the reaction injection molding process, *AIChE J.*, 28, 250–260.

Chamis, C.C. and Sinclair, J.H., 1977. Ten-degree off-axis test for shear properties in fiber composites, *Exp. Mech.*, 17, 339–346.

Chatterjee, S.N., Adams, D.F., and Oplinger, D.W., 1993a. *Test Methods for Composites—A Status Report, Vol. 1, Tension Test Methods*, DOT/FAA/CT-93/17-I, FAA Technical Center, Atlantic City International Airport, NJ.

Chatterjee, S.N., Adams, D.F., and Oplinger, D.W., 1993b. *Test Methods for Composites—A Status Report, Vol. 2: Compression Test Methods*, Report DOT/FAA/CT-93/17-II, FAA Technical Center, Atlantic City International Airport, NJ.

Chatterjee, S.N., Adams, D.F., and Oplinger, D.W., 1993c. *Test Methods for Composites—A Status Report, Vol. 3: Shear Test Methods*, Report DOT/FAA/CT-93/17-III, Federal Aviation Administration Technical Center, Atlantic City International Airport, NJ.

Choi, I. and Horgan, C.O., 1977. Saint-Venant's principle and end effects in anisotropic elasticity, *J. Appl. Mech.*, 44, 424–430.

Christensen, R.M., 1979. *Mechanics of Composite Materials*, Wiley, New York.

Coguill, S.L. and Adams, D.F., 1999a. Selection of the proper wedge grip surface for tensile testing composite materials, *Proceedings of the 44th International SAMPE Symposium*, Long Beach, CA, Society for the Advancement of Material and Process Engineering, Covina, CA, 2332–2345.

Coguill, S.L. and Adams, D.F., 1999b. Use of the Wyoming combined loading compression (CLC) fixture to test unidirectional composites, *Proceedings of the 44th International SAMPE Symposium*, Long Beach, CA, Society for the Advancement of Material and Process Engineering, Covina, CA, 2322–2331.

Coguill, S.L. and Adams, D.F., 2000. A comparison of open-hole compression fixtures by experimental evaluation, *Proceedings of the 45th International SAMPE Symposium and Exhibition*, Society for the Advancement of Material and Process Engineering, West Covina, CA, 1095–1105.

Conte, S.D. and de Boor, C., 1965. *Elementary Numerical Analysis: An Algorithmic Approach*, 2nd ed., McGraw-Hill, New York.

Corum, J.M., Simpson, W.A., Jr., Sun, C.T., Talreja, R., and Weitsman, Y.J., 1995, *Durability of Polymer-Matrix Composites for Automotive Structural Applications: A State of the Art Review*, Report ORNL-6869, Oak Ridge National Laboratory, Oak Ridge, TN.

Cox, B.N., Massabo, R., Mumm, D.R., Turrettini, A., and Kedward, K.B., 1997. Delamination fracture in the presence of through-thickness reinforcement, *Proceedings of the 11th International Conference on Composite Materials* (ICCM-11), M.L. Scott, Ed., Gold Coast, Australia, 1997, Technomic, Lancaster, PA, 159–177.

Crossman, F.W. and Wang, A.S.D., 1982. *The Dependence of Transverse Cracking and Delamination on Ply Thickness in Graphite/Epoxy Laminates*, ASTM Spec. Tech. Publ. 775, American Society for Testing and Materials, West Conshohoken, PA, 118–139.

Crossman, F.W., Warren, W.J., Wang, A.S.D., and Law, G.E., Jr., 1980. Initiation and growth of transverse cracks and edge delamination in composite laminates. Part 2. Experimental correlation, *J. Compos. Mater. Suppl.* 14, 88–108.

Cunningham, M.E., Schoultz, S.V., and Toth, J.M., Jr., 1985. *Effect of End-Tab Design on Tension Specimen Stress Concentrations*, ASTM Spec. Tech. Publ. 864, American Society for Testing and Materials, West Conshohoken, PA, 253–262.

Dale, M., Carlsson, L.A., and Acha, B.A., 2012a, Impact force analysis during low velocity impact of woven, carbon/vinylester, *J. Compos. Mater*, 46, 3163–3172.

Dale, M., Acha, B.A., and Carlsson, L.A., 2012b, Low velocity impact and compression after impact characterization of woven carbon/vinylester at dry and water saturated conditions, *Compos. Struct.*, 94, 1582–1589.

Dale, M., 2013, Damage tolerance characterization of carbon/vinylester composites exposed to sea water, master's thesis, Florida Atlantic University, Boca Raton.

Dally, J.W. and Riley, W.F., 1991. *Experimental Stress Analysis*, 3rd ed., McGraw-Hill, New York.

Dang, M.-L. and Hyer, M.W., 1998. Thermally-induced deformation behavior of unsymmetric laminates, *Int. J. Solids Struct.*, 35, 2101–2120.

Daniel, I.M., 1982. Failure mechanisms and fracture of composite laminates with stress concentrations, *Proceedings of the 7th International Conference on Experimental Stress Analysis*, Technion, Haifa, Israel.

Daniel, I.M. and Ishai, O., 2006. *Engineering Mechanics of Composite Materials*, 2nd ed., Oxford University Press, New York.

Davies, P., Sims, G.D., Blackman, B.R.K., et. al., 1999. Comparison of test configurations for determination of mode II interlaminar fracture toughness results from international collaborative test programme, *Plast. Rubber Compos.*, 28, 432–437.

DiBenedetto, A.T., 1987. Prediction of the glass transition temperature of polymers: A model based on the principle of corresponding states, *J. Polym. Sci. Part B: Polym. Phys.*, 25, 1779–2038.

Drzal, L.T., Herrera-Franco, P.J., and Ho, H., 2000. Fiber-matrix interface tests, in *Comprehensive Composite Materials*, A. Kelly and C. Zweben, Eds., Elsevier, Oxford, UK, Vol. 5, 71–111.

Drzal, L.T., Rich, M.J., Camping, J.D., and Park, W.J., 1980. Interfacial shear strength and failure mechanisms in graphite fiber composites, *35th Annual Reinforced Plastics/Composites Conference*, Paper 20C.

Drzal, L.T., Rich, M.J., Koenig, M.F., and Lloyd, P.F., 1983. Adhesion of graphite fibers to epoxy matrices: II. The effect of fiber finish, *J. Adhesion*, 16, 133–152.

Drzal, L.T., Rich, M.J., and Lloyd, P.F., 1982. Adhesion of graphite fibers to epoxy matrices: I. the role of fiber surface treatment, *J. Adhesion*, 16, 1–30.

Evans, A.G. and Adler, W.F., 1978. Kinking as a mode of structural degradation in carbon fiber composites, *Acta Metall.*, 26, 725–738.

Ewins, P.D., 1971. *Tensile and Compressive Test Specimens for Unidirectional Carbon Fiber Reinforced Plastics*, Technical Report 71217, Royal Aircraft Establishment, Farnborough, UK.

Farooq, M., 2009. Degradation of the composite fiber/matrix interface in marine environment, PhD thesis, Florida Atlantic University, Boca Raton.

Feih, S., Wonsyld, K., Minzari, D., Westermann, P., and Liholt, H., 2004. *Testing Procedure for the Single Fiber Fragmentation Test*, Riso-R-1483(EN), Riso National Laboratory, Roskilde, Denmark.

Figliolini, A.M., 2011, Degradation of mechanical properties of vinylester and carbon fiber/vinylester composites due to environmental exposure, master thesis, Florida Atlantic University, Boca Raton.

Flaggs, D.L. and Kural, M.H., 1982. Experimental determination of the in-situ transverse lamina strength in graphite/epoxy laminates, *J. Compos. Mater.*, 16, 103–115.

Fleck, N.A., 1977. Compressive failure of fiber composites, *Adv. Appl. Mech.*, 33, 43–117.

Flores, F., 2000. Cure-dependent material response of vinyl ester resin: Experimental investigation and model evaluation, master thesis, University of Delaware, Newark.

Flory, P.J., 1953. *Principles of Polymer Chemistry*, Cornell University Press, Ithaca, NY.

Freeman, W. and Campbell, M.D., 1972. *Thermal Expansion Characteristics of Graphite Reinforced Composite Materials*, ASTM Spec. Tech. Publ. 497, American Society for Testing and Materials, West Conshohoken, PA, 121–142.

Gere, J.M. and Goodno, B.J. 2013, *Mechanics of Materials*, 8th ed., Cengage Learning, Stamford, CT.

Gere, J.M. and Timoshenko, S.P., 1997. *Mechanics of Materials*, 4th ed., PWS, Boston.

Gibson, L.J. and Ashby, M.F., 1997, *Cellular Solids: Structure and Properties*, 2nd ed., Cambridge University Press, Cambridge, UK.

Gibson, R.F., 2007. *Principles of Composite Materials Mechanics*, 2nd ed., CRC Press, Boca Raton.

Gillespie, J.W., Jr., and Carlsson, L.A., 1988. Influence of finite width on notched laminate strength predictions, *Compos. Sci. Technol.*, 32, 15–30.

Gillespie, J.W., Jr., Carlsson, L.A., and Pipes, R.B., 1986. Finite element analysis of the end notched flexure (ENF) specimen for measuring mode II fracture toughness, *Compos. Sci. Technol.*, 26, 177–197.

Greszczuk, L.G., 1969. *Theoretical Studies of the Mechanics of the Fiber/Matrix Interface of Composites*, ASTM Spec. Tech. Publ. 452, American Society for Testing and Materials, West Conshohocken, PA, 42–58.

Griffith, A.A., 1920. The phenomena of rupture and flow in solids, *Phil. Trans. R. Soc.*, A221, 163–198.

Hahn, H.T. and Pagano, N.J., 1975. Curing stresses in composite laminates, *J. Compos. Mater.*, 9, 91–106.

Halpin, J.C. and Pagano, N.J., 1968. Influence of end constraint in the testing of anisotropic bodies, *J. Compos. Mater.*, 2, 18–31.

Halpin, J.C. and Pagano, N.J., 1975. Consequences of environmentally induced dilatation in solids, *Recent Adv. Eng. Sci.*, 5, 33–46.

Halpin, J.C. and Kardos, J.L., 1976. The Halpin-Tsai equations: a review, *Polym. Eng. Sci.*, 16, 344–352.

Hart-Smith, L.J., 1991, Generation of higher composite materials allowables using improved test coupons, *Proceedings of the 36th International SAMPE Symposium*, Anaheim, CA, Society for the Advancement of Material and Process Engineering, Covina, CA, 1029–1044.

Hart-Smith, L.J., 2000. Backing-out composite lamina strengths from cross-ply testing, in *Comprehensive Composite Materials*, A. Kelly and C. Zweben, Eds., Elsevier, Oxford, UK, Vol. 5, 149–161.

Hashemi, S., Kinloch, A.J., and Williams, J.G., 1990a. The analysis of the interlaminar fracture in uniaxial fiber-polymer composites, *Proc. Math. Phys. Sci.*, 427, 173–199.

Hashemi, S., Kinloch, A.J., and Williams, J.G., 1990b. The effects of geometry, rate, and temperature on the mode I, mode II, and mixed mode I/II interlaminar fracture of carbon-fiber/poly (ether-ether ketone) composites, *J. Compos. Mater.*, 24, 918–956.

Hashin, Z., 1983. Analysis of composite materials—a survey, *J. Appl. Mech.*, 50, 481–505.

Hashin, Z., 1985. Analysis of cracked laminates—a variational approach, *Mech. Mater.*, 4, 121–136.

Hellier, C.J. 2001, Eddy current testing, in *Handbook of Nondestructive Evaluation*, McGraw-Hill, New York, 8.1–8.7.

Herakovich, C.T., 1998. *Mechanics of Fibrous Composites*, Wiley, New York.

Highsmith, A.L. and Reifsnider, K.L., 1982. *Stiffness-Reduction Mechanisms in Composite Laminates*, ASTM Spec. Tech. Publ. 115, American Society for Testing and Materials, West Conshohocken, PA, 103–117.

Hinton, M.J., Soden, P.D., and Kaddour, A.S., 2004. *Failure Criteria in Fibre-Reinforced Polymer Composites*, Elsevier, Oxford, UK.

Ho, H., Morton, J., and Farley, G.L., 1994. Non-linear numerical analysis of the Iosipescu specimen for composite materials, *Compos. Sci. Technol.*, 50, 355–365.

Hodgkinson, J.M., Ed., 2000. *Mechanical Testing of Advanced Fibre Composites*, CRC Press, Boca Raton, FL.

Hofer, K.E, Jr. and Rao, P.N., 1977. A new static compression fixture for advanced composite materials, *J. Test. Eval.*, 5, 278–283.

Hoff, N.J. and Mautner, S.F., 1945. The buckling of sandwich-type panels, *J. Aeron. Sci.*, 12, 285–297.

Hojo, M., Sawada, Y., and Miyairi, H. 1994. Influence of clamping method on tensile properties of unidirectional CFRP in 0° and 90° directions—round robin activity for international standardization in Japan, *Composites*, 25, 786–796.

Holman, J.P. and Gajda, W.J., Jr., 1993. *Experimental Methods for Engineers*, 5th Ed., McGraw-Hill, New York.

Horgan, C.O., 1972. Some remarks on Saint-Venant's principle for transversely isotropic composites, *J. Elasticity*, 2, 335–339.

Horgan, C.O., 1982. Saint-Venant end effects in composites, *J. Compos. Mater.*, 16, 411–422.

Horgan, C.O. and Carlsson, L.A., 2000. Saint-Venant end effects for anisotropic materials, in *Comprehensive Composite Materials*, A. Kelly and C. Zweben, Eds., Elsevier, Oxford, UK, Vol. 5, 5–21.

Hornbeek, R.C., 1975. *Numerical Methods*, Quantum, New York.

Howarth, S.G. and Strong, A.B., 1990. Edge effects with waterjet and laser beam cutting of advanced composite materials, *Proceedings of the 35th International SAMPE Symposium*, Anaheim, CA, Society for the Advancement of Material and Process Engineering, Covina, CA, 1684–1697.

Hubert, P., 1996. Aspects of flow and compaction of laminated composite shapes during cure, PhD thesis, University of British Columbia, Vancouver.

Hubert, P. and Poursartip, A, 1998. A review of flow and compaction modelling relevant to thermoset matrix laminate processing, *J. Reinf. Plast. Compos.*, 17(4), 286–318.

Hussain, A.K. and Adams, D.F., 1998. *An Analytical and Experimental Evaluation of the Two-Rail Shear Test for Composite Materials*, Report UW-CMRG-R-98–105, Composite Materials Research Group, University of Wyoming, Laramie.

Hussain, A.K. and Adams, D.F., 1999. The Wyoming-modified two-rail shear test fixture for composite materials, *J. Compos. Technol. Res.* 21(4), 215–223.

Hyer, M.W., 1981. Calculations of the room temperature shape of unsymmetric laminates, *J. Compos. Mater.*, 15, 296–310.
Hyer, M.W., 2000. Laminated plate and shell theory, in *Comprehensive Composite Materials*, A. Kelly and C. Zweben, Eds., Elsevier, Oxford, UK, Vol. 1, 479–510.
Hyer, M.W., 2009. *Stress Analysis of Fiber-Reinforced Composite Materials*, updated edition, DES Tech, Lancaster, PA.
Hyer, M.W. and Waas, A.M., 2000 Micromechanics of linear elastic continuous fiber composite, in *Comprehensive Composite Materials*, A. Kelly and C. Zweben, Eds., Elsevier, Oxford, UK, Vol. 1, 345–375.
Instron Corporation, http://www.instron.com
Iosipescu, N., 1967. New accurate procedure for single shear testing of metals, *J. Mater.*, 2, 537–566.
Irwin, G.R., 1958. Fracture, in *Handbuch der Physik*, S. Flügge, Ed., Springer, Berlin, Vol. 6, 551–590.
Jackson, W.C. and Martin, R.H., 1993. *An Interlaminar Tensile Strength Test*, ASTM Spec. Tech. Pub. 1206, American Society for Testing and Materials, West Conshohocken, PA, 333–354.
Japan Industrial Standards, JIS 7086, 1993. *Testing Methods for Interlaminar Fracture Toughness of Carbon Fiber Reinforced Plastics*, Japanese Standards Association, Tokyo.
Jenkins., C.H., Ed., 1998. *Manual on Experimental Methods of Mechanical Testing of Composites*, 2nd ed., Society for Experimental Mechanics, Bethel, CT, Chap. 4–8.
Jeronimidis, G. and Parkyn, A.T., 1988. Residual stresses in carbon fibre-thermoplastic matrix laminates, *J. Compos. Mater.*, 22, 401–415.
Johnson, W.S. and Mangalgiri, P.D., 1987. *Influence of the Resin on Interlaminar Mixed Mode Fracture*, ASTM Spec. Tech. Publ. 937, American Society for Testing and Materials, West Conshohoken, PA, 295–315.
Johnston, A., Vaziri, R., and Poursartip, A., 2001. A plane strain model for process-induced deformation of laminated composite structures, *J. Compos. Mater.*, 35, 1435–1469.
Jones, R.M., 1999. *Mechanics of Composite Materials*, 2nd ed., Taylor & Francis, Philadelphia.
Jost, W., 1960. *Diffusion*, 3rd ed., Academic Press, New York.
Kamaya, M. and Kawakabo, M.A., 2011 A procedure for determining true stress-strain curve over a large range of strains using digital image correlation and finite element analysis. *Mech. Mater.*, 43, 243–253.
Karlak, R.F., 1977. Hole effects in a related series of symmetrical laminates, *Proceedings of 4th Joint ASM-Metallurgical Society of AIME Conference on Failure Modes in Composites (III)*, Chicago, 1977, American Society for Metals, Chicago.
Kedward, K.T., Wilson, R.S., and McLean, S.K., 1989. Flexure of simply curved composite shapes, *Composites*, 20, 527–536.
Kelly, A. and Tyson, W.R., 1965. Tensile properties of fiber-reinforced metals: copper/tungsten and copper/molybdenum, *J. Mech. Phys. Solids*, 13, 329–350.
Khoun, L., Centea, T., and Hubert, P., 2010. Characterization methodology of thermoset resins for the processing of composite materials—case study: CYCOM 890RTM epoxy resin, *J. Compos. Mater.*, 44, 1397–1415.
Kim, B.W. and Nairn, J.A., 2002. Observation of fiber fracture and interfacial debonding phenomena using the fragmentation test in single fiber composites, *J. Compos. Mater.*, 36, 1825–1858.
Kinloch, A.J., et al., 1993. The mixed-mode delamination of fiber composite materials, *Compos. Sci. Technol.*, 47, 225–237.

Kobayashi, A.S., Ed., 1993. *Handbook on Experimental Mechanics*, 2nd ed., Society for Experimental Mechanics, Bethel, CT.
Komanduri, R., 1993, April. Machining fiber-reinforced composites, *Mech. Eng.*, 58–63.
Konish, H.J. and Whitney, J.M., 1975. Approximate stresses in an orthotropic plate containing a circular hole, *J. Compos. Mater.*, 9, 157–166.
Kotha, S. and Adams, D.F., 1998. *Analytical Investigation of Crack Initiation and Propagation in Composite Materials*, Report UW-CMRG-R-98–119, Composite Materials Research Group, University of Wyoming, Laramie.
Kozey, V.V., Jiang, H., Mehta, V.R., and Kumar, S., 1995. Compressive behavior of materials: part II. High performance fibers. *J. Mater. Res.*, 10, 1044–1061.
Krueger, R., 2004. Virtual crack closure technique: History, approach and applications. *Appl. Mech. Rev.*, 57, 109–143.
Kural, M.H. and Flaggs, D.L., 1983. A finite element analysis of composite tension specimens, *J. Compos. Technol. Res.*, 5, 11–17.
Lal, K.M., 1983. Low velocity transverse impact behavior of 8-ply, graphite-epoxy laminates, *J. Reinf. Plast. Compos.*, 2, 226–238.
Lau, W.S. and Lee, W.B., 1991. A comparison between EDM wire-cut and laser cutting of carbon fibre composite materials, *Mater. Manuf. Process.*, 6, 331–342.
Lee, S.M., 1993. An edge crack torsion method for mode III delamination fracture testing, *J. Compos. Technol. Res.*, 15, 193–201.
Lee, W.I., Loos, A.C., and Springer, G.S., 1982. Heat of reaction, degree of cure, and viscosity of Hercules 35016 resin, *J. Compos. Mater.*, 16, 510–520.
Lekhnitskii, S.G., 1968. *Anisotropic Plates*, Gordon and Breach, New York.
Li, X., Davies, P., and Carlsson, L.A., 2004. Influence of fiber volume fraction on mode III interlaminar toughness of glass/epoxy composites. *Compos. Sci. Technol.*, 64, 1279–1286.
Loos, A.C. and Springer, G.S., 1983. Curing of epoxy matrix composites, *J. Compos. Mater.*, 17, 135–169.
Mahishi, J.M. and Adams, D.F., 1984. Analysis of neat resin cracking induced by rapid moisture loss, *Compos. Technol. Rev.*, 6, 159–163.
Mandell, J.F., Chen, J.H., and McGarry, J.F., 1980. A microdebonding test for in situ assessment of fibre/matrix bond strength in composite materials. *Int. J. Adhesion*, 1, 40–44.
Mandell, F., Wang, S.S., and McGarry, F.J., 1975. The extension of crack tip damage zones in fiber reinforced plastic laminates, *J. Compos. Mater.*, 9, 266–287.
Masters, J.E. and Ifju, P.G., 1997. Strain gage selection criteria for textile composite materials, *J. Compos. Technol. Res.*, 19, 152–167.
Measurements Group, 1982. *Errors Due to Transverse Sensitivity in Strain Gages*, Report TN-509, Measurements Group, Raleigh, NC.
Measurements Group, 1987. *Measurement of Thermal Expansion Coefficient Using Strain Gages*, Report TN-513, Measurements Group, Raleigh, NC.
Miller, B., Muri, P., and Rebenfeld, L., 1987. A microbond method for determination of the shear strength of a fiber/resin-interface, *Compos. Sci. Technol.*, 28, 17–32.
Miller, I., Freund, J.E., and Johnson, R.A., 1990. *Probability and Statistics for Engineers*, 4th ed., Prentice-Hall, Englewood Cliffs, NJ.
Moon, C.K. and McDonough, W.G., 1998. Multiple fiber technique for the single fiber fragmentation test, *J. Appl. Polym. Sci.*, 67, 1701–1709.
Morton, J., Ho, H., Tsai, M.Y., and Farley, G.L., 1992. An evaluation of the Iosipescu specimen for composite materials shear property measurement, *J. Compos. Mater.*, 26, 708–750.

MTS Systems Corporation, http://www.mts.com
Muller, D.E., 1956. A method for solving algebraic equations using an automatic computer, *Math. Tables Comput.*, 10, 208–215.
Nicholls, D.J. and Gallagher, J.P., 1983. Determination of G_{IC} in angle-ply composites using a cantilever beam test method, *J. Reinf. Plast. Compos.*, 2, 2–17.
Nordstrand, T.M. and Carlsson, L.A., 1994, Evaluation of transverse shear stiffness of structural core sandwich plates, *Compos. Struct.*, 37, 145–153.
Northrop Specification NAI-1504C, 1988. *Open-Hole Compression Test Method*, Revision B, Northrop, Hawthorne, CA.
Nuismer, R.J. and Whitney, J.M., 1975. *Uniaxial Failure of Composite Laminates Containing Stress Concentrations*, ASTM Spec. Tech. Publ. 593, American Society for Testing and Materials, West Conshohoken, PA., 117–142.
O'Brien, T.K. and Martin, R.H., 1993. Results of ASTM round robin testing for mode I interlaminar fracture toughness of composite materials, *J. Compos. Technol. Res.*, 15, 269–281.
O'Brien, T.K., Murri, G.B., and Salpekar, S.A., 1989. *Interlaminar Shear Fracture Toughness and Fatigue Thresholds for Composite Materials*, ASTM Spec. Tech. Publ. 1012, American Society for Testing and Materials, West Conshohoken, PA, 222–250.
Obst, A.W., Van Landingham, M.R., Eduljee, R.F., and Gillespie, J.W., Jr., 1996, November. The effect of hygrothermal cycling on the microcracking behavior of fabric laminates, *Proceedings of the 28th International SAMPE Technical Conference*, Seattle, WA, Society for the Advancement of Material and Process Engineering, Covina, CA, 994–1002.
Odom, E.M. and Adams, D.F., 1983. *A Study of Polymer Matrix Fatigue Properties*, Report No. NADC-83053-60, Naval Air Development Center, Warminster, PA.
Odom, E.M. and Adams, D.F., 1990. Failure modes of unidirectional carbon/epoxy composite compression specimens, *Composites*, 21, 289–296.
Ogonowski, J.M., 1980. *Analytical Study of Finite Geometry Plates with Stress Concentrations*, AIAA Paper 80-0778, American Institute of Aeronautics and Astronautics, Reston, VA.
Oplinger, D.W., Gandhi, K.R., and Parker, B.S., 1982. *Studies of Tension Test Specimens for Composite Material Testing*, AMMRC TR 82-27, Army Materials and Mechanics Research Center, Watertown, MA.
Outwater, M. and Murphy, M.C., 1969. On the fracture energy of unidirectional laminate, in *Proceedings of the 24th Annual Technology Conference, Society of the Plastics Industry*, Washington D.C.
Ozdil, F. and Carlsson, L.A., 1992. Mode I interlaminar fracture of interleaved graphite/epoxy, *J. Compos. Mater.*, 26, 432–459.
Ozisik, M.N., 1980. *Heat Conduction*, Wiley, New York.
Pagano, N.J. and Pipes, R.B., 1971. The influence of stacking sequence on laminate strength, *J. Compos. Mater.*, 5, 50–57.
Pagano, N.J. and Pipes, R.B., 1973. Some observations on the interlaminar strength of composite laminates, *Int. J. Mech. Sci.*, 15, 679–688.
Park, I.K., 1971. Tensile and compressive test methods for high-modulus graphite fibre-reinforced composites, *Proceedings of the International Conference on Carbon Fibres, Their Composites, and Applications*, Paper No. 23, Plastics Institute, London.
Perry, C.C., 1988. *Strain Gage Measurements on Plastics and Composites*, Basingstroke, Hants, Measurements Group, NC.
Peters, S.T., 2002. Composite materials and processes, in *Handbook of Plastics, Elastomers and Composites*, 4th ed., C.A. Harper, Ed., McGraw-Hill, New York, Chap. 4.
Peterson, R.E., 1974. *Stress Concentration Factors*, Wiley, New York.

Pickering, K.L. and Murray, T.L., 1999. Weak link scaling analysis of high strength carbon fiber, *Compos. Part A*, 30, 1017–1021.
Piggott, M.R., 1987. Debonding and friction at fiber-polymer interfaces. I: criteria for failure and sliding, *Compos. Sci. Technol.*, 30, 295–306.
Piggott, M.R. and Chua, P.S., 1987. Recent studies of the glass fiber-polymer interface, *Industrial Engineering and Chemistry*, 26, 672–677.
Pindera, M.J. and Herakovich, C.T., 1981. *An Endochronic Theory for Transversely Isotropic Fibrous Composites*, Report VPI-E-81–27, Virginia Polytechnic Institute and State University, Blacksburg, VA.
Pindera, M.J. and Herakovich, C.T., 1986. Shear characterization of unidirectional composites with the off-axis tension test, *Exp. Mech.*, 26, 103–112.
Pipes, R.B. and Cole, B.W., 1973. On the off-axis strength test of anisotropic materials, *J. Compos. Mater.*, 7, 246–256.
Pipes, R.B., Kaminski, B.E., and Pagano, N.J., 1973. *Influence of the Free-Edge upon the Strength of Angle-Ply Laminates*, ASTM Spec. Tech. Publ. 521, American Society for Testing and Materials, West Conshohoken, PA, 218–228.
Pipes, R.B. and Pagano, N.J., 1970. Interlaminar stresses in composite laminates under uniform axial extension, *J. Compos. Mater.*, 4, 538–548.
Pipes, R.B., Vinson, J.R., and Chou, T.W., 1976. On the hygrothermal response of laminated composite systems, *J. Compos. Mater.*, 10, 129–148.
Pipes, R.B., Wetherhold, R.C., and Gillespie, J.W., Jr., 1979. Notched strength of composite materials, *J. Compos. Mater.*, 13, 148–160.
Plantema, F.J., 1966. *Sandwich Construction*, Wiley, New York.
Ramirez, F.A., 2008. Evaluation of water degradation of polymer matrix composites by micromechanical and macromechanical tests, master thesis, Florida Atlantic University, Boca Raton.
Ramulu, M. and Arola, D., 1994. Traditional and non-traditional machining of fiber reinforced plastic composites, *Proceedings of the 39th International SAMPE Symposium*, Anaheim, CA, Society for the Advancement of Material and Process Engineering, Covina, CA, 1073–1087.
Rao, V., Herrera-Franco, P., Ozzello, A.D., and Drzal, L.T., 1991. A direct comparison of the fragmentation test and the microbond pull-out test for determining the interfacial shear strength, *J. Adhesion*, 34, 65–77.
Reddy, J.N., 1997. *Mechanics of Laminated Composite Materials—Theory and Analysis*, CRC Press, Boca Raton, FL.
Reeder, J.R. and Crews, J.H., Jr., 1990. Mixed mode bending method for delamination testing, *AIAA J.*, 28, 1270–1276.
Reeder, J.R. and Crews, J.H., Jr., 1991, July. Redesign of the mixed mode bending test for delamination toughness, *Proceedings of the 8th International Conference on Composite Materials*, Honolulu, HI, Society for the Advancement of Materials and Process Engineering (SAMPE), Covina, CA.
Reeder, J.R. and Crews, J.H., Jr., 1992. Redesign of the mixed mode bending delamination test to reduce nonlinear effects, *J. Compos. Technol. Res.*, 14, 12–19.
Rich, M.J., Drzal, L.T., Hunston, D., Holmes, G., and McDonough, W., 2002, October 21–23. Round robin assessment of the single fiber fragmentation test, *Proceedings of the American Society for Composits, 17th Techical Conference*, Purdue University, West Lafayette, IN.
Robinson, P. and Song, D.Q., 1992. A modified DCB specimen for mode I testing of multidirectional laminates, *J. Compos. Mater.*, 26, 1554–1577.

Rosen, B., 1972, A simple procedure for experimental determination of the longitudinal shear modulus of unidirectional composites, *J. Compos. Mater.*, 6, 552–554.

Rosen, B.W., 1965. Mechanics of composite strengthening, in *Fiber Composite Materials*, ASM International, Materials Park, OH.

Rosen, B.W. and Hashin, Z., 1987. Analysis of material properties, in *Engineered Materials Handbook, Vol. 1, Composites,* T.J. Reinhart, tech. chairman, ASM International, Materials Park, OH, 185–205.

Russell, A.J. and Street, K.N., 1982. Factors affecting the interlaminar fracture energy of graphite/epoxy laminates, in *Progress in Science and Engineering of Composites*, T. Hayashi, K. Kawata, and S. Umekawa, Eds., ICCM-IV, ASM International, Tokyo, 1982, 279–286.

Russell, A.J. and Street, K.N., 1985. *Moisture and Temperature Effects on the Mixed-Mode Delamination Fracture of Unidirectional Graphite/Epoxy*, ASTM Spec. Tech. Publ. 876, American Society for Testing and Materials, West Conshohoken, PA, 349–370.

Rybicki, E.F. and Kanninen, M.F., 1977. A finite element calculation of stress intensity factors by a modified crack closure integral, *Eng. Fract. Mech.*, 9, 931–938.

SACMA Recommended Method SRM 3R-94, 1994. *Open-Hole Compression Properties of Oriented Fiber-Resin Composites*, Suppliers of Advanced Composite Materials Association, Arlington, VA.

SACMA Recommended Method SRM 5R-94, 1994. *Open-Hole Tensile Properties of Oriented Fiber-Resin Composites*, Suppliers of Advanced Composite Materials Association, Arlington, VA.

Saenz, E., 2012. Fatigue and fracture of foam cores used in sandwich composites, master thesis, Florida Atlantic University, Boca Raton.

Sakuma, K. and Seto, M. 1983. Tool wear in cutting glass-fiber reinforced plastics: The relation between fiber orientation and tool wear, *Bull. JSME*, 26(218), 1420–1427.

Sawicki, A. and Minguet, P., 1999. Failure mechanisms in compression-loaded composite laminates containing open and filled holes, *J. Reinf. Plast. Compos.*, 18, 1706–1728.

Shen, C.W. and Springer, G.S., 1976. Moisture absorption and desorption of composite materials, *J. Compos. Mater.*, 10, 2–20.

Shi, Y.B., Hull, D., and Price, J.N., 1993. Mode II fracture of $+\theta/-\theta$ angled laminate interfaces, *Compos. Sci. Technol.*, 47, 173–184.

Sih, G.C., Paris, P.C., and Irwin, G.R., 1965. On cracks in rectilinearly anisotropic bodies, *Int. J. Fract. Mech.*, 1, 189–203.

Smoot, M.A., 1982. *Compressive Response of Hercules AS1/3501–6 Graphite/Epoxy Composites*, Report CCM-82–16, Center for Composite Materials, University of Delaware, Newark.

Springer, G.S., Ed., 1981. *Environmental Effects on Composite Materials*, Technomic, Lancaster, PA.

Srivatsan, T.S., Lane, C.T., and Bowden, D.M., Eds., 1994. *Machining of Composite Materials II*, ASM International, Materials Park, OH.

Stone, M.A., 1997. Thermo-chemical and thermo-mechanical response of reacting polymers, master thesis, University of Delaware, Newark.

Sullivan, J.L., Kao, B.G., and Oene, H., 1984. Shear properties and a stress analysis obtained from vinyl-ester Iosipescu specimens, *Exp. Mech.*, 24, 223–232.

Sun, C.T., 2000. Strength analysis of unidirectional composites and laminates, in *Comprehensive Composite Materials*, A. Kelly and C. Zweben, Eds., Elsevier, Oxford, UK, Vol. 1, 641–666.

Suo, Z., Bao, G., and Fan, B., 1992. Delamination R-curve phenomnena due to damage, *J. Mech. Phys. Solids*, 40, 1–16.

Sutton, M.A., Orteu, J.J., and Schreier, H.W., 2009, *Image Correlation for Shape, Motion and Deformation Measurements*, Springer: New York.

Swanson, S.R. and Qian, Y., 1992. Multiaxial characterization of T300/3900-2 carbon/epoxy composites, *Compos. Sci. Technol.*, 43, 197–203.

Swanson, S.R. and Trask, B.C., 1989. Strength of quasi-isotropic laminates under off axis loading, *Compos. Sci. Technol.*, 34, 19–34.

Taher, S.T., Thomsen, O.T., Dulieu-Barton, J.M., and Zhang, S., 2012, Determination of mechanical properties of PVC foam using a modified Arcan fixture, *Compos.* Part A, 43, 1698–1708.

Tan, T.M. and Sun, C.T., 1985. Use of statical indention laws in the impact analysis of laminated composite plates, *J. Appl. Mech.*, 52, 6–12.

Tarnopolskii, Y.M. and Kincis, T.Y., 1985. *Static Test Methods for Composites*, Van Nostrand Reinhold, New York.

Timoshenko, S.P., 1984. *Strength of Materials*, Part 1, 3rd. ed., Krieger, Malabar, FL.

Timoshenko, S.P. and Goodier, J.N., 1970. *Theory of Elasticity*, 3rd ed., McGraw-Hill, New York.

Tsai, S.W., 1968. Strength theories of filamentary structures, in *Fundamental Aspects of Fiber Reinforced Plastic Composites*, R.T. Schwartz and H.S. Schwartz, Eds., Wiley, New York, 3–11.

Tsai, S.W. and Hahn, H.T., 1980. *Introduction to Composite Materials*, Technomic, Lancaster, PA.

Tsai, S.W. and Wu, E.M., 1971. A general theory of strength for anisotropic materials, *J. Compos. Mater.*, 5, 58–80.

Tuttle, M.E. and Brinson, H.F., 1984. Resistance-foil strain gage technology as applied to composite materials, *Exp. Mech.*, 24, 54–65.

Vadakke, V. and Carlsson, L.A., 2004. Experimental investigation of compression failure mechanisms of composite faced foam core sandwich specimens, *J. Sandwich Struct. Mater.*, 6, 327–342.

van der Zwaag, S., 1989. The concept of filament strength and the Weibull modulus, *J. Testing Eval.*, 17, 292–295.

Velisaris, C.N. and Seferis, J.C., 1986. Crystallization kinetics of polyetheretherketone (PEEK) matrices, *Polym. Sci. Eng.*, 26, 1574–1581.

Viana, G. M. and Carlsson, L.A., 2002. Mechanical properties and fracture characterization of cross-linked PVC foams, *J. Sandwich Struct. Mater.*, 4, 99–113.

Vinson, J.R. 1999. *The Behavior of Sandwich Structures of Isotropic and Composite Materials*, Technomic, Lancaster, PA.

Vinson, J.R. and Sierakowsky, R.L., 2002. *The Behavior of Structures Composed of Composites Materials*, 2nd ed., Kluwer, Dordrecht.

Walrath, D.E. and Adams, D.F., 1984. *Verification and Application of the Iosipescu Shear Test Method*, Report No. UWME-DR-401-103-1, Department of Mechanical Engineering, University of Wyoming, Laramie.

Wang, A.S.D. and Crossman, F.W., 1980. Initiation and growth of transverse cracks and edge delamination in composite laminates. Part 1. An energy method, *J. Compos. Mater. Suppl.* 14, 71–87.

Waterbury, M.C. and Drzal, L.T., 1991. On the determination of fiber strengths by in-situ fiber strength testing. *J. Compos. Technol. Res.*, 13, 22–28.

Watson, A.S. and Smith, R.L., 1985. An examination of statistical theories for fibrous materials in the light of experimental data, *J. Mater. Sci.*, 20, 3260–3270.
Wegner, P.M. and Adams, D.F., 2000. *Verification of the Combined Load Compression (CLC) Test Method*, Report No. DOT/FAA/AR-00/26, Federal Aviation Administration Technical Center, Atlantic City, NJ.
Weibull, W.J., 1951. A statistical distribution function of wide applicability, *J. Appl. Mech.*, 18, 293–297.
Weitsman, Y., 1979. Residual thermal stresses due to cool-down of epoxy-resin composites, *J. Appl. Mech.*, 46, 563–567.
Weitsman, Y.J., 2012, *Fluid Effects in Polymers and Polymer Composites*, Springer, New York.
Welsh, J.S. and Adams, D.F., 1996. Unidirectional composite compression strengths obtained by testing cross-ply laminates, *J. Compos. Technol. Res.*, 18, 241–248.
Welsh, J.S. and Adams, D.F., 1997. Current status of compression test methods for composite materials, *SAMPE J.*, 33, 35–43.
Whitney, J.M., 1987. *Structural Analysis of Laminated Anisotropic Plates*, Technomic, Lancaster, PA.
Whitney, J.M. and Browning, C.E., 1985. On short-beam shear tests for composite materials, *Exp. Mech.*, 25, 294–300.
Whitney, J.M., Daniel, I.M., and Pipes, R.B., 1984. *Experimental Mechanics of Fiber Reinforced Composite Materials*, rev. ed., Society for Experimental Mechanics, Prentice-Hall, Englewood Cliffs, NJ.
Whitney, J.M. and Nuismer, R.J., 1974. Stress fracture criteria for laminated composites containing stress concentrations, *J. Compos. Mater.*, 8, 253–265.
Whitney, J.M., Pagano, N.J., and Pipes, R.B., 1971. *Design and Fabrication of Tubular Specimens for Composite Characterization*, ASTM Special Tech. Publ. 497, American Society for Testing and Materials, West Conshohocken, PA, 52–67.
Wilkins, D.J., Eisenmann, J.R., Camin, R.A., Margolis, W.S., and Benson, R.A., 1982. *Characterizing Delamination Growth in Graphite-Epoxy*, ASTM Spec. Tech. Publ. 775, American Society for Testing and Materials, West Conshohocken, PA, 168–183.
Wilson, D.W. and Carlsson, L.A., 1991. Mechanical characterization of composite materials, in *Physical Methods of Chemistry*, 2nd ed., Wiley, New York, 139–221.
Wimolkiatisak, A.S. and Bell, J.P., 1989. Interfacial shear strength and failure modes of interphase-modified graphite-epoxy composites, *Polym. Compos.*, 4, 238–248.
Wyoming Test Fixtures, http://www.wyomingtestfixtures.com.
Xie, M. and Adams, D.F., 1995. Effect of specimen tab configuration on compression testing of composite materials, *J. Compos. Technol. Res.*, 17, 77–83.
Yaniv, G., Peimanidis, G., and Daniel, I.M., 1987. Method for hygromechanical characterization of graphite/epoxy composite, *J. Compos. Technol. Res.*, 19, 21–25.
Zenkert, D., 1997, *An Introduction to Sandwich Construction*, EMAS, London.
Zweben, C., 1990. Static strength and elastic properties, in *Delaware Composites Design Encyclopedia*, J.W. Gillespie, Jr. and L.A. Carlsson, Review Eds., Vol. 1, Technomic, Lancaster, PA, 49–70.

Index

Experimental Characterization of Advanced Composite Materials, Fourth Edition

M

T

CPSIA information can be obtained
at www.ICGtesting.com
Printed in the USA
BVHW04*0313280818
525762BV00005B/97/P